Metal Complexes in
Aqueous Solutions

MODERN INORGANIC CHEMISTRY

Series Editor: John P. Fackler, Jr., *Texas A&M University*

CARBON-FUNCTIONAL ORGANOSILICON COMPOUNDS
Edited by Václav Chvalovský and Jon M. Bellama

COOPERATIVE PHENOMENA IN JAHN–TELLER CRYSTALS
Michael D. Kaplan and Benjamin G. Vekhter

GAS PHASE INORGANIC CHEMISTRY
Edited by David H. Russell

HOMOGENEOUS CATALYSIS WITH METAL PHOSPHINE COMPLEXES
Edited by Louis H. Pignolet

INORGANOMETALLIC CHEMISTRY
Edited by Thomas P. Fehlner

THE JAHN–TELLER EFFECT AND
VIBRONIC INTERACTIONS IN MODERN CHEMISTRY
I. B. Bersuker

METAL COMPLEXES IN AQUEOUS SOLUTIONS
Arthur E. Martell and Robert D. Hancock

METAL INTERACTIONS WITH BORON CLUSTERS
Edited by Russell N. Grimes

MÖSSBAUER SPECTROSCOPY
APPLIED TO INORGANIC CHEMISTRY
Volumes 1 and 2 • Edited by Gary J. Long
Volume 3 • Edited by Gary J. Long and Fernande Grandjean

MÖSSBAUER SPECTROSCOPY
APPLIED TO MAGNETISM AND MATERIALS SCIENCE
Volume 1 • Edited by Gary J. Long and Fernande Grandjean

ORGANOMETALLIC CHEMISTRY OF THE TRANSITION ELEMENTS
Florian P. Pruchnik
Translated from Polish by Stan A. Duraj

PHOTOCHEMISTRY AND PHOTOPHYSICS OF METAL COMPLEXES
D. M. Roundhill

A Continuation Order Plan is available for this series. A continuation order will bring delivery of each new volume immediately upon publication. Volumes are billed only upon actual shipment. For further information please contact the publisher.

Metal Complexes in Aqueous Solutions

Arthur E. Martell

Texas A&M University
College Station, Texas

Robert D. Hancock

IBC Advanced Technologies
American Fork, Utah

PLENUM PRESS • NEW YORK AND LONDON

Library of Congress Cataloging-in-Publication Data

On file

ISBN 0-306-45248-0

© 1996 Plenum Press, New York
A Division of Plenum Publishing Corporation
233 Spring Street, New York, N. Y. 10013

Printed in the United States of America

Preface

Stability constants are fundamental to understanding the behavior of metal ions in aqueous solution. Such understanding is important in a wide variety of areas, such as metal ions in biology, biomedical applications, metal ions in the environment, extraction metallurgy, food chemistry, and metal ions in many industrial processes. In spite of this importance, it appears that many inorganic chemists have lost an appreciation for the importance of stability constants, and the thermodynamic aspects of complex formation, with attention focused over the last thirty years on newer areas, such as organometallic chemistry. This book is an attempt to show the richness of chemistry that can be revealed by stability constants, when measured as part of an overall strategy aimed at understanding the complexing properties of a particular ligand or metal ion. Thus, for example, there are numerous crystal structures of the Li^+ ion with crown ethers. What do these indicate to us about the chemistry of Li^+ with crown ethers? In fact, most of these crystal structures are in a sense misleading, in that the Li^+ ion forms no complexes, or at best very weak complexes, with familiar crown ethers such as 12-crown-4, in any known solvent. Thus, without the stability constants, our understanding of the chemistry of a metal ion with any particular ligand must be regarded as incomplete.

In this book we attempt to show how stability constants can reveal factors in ligand design which could not readily be deduced from any other physical technique. These range from the affinity of metal ions for donor atoms of different types, to a detailed consideration of the effect of ligand architecture on metal ion selectivity. The modern approach to the study of complexation of metal ions should include a comprehensive attack including stability constant studies, crystallography, NMR, UV-visible spectroscopy, theoretical calculations, and techniques such as ESR or polarimetry where appropriate. The first four chapters of this book consider the basic chemistry of metal ions in solution as revealed by stability constants. Thereafter, two chapters are devoted to solution chemistry, and metal ions in biomedical applications and biology, with a final chapter on the determination of stability constants.

The authors thank Ramunas J. Motekaitis for a review of this book, and for many helpful discussions. We are also indebted to Mary Martell for preparing several preliminary drafts as well as the final photo-ready copy.

Contents

Chapter 7   Stability Constants and Their Measurement

CHAPTER 1

INTRODUCTORY OVERVIEW

1.1   Background to the Study of Complex-Formation in Aqueous Solution

The idea of coordination of ligands to metal ions goes back to the theories of Alfred Werner, and his painstaking accumulation of evidence[1,2] of the coordination numbers and geometries of complexes of metal ions such as Co(III) and Pd(II). The achievement of Werner was the realization that metal ions have typical numbers of ligands (see Appendix for Glossary of terms) coordinated to them, such as the Co(III) and Pd(II) ions with their coordination numbers of six and four respectively, and that these numbers were not necessarily the same as the oxidation state of the metal ion. Werner further realized that ligands were arranged in space in definite geometries, such as the octahedral coordination geometry of Co(III) complexes, or square-planar geometry of Pd(II) complexes. Specific coordination geometries accounted for the occurrence of optical and geometrical isomers in these complexes, as summarized in Figure 1.1.

Coordination chemistry bloomed in the 1950's, with the study of equilibria in solution by workers such as Bjerrum[3] and Schwarzenbach.[4] A particular achievement was the recognition of the *chelate effect*.[5] The result of the chelate effect is that ligands with many donor atoms (multidentate ligands), such as the polyamine ligands in Table 1.1, form thermodynamically more stable complexes than analogous complexes containing unidentate ligands (For discussion of the determination and significance of formation constants, see Chapter 7):

Table 1.1. Formation constants of complexes of copper(II) with nitrogen donor ligands of differing denticity.

ligand:	ammonia	ethylenediamine	triethylenetetramine.
Denticity:	1	2	4
	$\log \beta_4 = 13.0$	$\log \beta_2 = 19.6$	$\log K_1 = 20.1$

Data from Ref. 6.

1

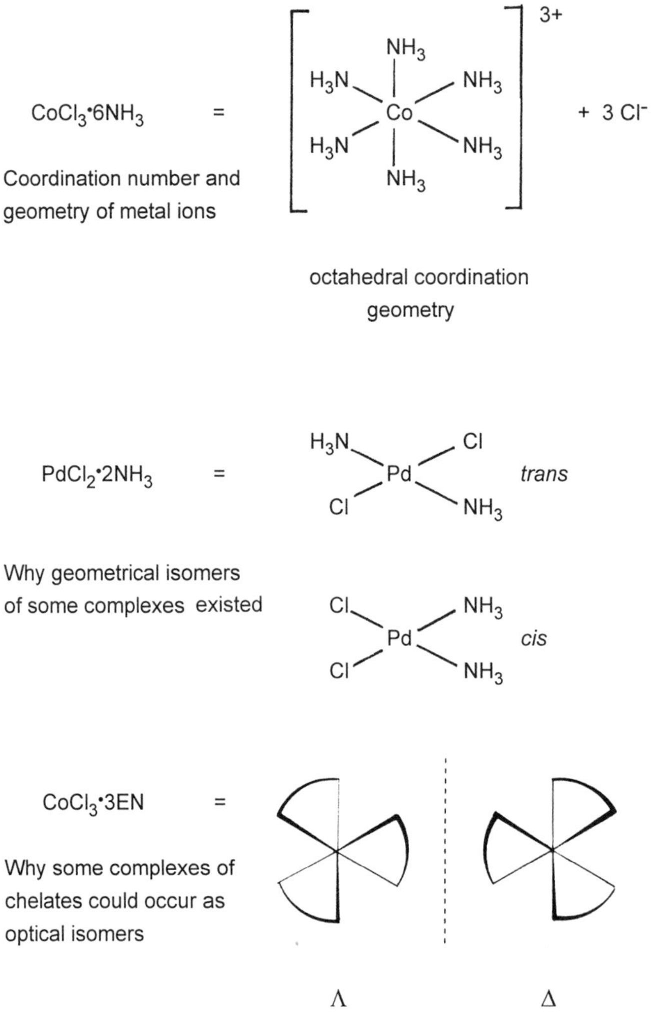

$CoCl_3 \cdot 6NH_3$ = [octahedral Co complex with 6 NH_3] $3+$ + 3 Cl$^-$

Coordination number and geometry of metal ions

octahedral coordination geometry

$PdCl_2 \cdot 2NH_3$ = [Pd complex] *trans*

Why geometrical isomers of some complexes existed

[Pd complex] *cis*

$CoCl_3 \cdot 3EN$ = [optical isomers]

Why some complexes of chelates could occur as optical isomers

Λ   Δ

Figure 1.1. A summary of facts explained by Werner's theory regarding the coordination number and geometry of metal ions.

There was considerable debate[7] over the cause of the chelate effect, as is discussed in Chapter 3. The chelate effect is important to one of the themes of this book, namely "pre-organization". The more nearly the donor atoms of the free ligand are arranged spatially as required for complexation, the more "preorganized" is the ligand, a term coined by Cram.[8] Chelates are generally more preorganized than unidentate ligands, which was recognized by early workers,[5] although they did not use this terminology.

A further achievement was the recognition of patterns present in preferences of individual metal ions for ligands with either more or less electronegative donor atoms[9]. These observations were formulated as the A and B type classification by Schwarzenbach,[4] the a and b type by Ahrland, Chatt, and Davies,[10] and the Hard and Soft Acids and Bases Principle (HSAB) by Pearson.[11] The patterns present are illustrated by the formation constants with halide ions of Ag(I), Pb(II), and In(III) in Table 1.2.

Table 1.2. Formation constants with halide ions of a representative metal ion from each group in the classifications of Schwarzenbach[4], Ahrland, Chatt and Davies[10], and Pearson.[11]

Log K_1	F⁻	Cl⁻	Br⁻	I⁻	Classification
Ag^+	0.4	3.31	4.68	6.58	B, b, or soft
Pb^{2+}	1.26	0.90	1.06	1.30	intermediate
In^{3+}	4.6	2.32	2.04	1.64	A, a, or hard

Data from Ref. 6.

Metal ions classified as soft, B, or b type, prefer to complex with less electronegative donor atoms such as iodide, whereas metal ions classified as hard, A, or a, prefer to complex with more electronegative donor atoms such as fluoride. Metal ions classified as intermediate show no strong preferences.

Linear Free Energy Relationships (LFER) came to be used to analyze formation constants, which further highlighted patterns and regularities in complex-formation in aqueous solution[12]. In Figure 1.2 is shown an example of a LFER between the log K_1(ML) values of ligands with Ni(II) and with Cu(II). The mechanisms of ligand substitution[13] and electron transfer[14] were also elucidated at this time.

There followed a lean period in the study of complexes in aqueous solution. The metal-carbon bond[15] and bonds between very soft ligands and low-valent metal ions, and metal-metal bonding[16] occupied the attention of the greater number of Inorganic Chemists. Recently there has been a resurgence in coordination chemistry, brought about in part by synthesis of new types of more highly preorganized ligands, such as the crown ethers,[17] cryptands,[18] sepulchrates,[19] and spherands.[8] The resurgence has also been driven by the use of complexes of metal ions in biomedical applications, as detailed in Chapter 5.

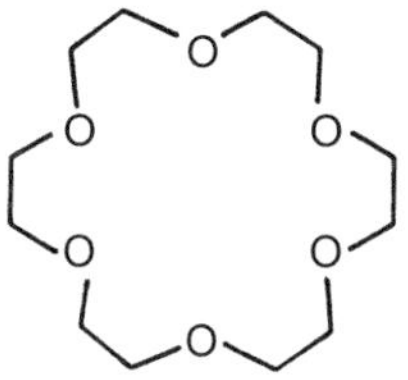

Crown ether

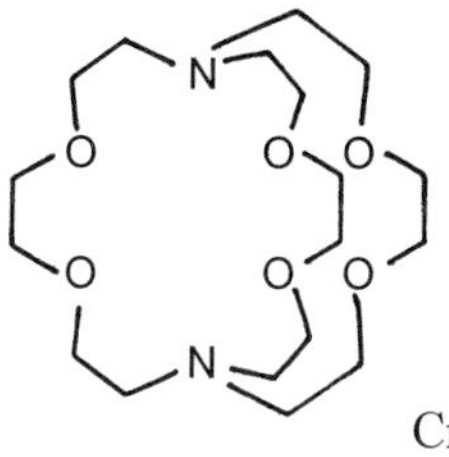

Cryptand

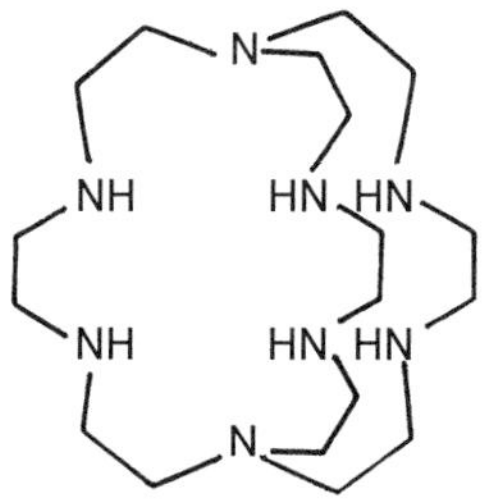

Sepulchrate

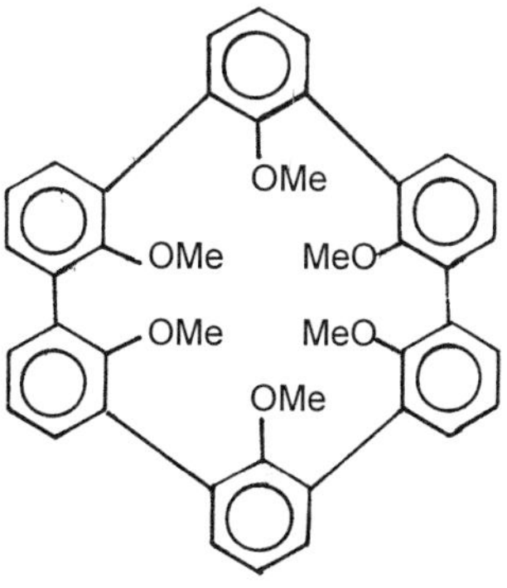

Spherand (Me = methyl)

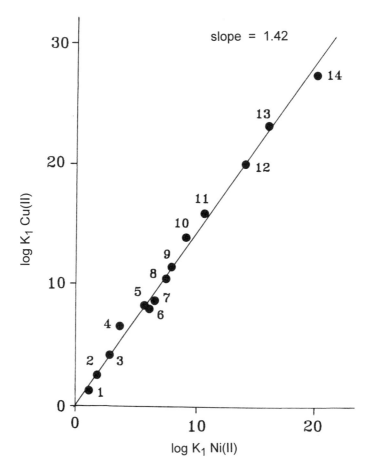

Figure 1.2. A Linear Free Energy Relationship (LFER.) of formation constants ($\log K_1$) of complexes of Cu(II) *vs* $\log K_1$ for the analogous complexes of Ni(II). Note that only a selection of ligands has been shown, because the number of ligands for which formation constants have been determined for Cu(II) and Ni(II) is now too large[6] to make display of such a correlation containing all of them practical. Ligands containing more than four donor atoms have been omitted, as in these the Jahn-Teller distortion of the Cu(II) leads to breakdown of the correlation. The slope of the correlation is 1.42. The ligands are 1, F^-; 2, pyridine; 3, NH_3; 4, OH^-; 5, acetohydroxamate; 6, acetylacetonate; 7, N,N,N',N'-tetrakis(2-hydroxyethyl)ethylenediamine; 8, ethylenediamine; 9, salicylate; 10, catecholate; 11, DIEN (1,4,7-triazaheptane); 12, TRIEN (1,4,7,10-tetraazadecane); 13, 2,3,2-tet (1,4,8,11-tetraazaundecane); 14, cyclam (1,4,8,11-tetraazacyclotetradecane). Data from Ref. 6.

Extensions of the chelate effect, such as the macrocyclic effect[20] and cryptate effect,[18] were recognized. Study of the new ligands shows that preorganization has a great role yet to play[21] in designing remarkable ligands and complexes. Study of coordination complexes in aqueous solution is being spurred on by the importance of this chemistry to biochemistry, medicine, industry, and the environment. In addition, as will be emphasized in this work, studies in solution are now carried out as part of an overall attack involving other techniques such as NMR or X-ray crystallography. These techniques greatly increase the level of understanding of factors controlling complex formation processes, and lead to improved ability to design ligands with specified complexing properties. A further important development has been the advent of computer programs[22-25] which have produced a shift of emphasis in studying complex formation equilibria. Originally the problem of how to determine what species are present in solution most exercised the chemist's mind, and books on the study of equilibria[26,27] were devoted almost entirely to this question. Modern computers and software have greatly simplified this problem. That is not to say that computers can now take over the role of providing insight in deciding on the correct model for the species in solution. Understanding and good chemical intuition are still essential. However, the interest in formation constants now lies in the constants themselves, rather than on how to determine them. Formation constants are essential as information needed for ligand design, and in understanding complex equilibria in living systems and the environment.

1.2   Metal Ions in Solution, and Metal Complexes

Metal ions in solution are usually indicated as M^{n+} ions, as though they were bare ions suspended in a structureless medium, whereas they are complexes of water, similar to complexes of other ligands. Water is both a good Lewis base, coordinating as a ligand via its oxygen to the metal ion, and a good Lewis acid, coordinating to anions via its hydrogen atoms. One may compute[28,29] heats of solvation of cations and anions in passing from the gas-phase to water, and (Table 1.3), these range from tens to hundreds of kcal.mol^{-1}. Such large energies of hydration are clearly necessary for solubility in water, to overcome the large lattice energies of ionic solids.

Table 1.3. Heats of solvation of some metal ions in aqueous solution.^{a}

Ion	- ΔG	- ΔH	- ΔS
H^+	260.5	269.8	31.3
Li^+	123.5	133.5	33.7
Na^+	98.3	106.1	35.3
K^+	80.8	86.1	17.7
Rb^+	76.6	81.0	14.8
Cs^+	71.0	75.2	14.1
Cu^+	136.2	151.1	42.9
Ag^+	114.5	122.7	27.6
Tl^+	82.0	87.0	16.7
Be^{2+}	582.3	612.6	101.6
Mg^{2+}	455.5	477.6	74.3
Ca^{2+}	380.8	398.8	60.8
Sr^{2+}	345.9	363.5	59.2
Ba^{2+}	315.1	329.5	48.5
Ra^{2+}	306.0	319.6	45.6
Cr^{2+}	444.8	460.3	53.0
Mn^{2+}	437.8	459.2	72.1
Fe^{2+}	456.4	480.2	79.8

Table 1.3. Continued

Ion	- ΔG	- ΔH	- ΔS
Co^{2+}	479.5	503.3	80
Ni^{2+}	494.2	518.8	82.4
Cu^{2+}	498.7	519.7	73.9
Zn^{2+}	484.6	506.8	74.5
Cd^{2+}	430.5	449.8	65.2
Hg^{2+}	436.3	453.7	58.4
Sn^{2+}	371.4	389.5	60.7
Pb^{2+}	357.8	371.9	47.4
Al^{3+}	1103.3	1141.0	126.6
Sc^{3+}	929.3	962.7	112.5
Y^{3+}	863.4	896.6	111.3
La^{3+}	791.1	820.2	97.5
Gd^{3+}	847.9	880.4	109.1
Lu^{3+}	891.8	925.6	113.3
Ga^{3+}	1106.0	1147.0	137.9
In^{3+}	973.2	1009.3	117.9
Tl^{3+}	975.9	1028.0	174.7
Fe^{3+}	1035.5	1073.4	127.5
Cr^{3+}	1037	1099.9	143.9
F^-	103.8	113.3	31.8
Cl^-	75.8	81.3	18.2
Br^-	72.5	77.9	14.5
I^-	61.4	64.1	9.0
S^{2-}	303.6	309.8	20.5
OH^-	90.6	101.2	35.6

[a] Data from Friedman, H. L.; Krishnan, C. V. in *Water: A Comprehensive Treatise*, Franks, F., Ed.; Plenum Press: New York, 1973, Vol.3, p.55. Units for ΔG and ΔH are kcal.mol^{-1}, and units for ΔS are cal. deg^{-1}.mol^{-1}. The energies refer to the process (at 298.16 K) M^{n+} (g) $\rightarrow$ M^{n+} (aq), and are based on the convention that for the reaction H^+ (g) $\rightarrow$ H^+ (aq) - ΔG is 260.5 kcal.mol^{-1}, and -ΔS is 31.3 cal. deg^{-1}.mol^{-1}.

The number of water molecules bound directly to the metal ion, its coordination number, is similar to the number of donor atoms found for the metal ion in its complexes in the solid state. In solution there is an inner sphere of water molecules bound directly to the metal ion, and an outer-sphere of more loosely held water molecules. Complex formation involves replacement of water molecules coordinated to the metal ion by ligands, which may replace either water molecules from the inner-sphere to form inner-sphere complexes, or water molecules from the outer-sphere to form outer-sphere complexes.

1.2.1 Metal ions in aqueous solution

Coordination numbers of metal ions in aqueous solution have been determined[28,30,31] by uv-visible spectroscopy, NMR spectrometry, X-ray diffraction, neutron diffraction, and isotope dilution. These methods are not always in good agreement, but roughly conform to the idea that metal ions in solution have the same coordination number and geometry as aquo ions in the solid state. The structures, as determined by X-ray diffraction,[32,33] of the hexaaquo ion of Ni(II) and the nonaaquo ion of Ho(III), which appear to be present in aqueous solution, are seen in Figure 1.3. Although different experimental techniques do not always give consistent coordination numbers in solution, the best values (Table 1.4) are similar to those found in the solid state.

Table 1.4. Approximate numbers of water molecules in the primary coordination sphere (N_p), the approximate total numbers of waters in the primary and secondary coordination spheres (N_T), and the M-O distance (R), for a selection of metal ions in aqueous solution.[a] Also shown are metal-oxygen distances from crystal hydrates (R_c).

Ion	N_p	N_T	R	R_c
Li^+	6	13.22	2.08	1.93-1.98
Na^+	6	7-13	2.35	2.35-2.52
K^+	6-8		2.79	2.67-3.22
Rb^+	8			
Cs^+	8		2.95	
H^+	4			
Be^{2+}	4		1.67	
Mg^{2+}	6	12-14	2.09	2.01-2.14
Ca^{2+}	6-10	8-12	2.42	2.30-2.49
Sr^{2+}	8		2.64	
Ba^{2+}	9.5	3-5	2.6	
Mn^{2+}	6		2.19	2.00-2.18
Fe^{2+}	6	10-13	2.11	1.99-2.08
Co^{2+}	6		2.10	1.93-2.12
Ni^{2+}	6	21	2.06	2.02-2.11
Cu^{2+}	6		1.96	1.93-2.00
Zn^{2+}	6	10-13	2.40 / 1.09	2.08-2.14
Cd^{2+}	6		2.30	2.24-2.31
Hg^{2+}	6		2.42	2.24-2.34
Al^{3+}	6	13	1.88	1.87
Y^{3+}	8		2.36	
La^{3+}	8-9		2.51	
Nd^{3+}	8.6		2.49	2.48-2.60
Lu^{3+}	8		2.34	
Cr^{3+}	6	17	1.96	2.02
Fe^{3+}	6		2.03	2.09-2.20
Rh^{3+}	6		2.04	2.09-2.10

Table 1.4. Continued

Ion	N_p	N_T	R	R_c
In^{3+}	6		2.15	2.23
Tl^{3+}	5 ± 1		2.23	
Th^{4+}	9		2.53	
U^{4+}			2.42	2.36
F^-	2-7	2.63		
Cl^-	6-8.5		3.18	
Br^-	6		3.37	
I^-	6-7		3.64	

^{a} The distances are in Å. The number of water molecules in the inner coordination sphere are largely from Ref. 30 and 31. The ion-water oxygen distances are mean values given in Ref. 31. The number of water molecules (inner plus outer sphere) are from Ref. 28. The ion-water oxygen distances from solid state structures are from Ref. 28.

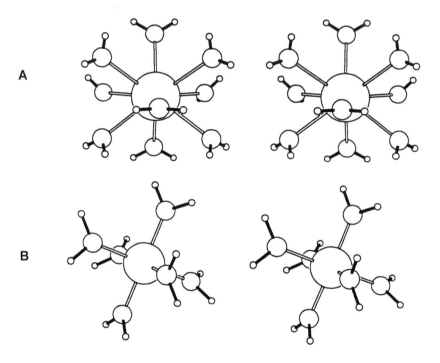

Figure 1.3. Stereoviews of the aquo ions A, $[Ho(H_2O)_9]^{3+}$ and B, $[Ni(H_2O)_6]^{2+}$. (Redrawn using atomic coordinates in Refs. 32 and 33).

A metal ion such as Ni(II) has (Figure 1.4) the six waters in its inner coordination sphere expected for high-spin Ni(II). In addition, solvent surrounding the six directly coordinated waters has a more ordered structure compared to bulk solvent. Water molecules in this outer sphere are held to waters in the inner sphere by hydrogen bonding. The number of these is[30] less certain, but about fifteen water molecules are present in the outer sphere of the Ni(II) cation. What is usually thought of as complex-formation is formation of inner sphere complexes, where ligands replace water molecules from the inner coordination sphere, and form bonds directly to the metal ion. In outer-sphere complexes the ligand is situated in the outer-sphere, forming no bonds directly to the metal ion. In a typical outer-sphere complex, a ligand such as sulfate is held to the metal ion by a combination of hydrogen bonding and electrostatic attraction. Outer sphere complexes are important[34] in the kinetics of complex-formation, since all complex-formation processes appear to involve initial formation of an outer-sphere complex, followed by entry of the ligand into the inner sphere. The formation constants of outer-sphere complexes may be predicted with reasonable accuracy using equations such as the Fuoss equation,[35] which treats complexes in terms of a simple electrostatic model.

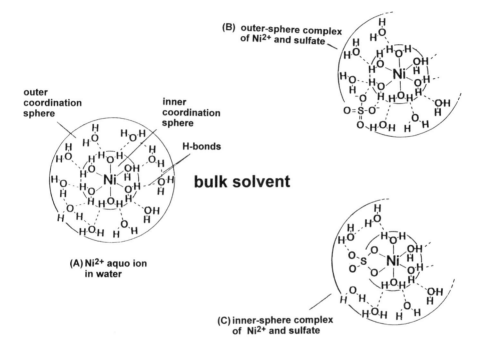

Figure 1.4. Diagrammatic representation of a metal ion, Ni(II), in water, showing its inner and outer sphere of coordinated water (A), the formation of an outer-sphere complex with sulfate (B), and an inner-sphere complex with sulfate where the sulfate ligand is now directly bonded to the Ni^{2+} ion (C).

1.2.2 Trends in the periodic table

It is important to develop an intuitive feeling for the ligand preferences and rates of complex formation of individual metal ions. The donor atoms that metal ions prefer can be summarized in the Hard and Soft Acid and Base (HSAB) ideas of Pearson[11] discussed in section 1.1.1 above. Soft metal ions prefer ligands with soft donor atoms, and hard metal ions prefer ligands with hard donor atoms, as shown in Table 1.2. The distribution of hard and soft metal ions in the periodic table is seen in Figure 1.5. Soft metal ions

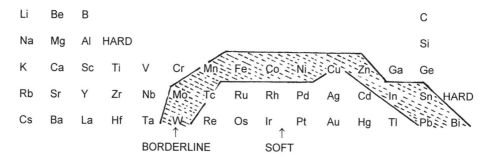

Figure 1.5. The distribution of hard, soft, and borderline metal ions in the periodic Table, according to the classification of Pearson.[11]

occur in a triangle in the center of the periodic Table, surrounded by metal ions of intermediate hardness, getting harder as one moves further from the soft metal ions. The most electronegative elements form the hardest donor atoms, as in F^- or H_2O, and less electronegative elements such as C, P, or I are soft donors, as shown in Figure 1.6. The

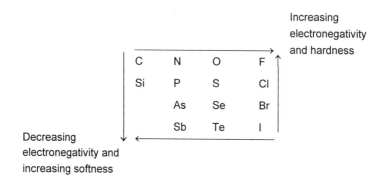

Figure 1.6. The distribution of hardness and softness in the Periodic Table as a function of the donor atom of the ligand, according to the Hard and Soft Acids and Bases classification of Pearson.[11]

HSAB approach provides an indication of the type of ligand a metal ion is likely to prefer, but (section 2.3) a more complete picture can be obtained by more detailed approaches. The rates at which ligands substitute for water molecules or other ligands on a metal ion fall into a typical range for each metal ion. The rates at which water in the inner coordination sphere of the metal ion exchanges with water in the bulk solvent is typical of the rates at which the metal ion forms complexes with other ligands, and may thus be taken as[34] representative of reaction rates for that metal ion, as summarized in Figure 1.7. When one ligand, such as ammonia, takes the place of another already on the metal ion, such as water, in general two possible mechanisms may be found.[34] In both mechanisms the incoming ligand must first form an outer-sphere complex so that it is correctly positioned to move into the inner sphere. In the first type of mechanism, (S_N1) the metal ion must lose a ligand from its coordination sphere, dropping its coordination number by one, to create a gap which is then filled by the incoming ligand, as seen in Figure 1.8. In the second type of mechanism (S_N2, Figure 1.8) the incoming ligand moves

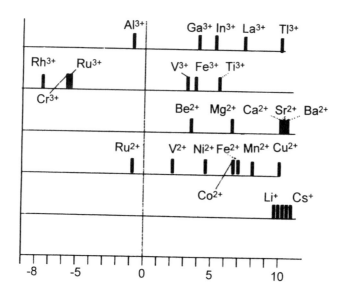

Figure 1.7. Logarithms of characteristic rate constants (s^{-1}) for substitution of inner-sphere water molecules on various metal ions. Redrawn after Ref. 34.

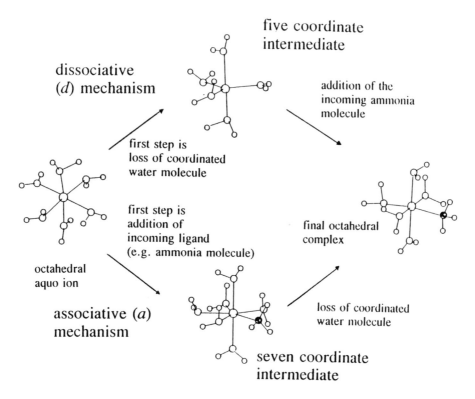

Figure 1.8. The dissociative mechanism (A) and the associative mechanism (B) of ligand substitution on an octahedral complex.

into the coordination sphere first, raising the coordination number of the metal ion, and then a departing ligand is lost. In general, ability to change coordination number from that usually favored by the metal ion is related to how tightly ligands are bound to the metal ion. The small Be(II) ion with its strong covalent M-L bonds strongly favors tetrahedral four coordination, being reluctant to form either three or five coordinate complexes, and so typically reacts only very slowly with all ligands. A large metal ion such as Sr(II) has very ionic M-L bonds, and a very variable coordination number. It thus changes coordination number with ease, and so its ligand substitution reactions are very rapid. It is possible to draw up three rules which give an indication of whether a metal ion will undergo rapid or slow ligand substitution reactions. These rules relate to factors that cause stronger M-L bonds and more strongly defined coordination geometries:

1. For metal ions of the same charge, substitution rates with the same ligand will increase with increasing metal ion size. Thus, typical reaction rates (Figure 1.7) increase Be(II) < Mg(II) < Ca(II) < Sr(II) < Ba(II) or Al(III) < Ga(III) < In(III) < Tl(III).

2. For metal ions of about the same size, rates of reaction will increase with decreasing charge on the cation, e.g. (Figure 1.7) reaction rates are typically In(III) < Mg(II) < Li(I)

3. Electrons in d-orbitals in arrangements that lead to ligand field stabilization energy (LFSE) cause slower rates of ligand substitution in direct relation to the extent of the LFSE.

The Cr(III) aquo ion has three electrons in the t_{2g} orbital of the d- shell, leading to considerable LFSE, and so undergoes ligand substitution reactions more slowly (Figure 1.7) than the Ga(III) ion, even though Ga(III) and Cr(III) are about the same size. Rule 3 is generally able to override rules 1 and 2. Thus, the M-L bond lengths for the trivalent ions of group VIII tend to increase Co(III) < Rh(III) < Ir(III), so that rule 1 would suggest that Co(III) would react the most slowly out of this group. However, LFSE increases Co(III) < Rh(III) < Ir(III), and the typical order of reaction rate for ligand substitution is Co(III) > Rh(III) > Ir(III). Rule 3 overcomes rule 2, in that Ru(II) reacts more slowly than most trivalent non-Transition metal ions. An important aspect of rule 3 is that when a metal ion changes spin state, there may be a large change in lability. Thus, high-spin complexes of Fe(II) are labile, but the low-spin complexes are classified as inert.

1.3 Steric Strain in Complex Formation

The formation of complexes in aqueous solution is controlled by a combination of steric and electronic effects. An attempt to unravel the role of electronic effects in the formation of the metal to ligand bond is made in Chapter 2. To a large extent, Molecular Mechanics (MM) are used in this book to attempt to understand steric effects. The subject of MM is discussed only briefly here, and the interested reader should consult references 36 and 37. In brief, bonds and angles are considered to have ideal values, and distortion away from these values gives rise to strain energy, which can be calculated using a simple Hooke's Law expression. The energy produced by bond length distortion, U_B, is calculated from the Hooke's Law expression:

$$U_B \quad = \quad 1/2 \, K_B(r - r^o)^2 \qquad\qquad\qquad 1.1$$

In equation 1.1, K_B is the force constant for restoring the bond to its ideal length, r^o, and r is the length to which the bond has been distorted. A similar expression applies to the distortion of bond angles away from the ideal value, θ^o, to the observed bond angle θ. In equation 1.2, K_θ is the force constant for restoring the bond angle to its ideal value.

$$U_\theta \quad = \quad 1/2\ K_\theta(\theta - \theta^o)^2 \qquad\qquad 1.2$$

Another important contribution in MM calculations is the torsional contribution. This models repulsion between the electrons in neighboring chemical bonds to atoms which are bonded to each other. An example of this is the C-H bonds in ethane, for which torsional contributions are the bulk of the barrier to rotation about the C-C bond in the ethane molecule. A fourth contribution to steric strain, which is probably the most important to consider, is the van der Waals repulsion between atoms that are not directly bonded to each other. The van der Waals (U_{NB}) energy is given by equation 1.3.

$$U_{NB} \quad = \quad A.e^{(-B.r)} - C/r^6 \qquad\qquad 1.3$$

In equation 1.3, r is the internuclear separation of the two interacting non-bonded atoms, and A, B, and C are empirical constants pertaining to the non-bonded interaction of the two particular atoms. At short internuclear separations, strong repulsive energies are produced by equation 1.3, while at distances close to the sum of the van der Waals radii, weak attractive forces are produced, with attractive energies of only about 0.1 kcal.mol^{-1}.

Other types of potential can also be included in the MM model. Thus, hydrogen bonding can be modeled electrostatically, and very ionic bonds such as those between alkali metal ions and oxygen donor atoms can be modeled electrostatically[38] rather than by use of equation 1.3. More sophisticated programs such as MM2[39] can also use a Huckel Molecular Orbital calculation to model bond length variation in aromatic systems.

Essentially, the MM calculation involves providing the program with trial coordinates for the molecule, which the program then perturbs to find the energy minimum. For small molecules this can be quite straightforward, although for large molecules such as proteins the enormous number of possible trial structures presents a problem in finding the global energy minimum which has at the time of writing not yet been overcome. Modern MM programs can be obtained which are very user friendly, and allow for easy use, such as SYBYL or ALCHEMY which can be obtained from TRIPOS associates in St. Louis, Mo. Such programs allow one to build up trial structures selecting atoms from a menu of atoms in the program, and set up the MM calculation themselves. Atoms not present in the library of the program can be easily added, although the user must then provide the appropriate parameters which apply to the added atoms.

The MM calculation gives an energy minimized structure in which the bond lengths and angles generally agree[37] to within 0.1 Å (lengths) and 1.5^o (angles) of the crystallographically observed structure. One also obtains the total strain energy of the molecule, broken down into types of contribution. These strain energies are an important indicator of how well a ligand is sterically suited for complexing a particular metal ion, and the specific contributions allow one to determine the origin of any particularly unfavorable steric effects in complex formation. Examples of this will be presented in the discussion of factors controlling complex formation throughout this book.

References

1. G. B. Kauffman, *Coord. Chem. Rev.*, **1972**, *9*, 339; **1973**, *11*, 161.
2. A. E. Martell and M. Calvin, *Chemistry of the Metal Chelate Compounds*, Prentice-Hall, New York, 1952.
3. J. Bjerrum, *Metal Ammine Formation in Aqueous Solution*, Thesis 1941, reprinted P. Haase & Son, Copenhagen, 1957.
4. G. Schwarzenbach, *Adv. Inorg. Radiochem.*, **1961**, *3*, 257.
5. G. Schwarzenbach, *Helv. Chim. Acta*, **1952**, *35*, 2344.
6. A. E. Martell and R. M. Smith, *Critical Stability Constants*, Plenum Press, New York, 1974-1989, Vols 1-6.
7. A. E. Martell, in *Essays in Coordination Chemistry*, Eds. W. Schneider, G. Anderegg, and R. Gut, Berkhauser Verlag, Basel, 1964, pp 52-64.
8. D. J. Cram, T. Kaneda, R. C. Helgeson, S. B. Brown, C. B. Knobler, E. Maverick, and K. N. Trueblood, *J. Am. Chem. Soc.*, **1985**, *107*, 3645.
9. J. O. Edwards, *J. Am. Chem. Soc.*, **1954**, *76*, 1540.
10. S. Ahrland, J. Chatt, and N. R. Davies, *Q. Rev., Chem. Soc.*, **1958**, *12* 265.
11. R. G. Pearson, *Chem. Br.*, **1967**, *3*, 103.
12. F. J. C. Rossotti, in *Modern Coordination Chemistry*, Eds. J. Lewis and R. G. Wilkins, Interscience, London, 1960, pp. 1-77.
13. See for example, R. G. Pearson and F. Basolo, *Mechanisms of Inorganic Reactions*, John Wiley and Sons, New York, 1967.
14. a) H. Taube and E. S. Gould, *Acc. Chem. Res.*, **1969**, *2*, 321. b) A. McAuley, *Coord. Chem. Rev.*, **1970**, *5*, 245.
15. F. R. Hartley and S. Patai, Eds., *The Chemistry of the Metal-Carbon Bond.*, Vols. 1-3, Wiley, New York, 1985.
16. a) F. A. Cotton, *Acc. Chem. Res.*, **1978**, *11*, 226. b) F. A. Cotton and R. A. Walton, *Multiple Bonds between Metal Atoms*, Wiley, New York, 1982.
17. C. J. Pedersen, *J. Am. Chem. Soc.*, **1967**, *89*, 2459.
18. J. M. Lehn, *Acc. Chem. Res.*, **1978**, *11*, 49.
19. A. M. Sargeson, *Pure Appl. Chem.*, **1984**, *56*, 1603.
20. D. K. Cabbiness and D. W. Margerum, *J. Am. Chem. Soc.*, **1969**, *91*, 6540.
21. R. D. Hancock and A. E. Martell, *Comments Inorg. Chem.*, **1988**, *6*, 237-284.
22. P. Brauner, L. G. Sillen, and R. Whiteker, *Arkiv. Kemi*, **1969**, *31*, 365.
23. A. Sabatini, A. Vacca, and P. Gans, *Talanta*, **1974**, *21*, 53.
24. D. J. Leggett (Ed.), *Computational Methods for the Determination of Formation Constants*, Pienum Press, New York, 1985.
25. R. J. Motekaitis and A. E. Martell, *The Determination and Use of Stability Constants*, VCH Publishers, New York, 1988.
26. F. J. C. Rossotti and H. Rossotti, *The Determination of Stability Constants*, McGraw Hill, New York, 1961.
27. M. T. Beck, *Chemistry of Complex Equilibria*, Van Nostrand London, 1970.
28. J. Burgess, *Metal Ions in Solution*, Ellis Horwood, Chichester, 1979.
29. F. Franks, *Water: A Comprehensive Treatise*, Plenum Press, New York, 1973, pp. 1-113.
30. J. P. Hunt and H. L. Friedman, *Prog. Inorg. Chem.*, **1983**, *30*, 359.
31. Y. Marcus, *Chem. Rev.*, **1988**, *88*, 1475.
32. F. Bigoli, A. Braibanti, A. Tiripicchio, and M. Tiripicchio-Camellini, *Acta Crystallogr., Section B.*, **1971**, *B27*, 1427.
33. C. R. Hubbard, C. O. Quicksall, and R. A. Jacobsen, *Acta Cryst, Sect. B*, **1974**, *B30*, 2613.
34. M. Eigen, *Pure Appl. Chem.*, **1963**, *6*, 105.
35. R. M. Fuoss, *J. Am. Chem. Soc.*, **1958**, *80*, 5059.
36. G. R. Brubaker and D. W. Johnson, *Coord. Chem. Rev.*, **1984**, *53*, 14.
37. R. D. Hancock, *Prog. Inorg. Chem.*, **1989**, *36*, 187.
38. G. Wipff, P. Wiener, and P. A. Kollman, *J. Am. Chem. Soc.*, **1982**, *104*, 3249.
39. a) N. L. Allinger, *Molecular Mechanics 1987 Force Field*, Quantum Chemistry Program Exchange, c/o Department of Chemistry, Indiana University, Bloomington, Indiana. b) N. L. Allinger, *J. Am. Chem. Soc.*, **1977**, *99*, 8127.

CHAPTER 2

FACTORS GOVERNING THE FORMATION OF COMPLEXES WITH UNIDENTATE LIGANDS IN AQUEOUS SOLUTION. SOME GENERAL CONSIDERATIONS

When a unidentate ligand such as NH_3 or SCN^- replaces a water molecule on a metal ion to form a complex in aqueous solution, several factors contribute to the final free energy of complex formation. This section summarizes these factors as they are presently understood.

2.1   The Role Of The Solvent

The solvent in complex-formation equilibria is usually water. Water is both an excellent Lewis acid and Lewis base, and both the metal ion and ligand in a complex-formation reaction are extensively solvated. As a starting point for understanding complex formation in water, acid-base reactions in the gas-phase, where no solvent is present, are examined.

2.1.1   Protonation reactions in the gas-phase

In the last twenty five years a large amount of data on energetics of protonation reactions in the gas-phase[1-10] has accumulated. These reactions have been studied, following kinetics and equilibria, using mass and ion cyclotron resonance spectrometry. The first results on gas phase protonation equilibria[1] were for saturated amines. These results caused a stir, because they contradicted many intuitive ideas derived previously from studies of protonation constants in aqueous solution. The enthalpies (and also free energies*) of protonation of amines in the gas phase (Figure 2.1) increase along the series $NH_3 < NH_2CH_3 < NH(CH_3)_2 < N(CH_3)_3$, in contrast to the protonation constants (pK), which in water are as follows:[11] NH_3, 9.22; NH_2CH_3, 10.6; $NH(CH_3)_2$, 10.6; $N(CH_3)_3$,

---

*   In the gas-phase there are not large and differing numbers of water molecules released in acid-base reactions, which in water may cause entropy to be the major contributor to the free energy of reaction. In the gas-phase enthalpy is the driving force of reactions, and entropy contributions to the free energy are small and differ little from one reaction to another, so that discussion of trends based on enthalpies of reaction in the gas-phase would also apply to results on free energy changes in the gas-phase.

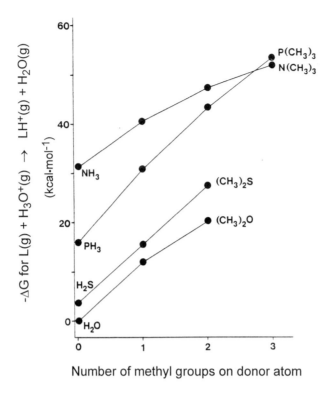

Figure 2.1. Variation of free energy of protonation in the gas phase for various series of bases as the number of methyl groups attached to the donor atom is varied. All free energies are for the removal by the base in the gas phase of a proton from the H_3O^+ ion as indicated. Reproduced with permission from Ref. 12.

9.9. The pK values in water give the impression that the basicity of nitrogen is little affected by added alkyl groups.

The differences in energetics of protonation of amines in the gas-phase[1] and water are explicable in terms of two factors. The first factor is the role of the solvent in dispersing charge[13] from the cation to the solvent by hydrogen-bonding. Dispersal of charge via hydrogen bonds stabilizes the cation, and as hydrogens on the ammonium cation are replaced by methyl groups, so the number of hydrogen bonding sites is reduced (Figure 2.2(a)). Thus, the basicity of trimethylamine in water is only slightly higher than that of ammonia, in contrast to the large difference anticipated from gas-phase basicities. This arises from the lower ability of the trimethylammonium cation to stabilize itself by hydrogen bonding, having fewer N-H hydrogens than the ammonium cation.

The second factor causing lowered basicity in trimethylamine is steric hindrance to solvation (Figure 2.2(b)). The proton on the trimethylammonium cation is unsolvated in the gas-phase, so that there is only a small steric clash between the coordinated proton and the methyl groups. In the gas phase the inductive (electron releasing) effects of the methyl groups are dominant, and the order of basicity is $NH_3 \ll NH_2CH_3 \ll NH(CH_3)_2 \ll N(CH_3)_3$. In general, then, addition of methyl or other alkyl groups to nitrogen donors has little effect on their aqueous phase basicity, and, except in special cases, inductive effects are not apparent. Figure 2.1 shows gas-phase basicity of bases with other donor atoms as hydrogens are replaced by methyl groups. All of the ligand series show the same response to added methyl groups, with strong increases in proton basicity. The curvature in the relationships for the R_3N and R_2O series with increasing methylation is attributed[2,14] to steric hindrance to the coordinated Lewis acid by methyl groups even in the gas-phase. The more nearly linear relationships in Figure 2.1 for the S and P donor series relate to the lower steric hindrance between the coordinated proton and the methyl groups because the longer P-C and S-C bonds, compared to N-C and O-C bonds, cause the methyl groups to be held further away from the proton in PR_3 and SR_2 bases.

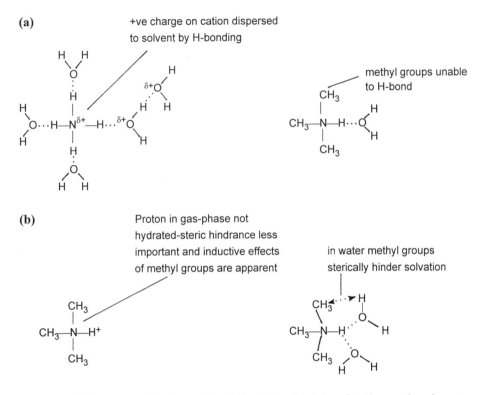

Figure 2.2. Factors contributing to the relatively low basicity of tertiary amines in water: **(a)** dispersal of charge to the solvent by hydrogen bonding in ammonia, which the methyl groups on trimethylamine cannot do, and **(b)** steric hindrance from the methyl groups, which is not so severe to the proton in the gas phase, but causes considerable steric hindrance to the solvated proton on trimethylamine.

The proton is classified as "hard" in the HSAB classification (See section 1.1), so that it may come as a surprise that R_2S bases have a higher affinity for the proton than do R_2O bases in the gas-phase, or that $P(CH_3)_3$ is a stronger proton base in the gas phase than is $N(CH_3)_3$. This effect has been analyzed,[9] and it is found that the fact that R_2S is a poor base in water derives from the inability of the hydrogens on S to hydrogen bond to the solvent. The same effect is true for the phosphines. Because hydrogen bonding is less important in stabilizing the trimethylammonium cation, one finds that the basicity is very similar to that of the trimethylphosphonium ion. However, the ammonium ion is greatly stabilized relative to the phosphonium ion by the ability of the ammonium ion to hydrogen bond:

BASES:	XH_3	$XH_2(CH_3)$	$XH(CH_3)_2$	$X(CH_3)_3$
pK, X = N	9.2	10.6	10.6	9.9
pK, X = P	-14	0.0	3.9	8.7

The protons on the phosphonium cation are only weakly hydrogen bonded to the solvent, even though the cation carries a positive charge. One assumes that P-H and S-H hydrogens on phosphonium and sulfonium cations are, because of the greater covalence in these bonds than in the N-H or O-H bond, more like C-H bonds, and therefore less able to form H-bonds. The effect of hydrogen bonding on proton basicity has been demonstrated by Taft[9] for $(CH_3)_2S$ relative to $(CH_3)_2O$. In the gas-phase the proton exchange reaction below proceeds to the right with an enthalpy of -8 kcal.mol^{-1}, indicating that the thioether is a stronger proton base than the oxygen ether, but with a single water molecule hydrogen-bonded to the proton, the enthalpy of the reaction is +6.0 kcal.mol^{-1}. This remarkable change in relative basicity reflects the poor ability of the $(CH_3)_2SH^+$ cation to stabilize itself by hydrogen-bonding.

$$(CH_3)_2S + (CH_3)_2OH^+ \longrightarrow (CH_3)_2SH^+ + (CH_3)_2O \qquad\qquad 2.1$$
$$\Delta H = -8.0 \text{ kcal.mol}^{-1}$$

$$(CH_3)_2S + (CH_3)_2OH^+\text{---}OH_2 \longrightarrow (CH_3)_2SH^+\text{---}OH_2 + (CH_3)_2O \qquad 2.2$$
$$\uparrow \qquad\qquad\qquad\qquad \uparrow$$

H-bond   H-bond

$$\Delta H = +6.0 \text{ kcal.mol}^{-1}$$

Another important series of amines is the series $MeNH_2$, $EtNH_2$, $iso\text{-}PrNH_2$, and $t\text{-}BuNH_2$ (Me = methyl, Et = ethyl, iso-Pr = isopropyl, and t-Bu = t-butyl), where methyl groups are added not to the nitrogen but to the α carbon atom. Here the protonation constants in water are almost constant at 10.6,[11] which also suggests that the basicity of nitrogen is unaffected by substitution of methyl groups onto the α carbon atom. Again, in the gas-phase, there is a strong increase in basicity along this series,[9] which is attributed to inductive and polarizability effects. In this series, the number of protons on the RNH_3^+ capable of hydrogen bonding to the solvent is a constant three. Therefore only steric hindrance to solvation is involved in preventing the increase in inductive effects as R is changed along the series from methyl through t-butyl from manifesting itself as an increase in pK$_a$. If, however, the R group is changed from methyl through t-butyl along the series $RN(CH_3)_2$, the inductive effects are observable:[9]

ligand $RN(CH_3)_2$ R =	Me	Et	iso-Pr	t-butyl
pK in water	9.9	10.3	10.5	10.7

The interpretation here[9] is that the extent of solvation of the proton in the dimethylalkylammonium cation is so diminished by steric hindrance that the steric hindrance to solvation produced by the alkyl groups along the series Me, Et, iso-Pr, t-Bu is less important than the increased inductive effects. One thus generally sees that in tertiary amines proton basicity in water increases as the alkyl groups are exchanged along the series Me < Et < iso-Pr < t-Bu. This accounts for such effects as the pK values of Me_3N (9.9) and Et_3N (10.7).

Taft[15] has produced an empirical equation, which separates the steric and electronic effects controlling the basicity of saturated bases in aqueous solution. This is similar to the Hammett σ-function[16] for the basicity (nucleophilicity) of aromatic bases in water. It is found, for example, that in a series of bases such as the carboxylic acids, RCOO⁻, the variation in proton basicity in water is controlled by the inductive strength of R. The constant steric environment provided by the carboxylate group interposed between the R group and the proton coordinated to the carboxylate means that steric effects of the R group are unimportant. The protonation constants of RCOO⁻ bases thus correlate well with the Taft σ-function, which function is in fact derived from the rates of hydrolysis of organic esters. Where the R groups are attached directly to the donor atom, as in the RNH_2 series, steric effects become important, and a second term must be added, the E_s parameter, which takes steric effects into account. The Taft equation 2.3 thus correlates the basicity of many saturated organic bases towards the proton (as well as other Lewis Acids, and also the rates of many organic reactions involving saturated bases.)

$$\log K \quad = \quad \log K^0 + \sigma^* \cdot \rho^* + \delta \cdot E_s \qquad\qquad 2.3$$

Here log K is the equilibrium constant, or rate constant, of the base with R as an alkyl substituent, K^0 is K for R = methyl, σ∗ is the inductive effect parameter for the substituent R, ρ is the responsiveness of the Lewis acid, or substrate, to σ∗, E_s is the tendency of R to sterically hinder complex-formation (or formation of the reaction intermediate), and δ is the susceptibility of the complex formed (or the substrate) to steric hindrance. In Table 2.1 are seen σ∗ and E_s values for a selection of alkyl substituents.[17] It is seen that the σ∗ values increase linearly along the series Me, Et, iso-Pr, t-Bu, but that the increase in E_s is more nearly exponential. This means that steric effects tend in general to become serious more rapidly than inductive effects.

Table 2.1. Some E_s and σ^* values for alkyl substituents.[a]

Substituent	σ^*	E_s	Substituent	σ^*	E_s
CH_3 - (methyl)	0.00	0.00	CH_3OCH_2 - (methoxymethyl)	+0.52	-0.2
CH_3CH_2 - (ethyl)	-0.10	-0.07	$NCCH_2$ - (cyanomethyl)	+1.30	-0.94
$(CH_3)_2CH$ - (iso-propyl)	-0.19	-0.47	H - (hydrogen)	+0.49[b]	+1.24
$(CH_3)_3C$ - (t-butyl)	-0.30	-1.54	$(CH_3)_3CCH_2$ - (neopentyl)	-0.17	-1.7
$CH_3CH_2CH_2$ - (n-propyl)	-0.12	-0.36	$(CH_3)_2CHCH_2$ - (isobutyl)	-0.13	-0.9
$CH_3(CH_2)_4$ - (n-pentyl)	-0.13	-0.40	C_6H_{11} - (cyclohexyl)	-0.20	-0.79
$ClCH_2$ - (chloromethyl)	+1.05	-0.24	ICH_2 - (iodomethyl)	+0.85	-0.37
Cl_2CH - (dichloromethyl)	+1.94	-1.54	Cl_3C - (trichloromethyl)	+2.65	-2.0
$C_6H_5CH_2$ - (benzyl)	+0.22	-0.38	C_6H_5 - (phenyl)	+0.6[c]	-2.6

[a] The σ∗ values refer to the inductive effects of the substituent, and are more negative as the substituent becomes more electron donating. The E_s values indicate the tendency of the substituent to cause steric hindrance, and are more negative as they become more sterically hindering. Values are from Refs. 15, 18. [b] This value of σ∗ applies only if the hydrogen is attached to a carbon atom, as in a series of carboxylic acids. If attached, for example, to an amine, the nature of the hydrogen itself is very changed, being strongly hydrogen bonded to the solvent, which is not the case for hydrogen attached to a carbon atom. [c] The value of σ∗ for the phenyl group is strongly dependent on the extent of conjugation that may occur with the donor group, as happens, for example, with carboxylic acids.

The LFER in organic reactions[16] is by now a familiar phenomenon.[17] What is of interest in the gas-phase is that the slope of the LFER is much steeper[4] than that in water, as seen in Figure 2.3. The slope shows that the response of proton affinity to change in electron donating or withdrawing nature of the p-substituent on pyridines is in the gas phase over three times as strong as in water. This highlights another effect of the solvent on chemical reactions. Replacing a substituent on a base will increase the electron density on the donor atom, strengthening the bond to the proton, or a metal ion,[19] for example. However, in a solvent, particularly water, the increased electron density on the donor atom will also make it more difficult to remove the waters of solvation hydrogen bonded to the donor atom. Thus, the increase in complex stability on adding a more electron-releasing substituent will in a strongly solvating solvent such as water be much less than the increase in the gas-phase. A solvent such as water thus has a levelling effect in that the difference in energy of reactions produced by changing substituents will be much smaller in the solvent than in the gas-phase.

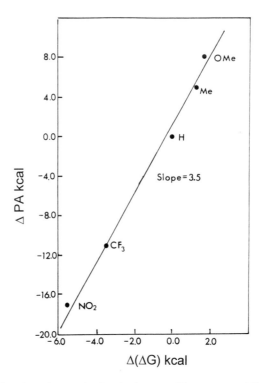

Figure 2.3. The damping out of substituent effects on pyridines by the solvent. The substituents shown for each point are for p-substituted pyridines. $\Delta(\Delta G)$ values are the change in free energy of protonation for each substituent in water relative to H as substituent. ΔPA values are the changes in free energy of protonation in the gas phase for each substituent relative to H as substituent. The slope of about 3.5 shows how much stronger the response to change of substituent is in the absence of the solvent, water. Redrawn after Ref. 4.

2.1.2 Complex-formation reactions of metal ions in the gas-phase

The kinetics (section 2.1.3) and thermodynamics of complex formation of a selection of metal ions with organic bases have been studied in the gas-phase. Metal ions studied include Li^+,[14,20] K^+,[21] Al^+,[22] Mn^+,[23] Ni^+,[24] Co^+,[25] $Ni(Cp)^+$ (Cp = cyclopentadienyl)[26] and $FeBr^+$.[27] The chemistry in the absence of a solvent does not at first sight correspond well with the typical coordination chemistry observed in solution. In Figure 2.4 are shown the enthalpies of formation of $Ni(I)L_2$ complexes with a variety of neutral bases in the gas phase.[24] These enthalpies of complex-formation are all relative to the complex of $Ni(I)L_2$ where L is acetylene, since the energies of complex-formation with the bare Ni^+ ion are too large to be determined directly. As with the proton (section 2.1), enthalpies of complex-formation for the amines RNH_2, and also other series of bases such as ROH, R_2O, $R_2C=O$, RSH, R_2S, and RCN, increase along the series Me < Et < i-Pr < (t-Bu). This type of response is rare in water, where there is considerable steric hindrance. However, the $Ni(I)L_2$ complexes should have linear coordination geometry, and in this arrangement steric effects would be minimized. In water most metal ions would have

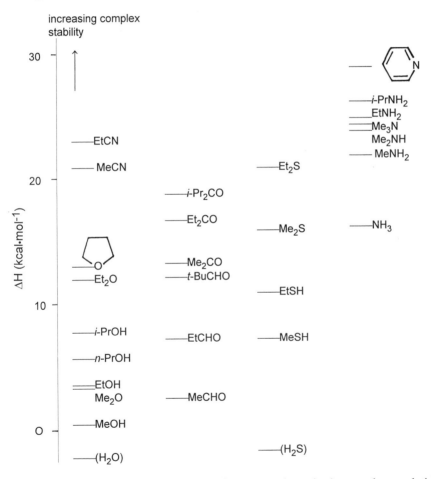

Figure 2.4. Enthalpies of formation of $Ni(I)L_2$ complexes in the gas phase, relative to L = acetylene. Redrawn after Ref. 28.

high coordination numbers, with solvent molecules occupying spare coordination sites, and under these sterically crowded conditions steric effects would outweigh inductive effects. It is of considerable interest to note[19] that the Ag^+ metal ion which has linear coordination geometry in its complexes with amines even in aqueous solution has responses to the series of ligands RNH_2 with R = Me through t-Bu like that of metal ions in the gas-phase:

Ligand	$MeNH_2$	$EtNH_2$	iso-$PrNH_2$	t-$BuNH_2$
log K_1 (Ag^+)	3.06	3.44	3.67	3.69

Virtually all other metal ions show sharp decreases in complex stability along this series of amines, and most do not even form complexes with the bulkier ligands because the metal ion reacts with hydroxide ion before a pH is attained where the complex with the amine is formed.

The enthalpies of complex formation for metal ions in the gas phase show patterns which can be interpreted[23-25,27] in terms of bonding ideas such as HSAB. In Figure 2.5(b) are shown the gas phase enthalpies of formation of MnL^+ complexes versus the gas phase enthalpies of protonation of the same ligands.[23] The ligands form separate LFER which each contain all of the ligands with a particular type of donor group. What is of particular interest[23] in Figure 2.5(b) is the fact that these LFER are displaced towards the proton in such a way that ligands with softer donor atoms (R_2S, R_3N) form relatively more stable complexes with the proton, while those with harder donor atoms (R_2O, $R_2C=O$) favor the Mn^+ ion. In all such diagrams involving any pair of metal ions out of the set of metal ions studied[23-25,27] preference of one of the pair of metal ions towards ligands with softer donor atoms is observed, and it can then be assumed that this is the softer of the pair of metal ions. Using this approach, one can order the metal ions in the gas-phase into a series indicating increasing softness, Al(I) < K(I) < Li(I) < Mn(I) < Co(I) < Ni(CP)$^+$ < H(I) < Ni(I) < Cu(I). This order conforms well with what one might have expected intuitively from a knowledge of HSAB trends in the periodic table derived from solution chemistry. One should point out that the nitriles are anomalous in Figure 2.5(b) in relation to being harder ligands than oxygen donor ligands, in spite of having a less electronegative nitrogen donor. The affinity for the proton is in effect anomalously low, and the nitriles appear to be reasonably soft on diagrams not involving the proton. The origin of the low affinity of the proton for nitriles in the gas-phase is not clear at this stage, but might, for example, involve side-on bonding to metal ions in the gas-phase to give π-bonding, which would not be possible for the proton. The observation of LFER diagrams in the gas-phase which accord with HSAB ideas on metal to ligand bonding is of importance to solution chemistry, as discussed in section 2.4.

The free energies of formation of metal halides (Table 2.2) in the gas phase[29] show that for all metal ions in the gas-phase, the order of affinity for halide ions is $F^- > Cl^- > Br^- > I^-$, so that all metal ions in the gas-phase are by this criterion "hard" in the HSAB classification. The reversals in this stability order, seen in Table 1.3, are brought about by the order of enthalpies of solvation of the halide ions which is $F^- > Cl^- > Br^- > I^-$. A metal ion is thus soft in water as solvent if the rate of increase in enthalpy of complex formation with the halide ions along the series from I^- to F^- is insufficient to overcome the increasing solvation energy of the halide ions from I^- to F^-. A metal ion such as Ag(I) which is "soft" in aqueous solution appears "hard" in a solvent such as acetone where the solvation of the halide ions is weaker.[30]

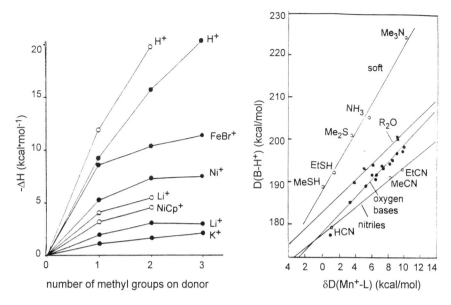

Figure 2.5. (a) Enthalpies of complex formation in the gas phase of a variety of Lewis acids with series of ligands NH_3, NH_2CH_3, $NH(CH_3)_2$, $N(CH_3)_3$ (●), and H_2O, CH_3OH, $(CH_3)_2O$ (o), as a function of the number of methyl groups attached to the donor amine. Energies relative to the NH_3 (o) or H_2O (●) complex. Reproduced with permission from Ref. 12. **(b)** The relationship between the enthalpies of formation of the complex with Mn^+ [$D(Mn^+$-L)], and the enthalpies of protonation [$D(B$-H)] in the gas phase of a variety of bases. The bases are as indicated, with (●) being aldehyde and ketone oxygen bases, and (◆) being ethers. Redrawn after Ref. 23.

Table 2.2. Total Coordinate Bond Energies (kcal.mol⁻¹) at 298.16 °C in the gas phase, and Pearson's Hardness Parameter, H_p, for cations.[a]

	F⁻	Cl⁻	Br⁻	I⁻	H_P
M⁺ Cations					
H	368	332	322	313	0.149
Li	184	153	147	138	0.250
Na	153	133	128	121	0.209
K	138	118	113	106	0.231
Rb	133	114	109	102	0.233
Cs	130	113	109	102	0.215
Cu	199	181	179	176	0.116
Ag	179	166	166	166	0.073
Au	206	194	195	197	0.044
Tl	172	146	142	135	0.215
M²⁺ Cations					
Be	777	688	669	643	0.172
Mg	598	545	531	507	0.152
Ca	523	464	449	430	0.178
Sr	492	437	425	406	0.175
Ba	466	413	402	380	0.185

Table 2.2. Continued[a]

	F⁻	Cl⁻	Br⁻	I⁻	H$_P$
M^{2+} Cations					
Zn	664	619	605	588	0.114
Cd	595	564	559	547	0.081
Hg	643	614	607	602	0.064
Sn	574	523	509	492	0.143
Pb	547	494	487	476	0.130
Mn	592	554	535	519	0.123
Fe	626	581	562	547	0.126
Co	638	588	574	557	0.127
Ni	657	605	588	574	0.126
Cu	667	621	621	598	0.103
M^{3+} Cations					
Al	1412	1281	1252	1219	0.137
Sc	1205	1097	1066	1035	0.141
Y	1114	1009	980	951	0.146
La	1066	951	918	884	0.170
Ga	1417	1324	1303	1277	0.099
In	1293	1200	1183	1164	0.099
Bi	1185	1087	1078	1052	0.112
Cr	1341	1252	1224	1197	0.107
Fe	1348	1257	1238	1217	0.097
Co	1410	1324	1298	1276	0.095
M^{4+} Cations					
Zr	2082	1917	1876	1831	0.121

[a] Data in kcal.mol^{-1}. The energies, ΔE, refer to the gas-phase process $MX_n(g) \rightarrow M^{n+}(g) + nX^-(g)$. For each metal ion M^{n+} the number of halide ions in the reaction is n. The hardness parameter, H$_P$, is calculated as $[\Delta E(F^-) - \Delta E(I^-)]/\Delta E(F^-)$ for each metal ion.[29]

2.1.3 Reaction rates in the gas-phase

Rates of reaction in the gas-phase highlight the role of the solvent in slowing down chemical reactions. A familiar example[10] from organic chemistry illustrates this most dramatically. The replacement of Br⁻ from CH_3Br by Cl⁻ (reaction 2.4) proceeds only slowly in water.

$$CH_3Br + Cl^- \longrightarrow [Cl\cdots\underset{\underset{H\,H}{/\backslash}}{\overset{\overset{H}{|}}{C}}\cdots Br] \longrightarrow CH_3Cl + Br^- \qquad\qquad 2.4$$

transition state

In acetone it is 3×10^9 times faster than in water[10] while in the gas-phase it is 10^{15} times faster. This once again comes down to the fact that solvating water molecules (Figure 2.6) must be removed from the Cl⁻ ion and the site of attack on the CH_3Br molecule before the activated complex can be formed. The reaction coordinate in water and in the gas phase for reaction 2.4 is shown in Figure 2.7. Without the waters solvating the Cl⁻ ion and the CH_3Br, the activation energy required in the gas phase to form the intermediate complexes is actually negative.

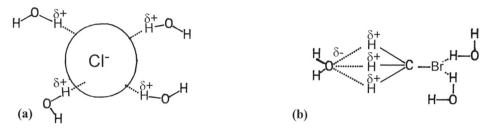

(a)

(b)

Figure 2.6. Diagrammatic representation of the solvation by hydrogen bonding of **(a)** the chloride ion, and **(b)** methyl bromide. As discussed in the text, this solvation of chloride and methyl bromide has to be at least partly removed before the reaction intermediate can form, accounting for the very high activation energy of the substitution of bromide by chloride on methyl bromide with water as solvent.

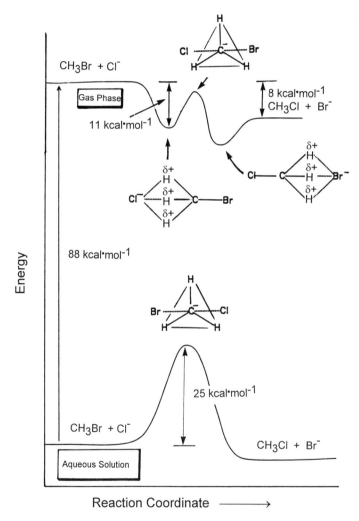

Figure 2.7. The reaction profiles for the reaction $CH_3Br + Cl^- = CH_3Cl + Br^-$ in aqueous solution, and the gas-phase. Redrawn after Ref. 10.

25

The first intermediate formed has the incoming chloride ion bound to the methyl bromide molecule. The transition state, at higher energy than this, is the activated complex with Cl and Br both bonded to the carbon atom. This is then followed by an even lower energy intermediate complex with Br⁻ coordinated to the methyl chloride product. In water, by contrast, (Figure 2.7) the activation energy required to desolvate the chloride ion and methyl bromide molecule leads to a positive activation energy of +25 kcal.mol⁻¹, and consequently much lower rates of reaction than in the gas-phase. These results once again illustrate the all-important effects of the solvent in chemistry. The greatly increased activity of "naked" anions has led to the use of crown ethers to generate weakly solvated anions such as F⁻ from the alkali metal salts in solvents of low dielectric constant for use as more reactive nucleophiles in organic synthesis.[31]

2.2   Linear Free Energy Relationships (LFER)

One can discover many relationships between the free energies or rates of complex formation of sets of complexes (LFER), and a variety of properties of the metal ions, ligands, or complexes. Such regularities are not derivable in any strict thermodynamic way, and are hence called extra-thermodynamic relationships. The correlations do, however, provide insights into the factors governing complex-formation, and in addition may allow for prediction of unknown formation or rate constants. The LFER has been known for a long time in Organic Chemistry[16] with correlations involving rates and proton basicity of series of organic aromatic bases. In coordination chemistry the first observations of LFER were[32] correlations between the protonation constant of the ligand and log K_{ML} with a variety of metal ions. This is still the most common type of correlation. An example of this type of correlation is seen in Figure 2.8, where log K_{ML}

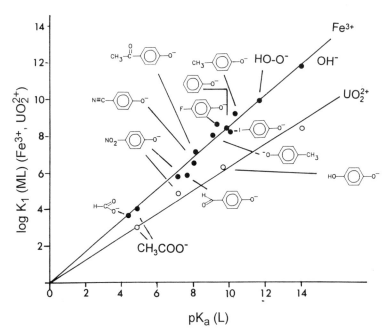

Figure 2.8. Linear free energy relationships for log K_1 (ionic strength = 0, t = 25 °C) for Fe^{3+} and UO_2^{2+} *versus* ligand pK_a for unidentate ligands with negatively charged oxygen donors. Data from Ref. 11.

for M = Fe^{3+} and UO_2^{2+} for a series of unidentate RO^- type ligands (mainly phenols and carboxylic acids) has been drawn up against the protonation constants of the ligands. The linearity of the relations demonstrates that the factors which increase or decrease the pK_a of the ligand by increasing or decreasing the electron density on the oxygen donor atoms of the RO^- ligand also affect the log $K_1(ML)$ values for Fe^{3+} and UO_2^{2+} in an exactly parallel manner. The slope of the relationship for Fe^{3+} is 0.85, and for UO_2^{2+} it is 0.63, indicating a smaller response on the part of the Lewis acids Fe^{3+} and UO_2^{2+} to variation in electron density than is found for the proton. It also follows from Figure 2.8, as is found for many pairs of metal ions, that an LFER for log $K_1(ML)$ for Fe^{3+} versus log $K_1(ML)$ for UO_2^{2+} would also be linear.

One may also draw up LFER of log K_1 values against non-thermodynamic properties of the metal ions or ligands. In Figure 2.9 is shown[33] a correlation of log $K_1(F^-)$ for a variety of metal ions against Z^2/r (Z = cationic charge, r = ionic radius) for the metal ions. Such a correlation is an indication that the M-F bonds are largely electrostatic, since they are reasonably well modeled in terms of simple electrostatic considerations. The LFER is a common phenomenon in coordination chemistry, and is used throughout this work to analyze metal to ligand bonding.

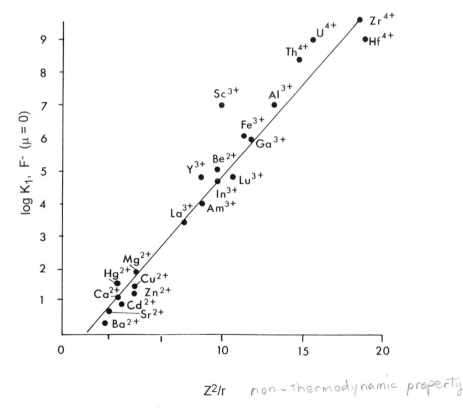

Figure 2.9. Relationship between log K_1 for fluoride complexes of metal ions versus Z^2/r, where Z is the cationic charge on the metal ion, and r the ionic radius. Ionic radii from Ref. 34, formation constant data at ionic strength = 0, t = 25 °C, from Ref. 11.

2.3 Ligand Field Theory And Metal To Ligand Bonding

For an extensive discussion of Ligand Field (LF) theory one should consult references such as 35-37. Our interest here is merely to highlight those aspects of the theory of direct importance to an understanding of solution chemistry. The presence of the ligands in an octahedral arrangement around the metal ion leads to a splitting of the d-shell into an e_g and t_{2g} level as seen in Figure 2.10. The splitting of the d-shell means that those electrons in the t_{2g} level will be lowered in energy, while those in the e_g level will be raised in energy. The stabilization of the complex that results from this splitting can be calculated from the empirical value of 10 Dq, the LF splitting parameter (Figure 2.10), and equation 2.5:

$$\Delta H(LFSE) = 10\ Dq[0.4n(t_{2g}) - 0.6(n_{eg})] \qquad\qquad 2.5$$

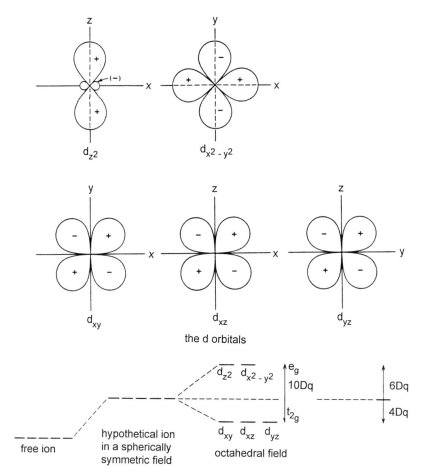

the d orbitals

Splitting of the d energy levels in an octahedral complex

Figure 2.10. The d-orbitals (above), and the splitting of the d energy levels in an octahedral complex (below).

The value of 10Dq varies for all metal ions in an approximately constant manner with the same series of ligands, and any one metal ion. This series of ligands, arranged in order of increasing ability to split the d shell, is called the spectrochemical series: (See Chart 2.1 for key to ligand abbreviations.)

$$I^- < Br^- < Cl^- \sim S\text{-donors} < F^- < H_2O \sim ox^{2-} < Acac^- < NH_3 < EN < BPY \sim PHEN \sim$$
$$9\text{-aneN}_3 < NO_2^- < CN^- \sim CO$$

The splitting of the d-shell is interpreted[37,38] in Molecular Orbital (MO) theory in terms of increasing overlap in the M-L bond, as seen in Figure 2.11. Larger splitting thus generally indicates increased covalence in the M-L bond. This appears to be true for the lighter donor atoms, F, O, N, and C, and for ligands containing these donor atoms LF splitting is primarily a function of the electronegativity of the donor atoms $C < N < O < F$, with increased electronegativity leading to decreased covalence. The heavier donor atoms, S, Se, P, As, Cl, Br, I, however, are lower in the spectrochemical series than would be expected from their low electronegativity, and here 10 Dq may not be a good guide to covalence. It may be that $L \rightarrow M$ π-bonding in the complexes of these heavier donor atoms leads to a lowering of 10 Dq, as seen in Figure 2.11(d).

The variation of formation constants for divalent metal ions, the "Irving-Williams"[39] stability order $Mn(II) < Fe(II) < Co(II) < Ni(II) < Cu(II) > Zn(II)$ has been explained in terms of the variation of LFSE in the same order along this series. In Figure 2.12(a) is seen the variation in log K_1 for EN and EDTA complexes of divalent metal ions of the first row of transition metal ions as a function of the number of electrons in the d-shell. The typical double humped curve is seen, with minima in log K_1 occurring for the d^0, d^5, and d^{10} metal ions with no LFSE. The ligands EN and EDTA are obviously displacing water molecules, and there would also be LFSE for the metal aquo ions from which the complexes are formed. The dependence of log K_1 on LFSE in Figure 2.12(a) arises because EN and EDTA are higher in the spectrochemical series than water, and so there is a net LFSE left over. One would thus expect for ligands which are lower in the spectrochemical series than water, that the variation of log K_1 with d-orbital population would be the opposite of what is normally found. In Figure 2.12(b) the variation of log $K_1(F^-)$ with d-orbital population for trivalent metal ions from the first row of transition metal ions is the inverse of the usual stability order found for these ions,[40] and the inverse of the order of LFSE usually calculated for these metal ions.

A change of spin state, particularly for d^6 and d^8 ions, can also lead to a disruption of normal stability orders. In Figure 2.12(c) is shown log β_3 for the BPY complexes of the series of transition metal ions Ca(II) through Zn(II). Here it is known[41] that the complex with Fe(II) is low-spin, which raises (eq 2.4) the LFSE from 4Dq to 24Dq. This greater LFSE causes log β_3 for the Fe(II) complex to be much higher than expected compared with normal trends seen in Figure 2.12(a). One might ask why all Fe(II) complexes are not low-spin if this leads to such great complex stabilization. The answer is that there is an unfavorable spin pairing energy, P, which must be overcome before spin-pairing can occur, and the stabilization of the complex is ΔLFSE - P. For Fe(II), ΔLFSE is LFSE (low spin) - LFSE (high-spin) = 20Dq, so that the actual stabilization of the complex is 20Dq - P. For d^6 ions P is given by $5B + 8C$, where B is the Racah interelectronic repulsion parameter, and C is the Racah spin-pairing parameter. For spin-pairing in other electronic arrangements, see reference 35.

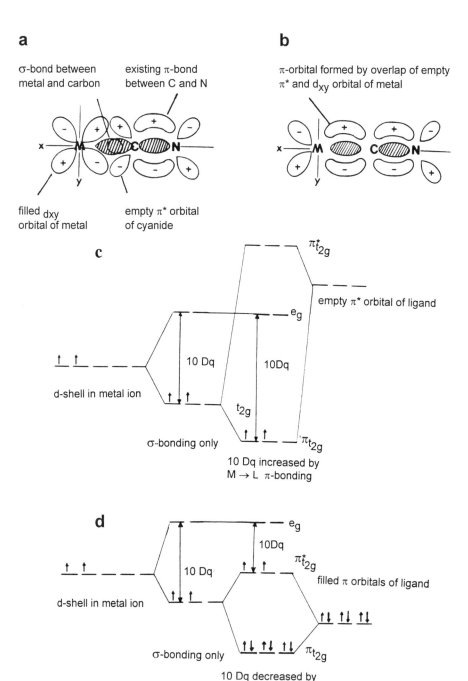

a

σ-bond between
metal and carbon

existing π-bond
between C and N

filled d_{xy}
orbital of metal

empty π* orbital
of cyanide

b

π-orbital formed by overlap of empty
π* and d_{xy} orbital of metal

c

$\pi^* t_{2g}$

empty π* orbital of ligand

e_g

10 Dq

10Dq

d-shell in metal ion

σ-bonding only

t_{2g}

πt_{2g}

10 Dq increased by
M → L π-bonding

d

e_g

10Dq

$\pi^* t_{2g}$

filled π orbitals of ligand

10 Dq

d-shell in metal ion

σ-bonding only

πt_{2g}

10 Dq decreased by
M → L π-bonding

Figure 2.11. The possible role of dπ-pπ π-bonding in the CN⁻ ligand to metal bond in increasing the ligand field (LF) strength of the CN⁻ ligand. At **a)** is shown the formation of a σ bond between a metal ion and cyanide, and **b)** is shown how the d_{xy} orbital of the metal and the empty π* orbitals of the cyanide overlap to form a π-bond. At **c)** is shown a MO diagrams illustrating the role of dπ-pπ π-bonding in increasing the LF strength through overlap with empty π orbitals on the ligand. At **d)** is shown how filled π orbitals on the ligand may overlap with the t_{2g} level of the metal ion to decrease LF strength.

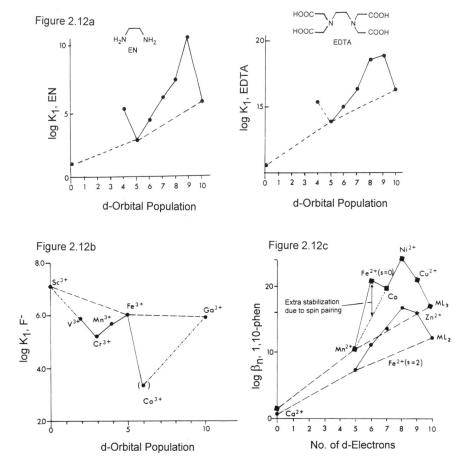

Figure 2.12. (a) Variation of log K_1(EN) and log K_1(EDTA) for divalent metal ions of the first row of transition metals with d-orbital population. Formation constant data from Ref. 11. **(b)** Variation of log K_1(F^-) with d-orbital population for the trivalent metal ions of the first row of transition metals. Formation constant data from Ref. 40, at ionic strength = 0, t = 25 °C. **(c)** Variation with d-orbital population of log β_2 ($\bullet$) and log β_3 ($\blacksquare$) for the 1,10-phenanthroline complexes of the divalent metal ions of the first row of transition metals. The diagram shows the extra stabilization in log β_3 due to spin pairing of Fe(II) (S=O), which is not present in log β_2 for high-spin Fe(II) (S = 2). Formation constant data from Ref. 11.

For spin pairing to occur, the condition ΔLFSE > P must be met, so that in general ligands with high LF strength produce spin-paired complexes, and those with low LF strength are high spin. Thus, Fe(II) forms high-spin complexes with F^- and all O-donor ligands, and EN and NH_3, but with stronger nitrogen donors such as 9-aneN_3 or BPY it forms low-spin complexes (see section 2.5). The heavier donor atoms P, As, S, Se, Cl, Br, and I generate weak ligand fields, but many of their complexes are low spin because the Racah parameters, and hence P, which oppose spin pairing, are also very small. Smaller B and C values are associated with more covalent M-L bonding. For ions such as Ni(II), spin-pairing is accompanied by a change in coordination number and geometry from octahedral (high-spin) to square planar (low-spin). Thus, complexes of Ni(II) with F^- and virtually all O-donor ligands are high-spin and octahedral, and also for less basic N-donors such as NH_3 and EN. Complexes with CN^- and more basic N-donors such as TMEEN or 13-aneN$_4$ are low-spin and square planar. Steric constraints prevent the complexes of Ni(II) with BPY, PHEN, and 9-aneN$_3$ from being low-spin, even though these ligands generate a strong LF. It is found that 10Dq tends to increase, and B and C to decrease, reflecting more covalent M-L bonding as one passes down a group in the periodic table, such as Co(III), Rh(III), Ir(III), or Ni(II), Pd(II), or Pt(II). Thus, while the complexes of many of the first row of transition metal ions are high spin, only PdF_2 is high spin amongst the heavier d-block elements. It is important to note that LFSE is at a maximum for d^3 ions and low-spin d^6 ions, so that the complexes of these ions are of particularly high stability, and because (section 1.2.2) of the relationship between LFSE and the rate of ligand exchange reactions, tend to form rather kinetically inert complexes.

A further effect of importance in LF theory is Jahn-Teller distortion, found in Cu(II) in particular. It is noted in Figure 2.12(a) that, although LFSE predicts the order of log K_1 such that Ni(II) should form more stable complexes than Cu(II), Cu(II) usually forms more stable complexes than does Ni(II). This is due to the fact that the Cu(II) ion is Jahn-Teller distorted. The presence of three electrons in the e_g level of the d shell leads to two being present in, usually, the d_{z^2} orbital, and one in the $d_{x^2-y^2}$ orbital, with uneven repulsion being experienced by the six ligands surrounding the metal ion. Thus, the four donor atoms forming bonds to the metal ion in the plane experience repulsion only from the single electron present in the $d_{x^2-y^2}$ orbital. These bonds are thus shorter, and stronger, than the two bonds to the axial coordination sites, where the two donor atoms are experiencing repulsion from the two electrons present in the d_{z^2} orbital. In the process of complex formation, bidentate ligands such as EN complex first in the plane utilizing the short strong bonds, and so form complexes that are more stable than those of Ni(II). Only where the complex formed involves covalent bonds to all six coordination sites, so that coordination to the two weak axial sites is included, does the normal LFSE order of complex stability Ni(II) > Cu(II) occasionally become apparent. This is found for the ML$_3$ complexes where L is PHEN or BPY (Figure 2.12c), and also for ligands such as 9-aneN$_3$ (log β_2 Ni(II) = 28, log β_2 Cu(II) = 27), where geometrical factors force octahedral coordination on the Cu(II). Even for ligands forcing octahedral coordination, such as EDTA, log K_1 for the Cu(II) complex is slightly higher than for the Ni(II) complex because of the Jahn-Teller effect and because of the rise in stability in the absence of LFSE effects, due to ionic contraction from Ca(II) to Zn(II) along the first row of transition metal ions. This increase in log K_1 with contraction of ionic radius in the absence of LFSE effects is indicated as broken lines in Figures 2.12(a)-2.12(c).

Fabbrizzi, et al.[42] have demonstrated the existence of a LFER involving the enthalpies of complex formation of Cu(II) with polyamines, and the LF strength of the complexes. Figure 2.13 shows an analogous correlation involving free energies of complex-formation instead. The free energies are given by log β_n - (n - 1) log 55.5, where the (n - 1) log 55.5 term corrects for the entropic contribution to the chelate effect from the asymmetry of the standard reference state, discussed in section 3.3. The correlation holds because increased overlap in the Cu-N bond leads to increased enthalpy of Cu-N bond-formation, and also increased energy of the d-d band. A requirement for the correlation to hold is that there should be no unusual energetic situations in the free ligand itself, since this will not be reflected in the Cu-N bond. Thus, there is considerable steric strain involved in getting the free ligands BAE-PIP [N,N′-bis(2-aminoethyl)-piperazine] and BAP-PIP [N,N′-bis(2-aminopropyl)-piperazine] from their low energy

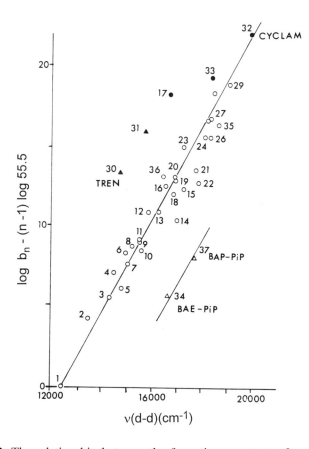

Figure 2.13. The relationship between the formation constants of complexes of Cu(II) with saturated polyamines, and the energy of the d-d band in the complex in solution, ν(d-d). The formation constants are corrected for the entropy contribution derived from the chelate effect[45,46] by subtraction of (n - 1) log 55.5, where n is the denticity of the ligand. Ligands attached to Cu(II), in which there are no unusual steric effects (O) are 1) H_2O 2) NH_3 3) N,N,N,',N'-tetramethylethylenediamine 4) N,N'-diethylethylenediamine 5) $2NH_3$ 6) N,N'-dimethylethylenediamine 7) N,N-dimethylethylenediamine 8) N-methylethylenediamine 9) ethylenediamine (EN) 10) 1,1-dimethyl-1,2-diaminoethane 11) C-methylethylenediamine (PN) 12) $3NH_3$ 13) 1,5,9-triazanonane (3,3-TRI) 14) bis-(N,N-diethylethylenediamine) 15) bis-(N,N'-diethylethylenediamine) 16) 1,4,7-triazaheptane (DIEN) 18) 1,5,9,13-tetraazatridecane 19) 1,4,8-triazaoctane (2,3-TRI) 20) $4NH_3$ 21) bis-(N,N'-dimethylethylenediamine) 22) bis-(N,N-dimethylethylenediamine) 23) 1,4,7,10-tetraazadecane (TRIEN, 2,2,2-tet) 24) bis-(N-methylethylenediamine) 25) bis-(C-methylethylenediamine) 26) bis-(1,1-dimethyl-1,2-diaminoethane) 27) bis-(ethylenediamine) 28) 1,5,8,12-tetraaza-dodecane (3,2,3-tet) (29) 1,4,8,11-tetraazaundecane (2,3,2-tet). Ligands that force at least one donor atom to occupy the unfavorable axial site on Cu(II) ($\triangle$) are 30) tris-(2-aminoethyl)amine (TREN) and 31) 1,4,7,10,13-pentaazatridecane (TETREN). Ligands ($\triangle$) with piperazine bridges that have to change conformation from the energetically favored chair to the unfavorable boat conformer before complex formation can occur are 34) N,N'-bis-(2-aminoethyl)piperazine (BAE-PIP) and 37) N,N'-bis-(3-aminopropyl)piperazine (BAP-PIP) while 36) is N,N'-bis(2-aminoethyl)homopiperazine. The tetraazamacrocycles ($\bullet$) are 17) 1,4,7,10-tetraazacyclododecane (12-aneN_4) 32) 1,4,8,11-tetraazacyclotetradecane (CYCLAM) 33) 1,4,8,12-tetraazacyclopentadecane. Formation constants from reference 11, d-d band energies from Ref. 42.

boat-conformers to the high energy chair conformers required for complex formation, so that log K_1 for their Cu(II) complexes is low (Fig 2.13) in relation to the high d-d band energies of the complexes. This demonstrates that the weakness of BAE-PIP and BAP-PIP as ligands is related mainly to very low levels of preorganization of the free ligand rather than steric problems in the complex itself. The points for ligands such as TREN [N,N,N–tris(2-aminoethyl)-amine] are displaced to the left, which appears[43] to relate to the fact that one of the amine donors must occupy an axial site in the copper complex. Occupation of the axial coordination site causes a lowering in d-d band energies. It is of considerable interest that the point for the macrocycle cyclam in Figure 2.13 is well-behaved, suggesting that perhaps the 'macrocyclic effect'[44] is due to the greater basicity of the secondary N-donors of the macrocycle rather than any preorganization effects (see section 2.1.1 for a discussion of basicity of saturated N-donors). The use of d-d band energies and ligand field splittings as pointers to factors controlling the strength of the M-L bond, as exemplified by Figure 2.13, are of importance to the theses developed in this book.

2.4   Patterns In Lewis Acid-Base Behavior In Aqueous Solution

The fact that Lewis acids and bases can be assigned as hard, soft, and intermediate in the HSAB classification[44] has already been mentioned in section 1.1. The pattern present in preferences of metal ions for donor atoms of different types can be investigated with LFER diagrams.[47,48] If all the log K_1 values for complexes of unidentate ligands with one metal ion are plotted against those of another metal ion, an LFER diagram is obtained. Figure 2.14 shows the LFER diagram for Ag(I) and Hg(II). This is similar to the LFER diagram for Mn(I) and the proton in the gas phase seen in Figure 2.5(b). Ligands with the

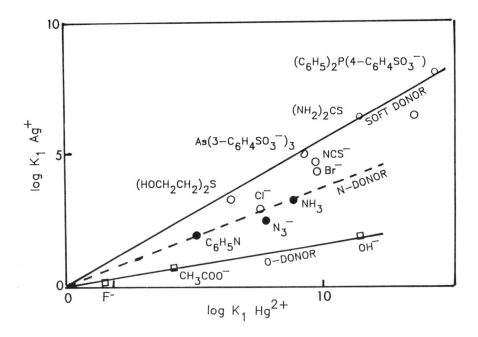

Figure 2.14. LFER (Linear Free Energy Relationship) diagram for Ag(I) *vs* Hg(II). Formation constants from Ref. 11. Redrawn after Ref. 47.

same type of donor atom tend to occur on the same LFER, e.g. all the amines occur on one LFER, or all the ligands with soft donor atoms on another. Normally, a LFER comparing the thermodynamics of complex-formation of one Lewis acid with another is restricted to a single type of donor atom. In cases where a single LFER is observed[30] (Figure 1.2) with a wide variety of donor types, this is because the two metal ions, Cu(II) and Ni(II), are very similar. The Ag(I) and Hg(II) ions are not sufficiently similar for a single order of stability with all unidentate ligands to be observed, and the ligands become displaced towards Ag(I) as they become softer, indicating that Ag(I) is softer than Hg(II). In Figure 2.15 the LFER diagram[44,47] for Bi(III) *versus* Hg(II) is like that in Figure 2.14, except that the softer ligands are now displaced towards Hg(II), indicating that Hg(II) is softer than Bi(III).

For a pair of metal ions, e.g. Cu(II) and Ni(II) (Fig 1.2), where the LFER diagram contains all ligands approximately on a single LFER, log K_1 for Cu(II)L complexes can be predicted using a two-parameter equation, equation 2.6.

$$\log K_1[Cu(II)] \sim 1.2 \log K_1[Ni(II)] \qquad\qquad 2.6$$

However, for most metal ions a single LFER is not obtained, and more pairs of parameters are required. The earliest equation of this type was the Edwards[49] equation (Eq 2.7):

$$\log K^0 = \alpha \cdot E + \beta \cdot H \qquad\qquad 2.7$$

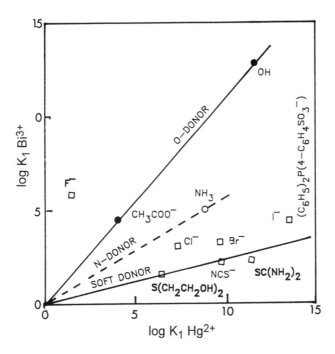

Figure 2.15. LFER (Linear Free Energy Relationship) diagram for Bi(III) *vs* Hg(II). Formation constants from Ref. 11. Log K_1 for NH_3 (O) with Bi(III) was estimated as described in section 2.5.3. Redrawn after Ref. 47.

where log K^o = log K(ML) + log 55.5, E is the oxidation potential of the ligand + 2.60, and H is the pK$_a$ of the ligand + 1.74. The parameters α and β were empirically adjusted to a best fit. Other four-parameter equations have been developed,[50] and an empirical order of hardness for metal ions and ligands has been derived[51] by an empirical fitting of the Edwards equation to log K_1 values in aqueous solution. Drago[52] has applied an equation with two pairs of parameters to enthalpies of adduct formation in solvents of low dielectric constant. Subsequent work[47,48] has shown that for equilibria in aqueous solution, two pairs of parameters as used in the Edwards equation are not sufficient. An equation containing three pairs of parameters was found[47,48] to give significantly better prediction of log K_1 values (Eq 2.8). The E and C parameters in equation 2.8 are identified as being the tendency of the Lewis acid A or Lewis base B to undergo either electrostatic ($E_A{}^{aq}$ and $E_B{}^{aq}$) or covalent ($C_A{}^{aq}$ and $C_B{}^{aq}$) bonding, and in this resembles the Drago[52] equation.

$$\log K_1 = E_A{}^{aq} \cdot E_B{}^{aq} + C_A{}^{aq} \cdot C_B{}^{aq} + D_A \cdot D_B \qquad 2.8$$

In addition to the parameters in the Drago equation, the parameters D_A and D_B corrected for what the authors[47] identified as the steric hindrance to solvation of the Lewis acid and base on complex formation. It may also be that there is a component in the D parameters from desolvation of the complex brought about by the formation of an extremely strong covalent bond. It was noted that the D parameters were strongly related to the size of the acid or base. The D parameters were large and made a big contribution to destabilizing the complex when the metal ion was small and the donor atom was large (e.g. Co(III) with I$^-$) but were zero for small donor atoms (e.g. F$^-$, OH$^-$ and NH$_3$) or large metal ions (e.g. Ag(I), La(III), Pb(II)). The low affinity of the proton for ligands with large donor atoms, e.g. R$_2$S, I$^-$, is almost entirely due to the uniquely large D_A parameter of the proton. The large D_A parameter for the proton is responsible for the fact that HI, unlike HF, is a strong acid in water. This arises because of the weak solvation of the HI molecule in solution,[47,48] which fits in with the idea of the bulky iodide ion disrupting the solvation of the proton. This is analogous to the low basicity of trimethylamine because of the disruption of the solvation of the proton by bulky methyl groups (section 2.1.1). In addition, the more covalent HI bond would reduce the charge on the proton, and reduce its ability to hydrogen bond with the solvent.

Evidence that the D parameters can be dramatically altered can be found in the complexes of [Cu(tetb)]$^{2+}$, where log K_1 values have been reported[53] for binding of unidentate ligands to the axial coordination site on the complex (for structure of tetb see Chart 2.1, page 59). Here, the formation constants are more like those for a soft metal ion, and the Cu(II) is no longer borderline as it is for the Cu^{2+} aquo ion:

	log K_1		
	Cl$^-$	Br$^-$	I$^-$
[Cu(tetb)]$^{2+}$	0.04	0.30	0.81
Cu^{2+}	0.4	0.03	no evidence complex

(data from Refs. 11 and 53)

It would seem possible that the solvation at the coordination site in [Cu(tetb)]$^{2+}$ is less extensive than in Cu^{2+}(aq), and so a large donor atom such as iodide is better accommodated. Of course, other interpretations are also possible, such as that the four nitrogen donors attached to the Cu in [Cu(tetb)]$^{2+}$ render the metal ion more covalent in its bonding, and so make it appear soft. However, Chung et al.[53b] have recently reported formation constants for binding of amines to the axial coordination site of [Cu(tetb)]$^{2+}$, which show the order of log K$_1$ NH$_3$ < NH$_2$CH$_3$ < NH(CH$_3$)$_2$, which is consistent with a very low level of steric hindrance, supporting the steric interpretation of the change in softness from [Cu(tetb)]$^{2+}$ to Cu^{2+} discussed above. It is of interest to note that Klopman[54] in an electron-perturbation model of complex-formation concluded that three factors were important in HSAB behavior. Two of these factors were the covalence and ionicity that have always been at least loosely associated with softness and hardness. The third factor identified by Klopman was steric hindrance to solvation on complex-formation, which accords well with the interpretation of the D parameters in terms of steric effects.

Equation 2.8 predicts formation constants for metal ions with unidentate ligands to an accuracy of about 0.2 log units. The E, C, and D parameters for the more common metal ions and bases are shown in Tables 2.3 and 2.4. The acids in Table 2.3 are placed in order of increasing I$_A$ value, where I$_A$ is the ratio E$_A$/C$_A$. The I$_A$ parameters can be shown by matrix algebra[47] to give a unique order of hardness if a certain condition is met in assigning the parameters. The condition is that the ligand for which C$_B$ is given a value of zero be the hardest ligand, and have effectively only electrostatic bonding. Figure 2.9 appears to support the idea that F$^-$ has very little covalence in its M-L bonds, at least relative to the water molecules which it must displace from the metal ion on coordination.

Table 2.3. The tendency to ionicity in bonding of Lewis Acids as measured by the I$_A$ parameter, and the electrostatic (E$_A$) and covalent (C$_A$) contributions to the formation constants of Lewis acid-base complexes in aqueous solution.[a] The D$_A$ parameters are a measure of the susceptibility to steric hindrance of the Lewis acid in formation of the metal to ligand bond. The parameters are for use in equation 4.1. The Lewis acids are arranged in order of increasing I$_A$ parameter, the order of increasing ionicity in the bond between the Lewis acid and base.

Lewis Acid	I$_A$	E$_A$	C$_A$	D$_A$
Au$^+$	-16	-3.0	0.190	0.0
Ag$^+$	-10.6	-1.52	0.143	0.0
Cu$^+$	-1.3	-0.56	0.43	2.5
Hg^{2+}	1.63	1.346	0.826	0.0
Pd^{2+}	1.85	1.72	0.929	6.0
CH3Hg$^+$	2.5	1.60	0.64	0.0
Tl^{3+}	2.66	2.55	0.96	0.0
Cu^{2+}	2.68	1.25	0.466	6.0
H$^+$	3.04	3.07	1.009	20.0
Cd^{2+}	3.31	0.99	0.300	0.6
Ni^{2+}	3.37	1.20	0.300	4.5
Co^{3+}	3.77	3.30	0.875	7.0
Zn^{2+}	4.26	1.43	0.312	4.0
Co^{2+}	4.34	1.33	0.276	3.0
Fe^{2+}	5.94	1.40	0.256	2.0
In^{3+}	6.30	4.49	0.714	0.5
Bi^{3+}	6.39	5.91	0.926	0.0
VO^{2+}	6.42	3.81	0.593	3.5
Pb^{2+}	6.69	2.76	0.413	0.0
Mn^{2+}	7.09	1.64	0.223	1.0
Ga^{3+}	7.07	5.72	0.809	1.5
Cr^{3+}	7.14	5.15	0.721	1.5
Am^{3+}	7.21	4.29	0.595	0.0

Table 2.3. Continued

Lewis Acid	I_A	E_A	C_A	D_A
Fe^{3+}	7.22	6.07	0.841	1.5
U^{4+}	7.80	7.55	0.968	3.0
Sn^{2+}	8.07	5.65	0.700	0.0
Pu^{4+}	8.31	7.9	0.950	
UO_2^{2+}	8.40	4.95	0.589	1.0
Be^{2+}	8.84	5.43	0.614	
Lu^{3+}	10.07	4.57	0.454	0.0
La^{3+}	10.30	3.90	0.379	0.0
Mg^{2+}	10.46	1.86	0.178	1.5
Sc^{3+}	10.49	7.03	0.671	0.0
Al^{3+}	10.50	6.90	0.657	2.0
Ca^{2+}	10.53	0.98	0.093	0.0
Y^{3+}	10.64	4.76	0.477	0.0
Th^{4+}	10.94	8.44	0.771	0.0
Ba^{2+}	12.7	0.54	0.043	0.0
Na^+	$(14.6)^b$	-0.20	-0.014	0.0
Li^+	22.2	0.57	0.026	0.0

[a] The parameters are from references 28, 47 and 48. [b] Estimated by comparison with Pearson's hardness parameter, HP, in Table 2.2.

Table 2.4. The softness of Lewis bases in aqueous solution as indicated by the I_B parameter. The electrostatic (E_B) and covalent (C_B) contributions to the formation constants of Lewis acid-base complexes in aqueous solution[a] are given. The D_B parameters are a measure of the susceptibility of steric hindrance of the Lewis base in formation of the metal to ligand bond. The parameters are for use in equation 2.8. The Lewis bases are arranged in order of decreasing I_B parameter, which is increasing softness.

Lewis base	I_B	E_B	C_B	D_B
F^-	∞	1.00	0.0	0.0
CH_3COO^-	0.0	0.0	4.76	0.0
OH^-	0.0	0.0	14.00	0.0
N_3^-	-0.064	-0.067	10.4	0.2
$S=C=N^-$	-0.082	-0.76	9.3	0.2
NH_3	-0.088	-1.08	12.34	0.0
C_5H_5N	-0.102	-0.74	7.0	0.0
Cl^-	-0.100	-1.04	10.4	0.6
SO_3^{2-}	-0.107	-1.94	18.2	0.4
Br^-	-0.108	-1.54	14.2	1.0
$S_2O_3^{2-}$	-0.119	-3.15	26.5	1.1
I^-	-0.122	-2.43	20.0	1.7
NCS^-	-0.128	-1.83	14.3	1.0
$(HOCH_2CH_2)_2S$	-0.135	-1.36	10.1	0.6
$PPh_2(4-C_6H_4SO_3^-)$	-0.132	-3.03	23.0	0.7
$As(3-C_6H_4SO_3^-)_3$	-0.135	-1.93	14.3	
$(NH_2)_2C=S$	-0.135	-2.46	18.2	0.6
$HOCH_2CH_2P(C_2H_5)_2$	-0.141	-4.89	34.7	0.6
CN^-	-0.148	-4.43	30.0	0.3

[a] The parameters are from Refs. 28, 47 and 48.

The parameters in Tables 2.3 and 2.4 thus give rise to the interpretation[47,48] that the free energy of complex formation with unidentate ligands in aqueous solution is governed by ionicity and covalence, which might be loosely equated with hardness and softness in the HSAB classification. However, many Lewis acids that are classified[44] as hard, such as Co(III) or the proton, form highly covalent bonds to the lighter donor atoms, such as the C present in CN⁻, or nitrogen donor ligands. The reluctance of Co(III) or the proton to bind to ligands with heavier donor atoms such as S or I is traceable to their very large D parameters, and may therefore reflect steric hindrance to coordination rather than any lack of ability to form covalent bonds. The steric properties of Co(III) complexes are further discussed in section 3. The role of steric hindrance in modifying HSAB behavior beyond what would be expected from contributions from ionicity and covalence in the M-L bond is summarized as follows:

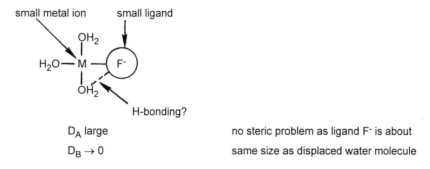

D_A large

$D_B \rightarrow 0$

no steric problem as ligand F⁻ is about
same size as displaced water molecule

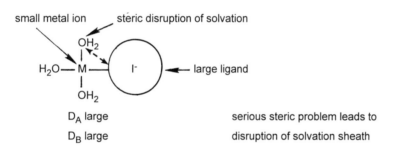

D_A large

D_B large

serious steric problem leads to
disruption of solvation sheath

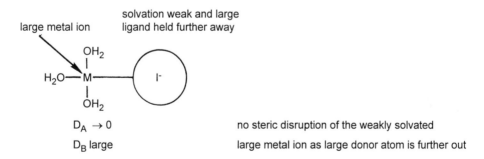

$D_A \rightarrow 0$

D_B large

no steric disruption of the weakly solvated
large metal ion as large donor atom is further out

Equation 2.8 can be used to decide on the suitability of donor atom types for complexing a metal ion for which little information is available on formation constants with that type of ligand in the literature. For example, one might be interested in promoting selectivity for In(III) over Fe(III) in designing imaging agents, where one did not wish the In(III) to be displaced from its ligand by the Fe(III). Would addition of soft donor atoms such as RS^-, R_2S, or PR_3 improve selectivity for In(III)? Substituting parameters for the triphenyl phosphine in Table 2.4 and Fe(III) and In(III) in Table 2.3 predicts a log K_{ML} for Fe(III) with phosphine of -0.1, but log $K_{ML} = 2.5$ for In(III). This calculation, and similar calculations for RS^- and R_2S suggests that In(III)/Fe(III) selectivity would be improved by addition of these donor groups to chelating ligands.

2.5   The Coordinating Tendencies of Different Donor Groups

In Chapter 3 it is shown how most chelating ligands behave in a fairly additive fashion, so that their coordinating tendencies are the sum of the coordinating tendencies of their unidentate analogues. In other words, once the chelate effect is taken into consideration as an entropy effect derived from a decrease in the number of particles coming together to form the complex, chelating ligands are not much different from their unidentate analogues. This simplicity is only altered if the steric situations in the free ligand or the final complex are made very demanding, and here the potential for breaking away from the limitations set by the affinities of the metal ions for different types of donor atom can be overcome, which is considered in Chapter 3. In this section the coordinating tendencies of the groups most commonly encountered in chelating ligands, namely those containing oxygen and nitrogen, are considered in more detail. The heavier donor atoms encountered in chelating ligands, P, As, S, and Se have been dealt with to some extent in Section 2.4.

2.5.1   The neutral oxygen donor

The neutral oxygen donor atom is of especial interest because of its occurrence in the solvent, water. It is also of great interest because of its occurrence in crown ethers.[55] Reference to Figures 2.4 and 2.5 shows that the inductive effects of the added alkyl groups mean that, for added methyl groups, for example, the order of increasing basicity is H_2O < MeOH < Me_2O. Steric problems, and the inability of the methyl group to disperse charge to the solvent by hydrogen bonding, however, mean that this is not necessarily the order of complex stability in water. It is not possible to investigate the coordinating properties of ligands which contain neutral oxygen donors only with a wide variety of metal ions, other than water, since such ligands tend to bind only to large metal ions. It is[56] possible, however, to investigate the effect on complex stability of addition of groups containing neutral oxygen donors to ligands such as EN which coordinate to a wider range of metal ions. One finds that addition of neutral oxygens may lead to stabilization for the complexes of some metal ions, or to destabilization for others.

It is found as a general observation[28,56] that addition of neutral oxygen donors leads to an increase in complex stability for larger metal ions relative to the stability of complexes of smaller metal ions. Thus Pb^{2+} has[34] an octahedral radius of 1.18 Å as against 0.69 Å for Ni^{2+}, and so Pb^{2+} will show increases in selectivity relative to Ni^{2+} on addition of neutral oxygen donors.

The relationship between change in complex stability (Δ log K) produced by adding neutral oxygen donors to a ligand, and ionic radius (r^+) is so close that plots of Δ log K versus r^+ lead to smooth relationships, as seen in Figures 2.16 and 2.17. In Figure 2.16 Δ log K for passing from alanine to DHE-AL and from oxalate to DETODA has been plotted against the octahedral radii of the metal ions. Octahedral radii appear satisfactory except for Cu(II) and metal ions such as low-spin Ni(II), whose size appears to be better represented by the radii for the square-planar ions, or Be(II) for which tetrahedral radii appear appropriate.

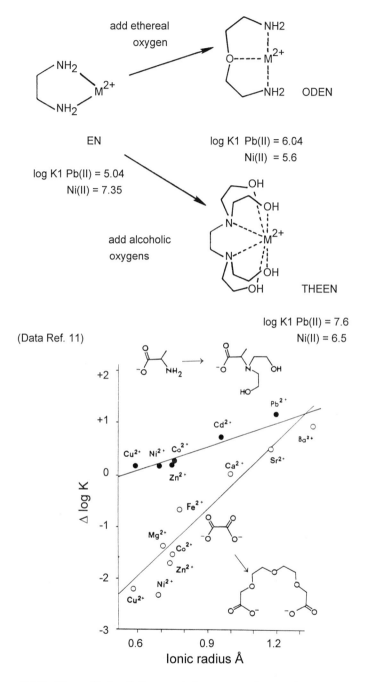

Figure 2.16. The effect of the neutral oxygen donor atom on complex stability. Relationship between the change in complex stability, Δ log K, that occurs on adding 2-hydroxyethyl groups to alanine (●) or ethereal oxygens to oxalate (○), and the ionic radius of the metal ion. Ionic radii are from Ref. 34 and are for octahedral coordination, except for Cu(II) which is square planar. Formation constants are from Ref. 11.

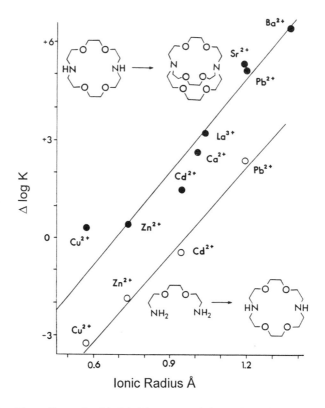

Figure 2.17. The effect of added bridges containing ethereal oxygen donors on the stability of complexes of macrocyclic ligands. Relationship between the change in formation constant, Δ log K, that occurs on adding groups containing ethereal oxygen donors, and ionic radius[34] of the metal ion. In the upper relationship ($\bullet$) cryptand-222 is formed from 18-aneN$_2$O$_4$, while in the lower relationship ($\circ$) 18-aneN$_2$O$_4$ is formed from BAEDOE (4,7- dioxa-1,10-diazadecane). Formation constants (log K$_1$) from Ref. 11.

In Figure 2.17 is shown the plot of Δ log K *versus* r$^+$ for passing from ethyleneoxabis(ethylamine) (EDODA) to 18-aneN$_2$O$_4$, and then from 18-aneN$_2$O$_4$ to cryptand-222. Figures 2.16 and 2.17 thus show that the tendency to coordinate best to large metal ions found in the crown ethers and cryptands is shared by neutral oxygen donors that are part of non-macrocyclic ligands. One may formulate here a rule of ligand design: "Addition of groups containing neutral oxygen donor atoms to an existing ligand leads to an increase in selectivity of the ligand for large metal ions over small metal ions". This rule applies to both macrocyclic and non-macrocyclic ligands, except for some more rigid small macrocycles (section 3.3) which may be able to override the effect of neutral oxygen donors with other selectivity determining factors.

The application of the rule is quite straightforward. For example, the ligand 18-aneN$_4$O$_2$ has poor selectivity for Pb(II) over Zn(II), as seen below. To improve this selectivity, one can add four hydroxyethyl groups to give THE-18-aneN$_4$O$_2$ as shown. Here no new macrocyclic structure is created in adding the four neutral oxygens. Conversely, the open chain ligand AMPY below has no selectivity for Pb(II) over Zn(II), but when four ethereal oxygens are added in such a way as to create a macrocyclic ring, strong Pb(II)/Zn(II) selectivity is generated:

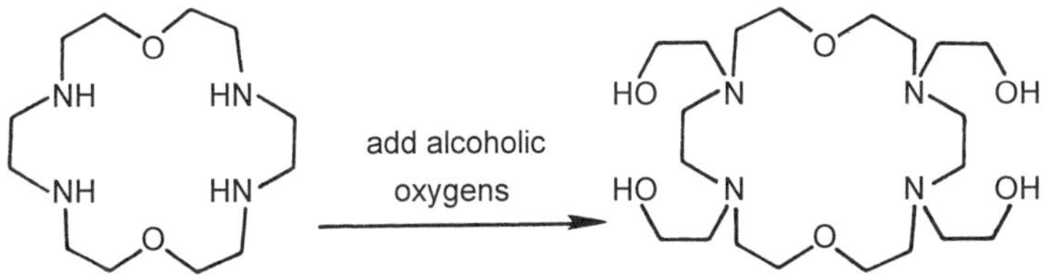

add alcoholic oxygens

18-aneN$_4$O$_2$

log K$_1$ Pb(II) = 9.0
log K$_1$ Zn(II) = 10.5
selectivity
Pb(II)/Zn(II) = -1.5

HO
HO
OH
OH

THE-18-aneN$_4$O$_2$

log K$_1$ Pb(II) = 10.7
log K$_1$ Zn(II) = 5.9
selectivity
Pb(II)/Zn(II) = +4.8

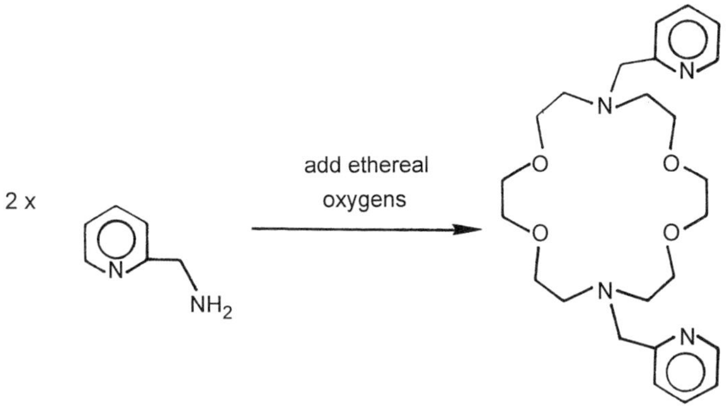

2 x

add ethereal oxygens

NH$_2$

AMPY

log β$_2$ Pb(II): 6.0
log β$_2$ Zn(II): 9.4
selectivity
Pb(II)/Zn(II): -3.4

py$_2$-18-aneN$_2$O$_4$

log K$_1$ Pb(II): 11.7
log K$_1$ Zn(II): 7.0
selectivity
Pb(II)/Zn(II): +4.7

The origin of the metal ion size related effect of the neutral oxygen donor on complex stability may have at least two causes. The more obvious cause must be that large metal ions have higher coordination numbers than small metal ions, and so are able to accommodate the larger numbers of donor atoms involved in ligands such as THE-18-aneN$_4$O$_2$, which small metal ions with their low coordination numbers cannot do. However, the effect holds as well even for passing from alanine to DHE-AL (Figure 2.16), an increase in coordination number from two to four, which can hardly be exceeding the coordination numbers of any of the metal ions involved. The inductive effects of the alkyl groups on ethereal and alcoholic oxygen donors produces a stronger base than water. However, these alkyl substituted oxygens are not so much more basic than water, and whether their addition will produce an increase in complex stability will always be subject to the steric effects involved. One might say that the effect on complex stability of neutral oxygen donors is steric strain controlled. A part of this steric strain should arise from steric crowding effects, which would be more serious for small metal ions than for large metal ions. Another consideration (section 3.1.3) is the size of the chelate ring formed. Large metal ions are favored by five-membered chelate rings for steric reasons, and it should be noted that in virtually all cases neutral oxygen donors occur in ligands where they will be part of five membered chelate rings. Where the oxygen donors (section 3.1.3) are part of six membered chelate rings, a preference for small metal ions is found.

It appears possible to improve the coordinating properties of the neutral oxygen donor by improved steric efficiency coupled with greater coordinating strength. Thus, THP (tetrahydropyranyl) groups as bearers of neutral oxygen donors on iminodiacetates[57] lead to improved affinity for large metal ions and greater selectivity over small metal ions. One thus finds that the THP side arms are better preorganized than is found for simple ethanolic groups: (data from reference 11).

		log K_1 for R	= H	-CH$_2$CH$_2$OH	-H$_2$C (THP)
N-substituted	Pb(II)	= 7.36		9.41	10.30
iminodiacetates	Zn(II)	= 7.24		8.45	9.06

R-N(CH$_2$COO$^-$)(CH$_2$COO$^-$)

It is seen that the selectivity of the large Pb(II) ion over the small Zn(II) ion is improved successively by the addition of a 2-hydroxyethyl group, and then the sterically efficient THP group, to iminodiacetate. The same trends can be seen with similar neutral O-donor groups added to the nitrogens of 18-aneN$_2$O$_4$.[58]

The neutral carbonyl oxygen donor atom occurs in groups such as the amide, keto, aldehydic, and ester group. These are important because of their occurrence in compounds such as valinomycin or enniatin-B, which owe their anti-microbial action to their ability to complex alkali or alkaline earth metal ions. The available evidence in the form of ligands with these groups present as donors suggests that the amide carbonyl oxygen is often a stronger donor than the neutral saturated oxygen as a donor, as in:

NH$_2$-R	R = -H	-CH$_2$CH$_2$OH	-CH$_2$CO·NH$_2$	-CH$_2$CO·N(Et)$_2$
log K_1 Cu(II)	4.1	4.1	5.4	6.2

The inductive effect of the N-ethyl groups is apparent in increased stability of the Cu(II) complex with the N,N-diethylglycinamide, because of the steric efficiency of the carbonyl group. The two N-ethyl groups are attached to the amide nitrogen, well away from the coordinated carbonyl group. In other situations the neutral alcoholic oxygen may lead to greater stabilization than does the amide group. In general the available stability constants[11] lead to the conclusion that the order of donor strength of neutral carbonyl oxygens is amide > ketone > aldehyde >> ester. However, it is clear that more work needs to be done to establish fully the coordinating properties of the neutral carbonyl oxygen.

2.5.2 The negatively charged oxygen donor

Many ligands contain groups with negatively charged oxygen donors, including the carboxylate, phenolate, hydroxamate, phosphonic acid, sulfonic acid, and alkoxide oxygens, of widely differing basicity. The simple dependence of log K_{ML} for unidentate

RO⁻ type ligands on ligand pK$_a$ has been demonstrated in the LFER in Figure 2.8. The values of log K$_{ML}$ for ligands containing RO⁻ donors only depends in a simple fashion on the affinity of M for HO⁻, the archetypal RO⁻ ligand, as shown in Figure 2.18. In Figure 2.18, LFER of log K₁ML for the ligands catecholate, 5-nitrosalicylate, kojate, and malonate, versus log K₁ for the hydroxide complexes have been drawn up. In all cases the affinity of the metal ion for the chelate with negatively charged oxygen donors is a simple function of log K₁(MOH). Such LFER can be drawn up even for such complex ligands as

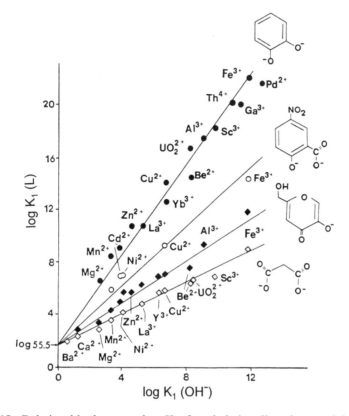

Figure 2.18. Relationship between log K₁ for chelating ligands containing negative oxygen donors, and log K₁ for the formation of the hydroxide complex, for a variety of metal ions. The ligands shown are catechol (●), 5-nitrosalicylic acid (O), kojate (■), and malonate (□). Formation constants at ionic strength zero and 25 °C are from reference 11. The intercept at log 55.5 is that expected from theories[45,46] of the chelate effect. Redrawn after Ref. 59.

the hexadentate BAMTPH (Chart 2.1) among other ligands, as seen in Figure 2.19. It is important to emphasize that log K₁ (OH⁻) values do not correlate well with Z^2/r, unlike the case with log K₁ (F⁻) (Figure 2.9), indicating that the bonding to OH⁻ is governed by more complex factors than simple electrostatic attraction. As an example, for the trivalent group III A metal ions log K₁ (F⁻) decreases with increasing ionic radius, as expected from electrostatic considerations. However, log K₁ (OH⁻) is more strongly dependent on covalence, and shows no simple dependence on metal ion radius:

Metal ion	Al(III)	Ga(III)	In(III)	Tl(III)
ionic radius (A)	0.54	0.62	0.80	0.89
log $K_1(F^-)$	7.0	5.9	4.6	(2.6)*
log $K_1(OH^-)$	9.1	11.4	10.0	13.4

* Estimated from equation 2.4.3 and parameters in Tables 2.4.1 and 2.4.2. In accord with this low formation constant, TlF_3 is at once decomposed in water. Other data from Ref. 11.

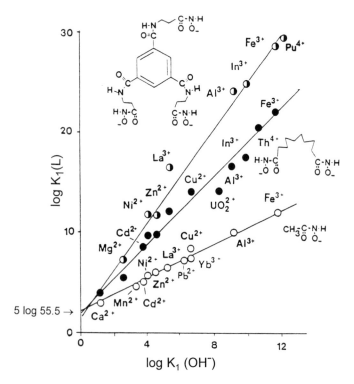

Figure 2.19. Relationship between log K_1 for chelating ligands containing hydroxamate groups, and log K_1 for the formation of the hydroxide complex, for a variety of metal ions. The ligands are BAMTPH (O), 1,8-dihydroxamatooctane (●), and acetohydroxamic acid (O). Formation constants at ionic strength zero and 25 °C are from Ref. 11. An intercept at 5 log 55.5 is that expected from theories[45,46] of the chelate effect for an hexadentate ligand such as BAMTPH. Redrawn after Ref. 59.

The behavior seen in Figure 2.18 is the simplest type of behavior found for negatively charged O-donor ligands. Numerous examples of such LFER can be drawn up for bidentate ligands with RO⁻ donors, including oxalate, salicylates, catecholates, dihydroxynaphthols, acetylacetonates, tropolonates, maltols, and polyphosphates. With substituents on the aromatic rings of ligands such as catecholates or salicylates, the slopes of the LFER decrease with electron- withdrawing substituents ($-NO_2, -CN, -I$) in line with their Hammett σ-functions.[16] What is of particular interest with the ligands in Figure 2.18 are the intercepts on the LFER, which all lie close to log 55.5, where 55.5 is the molarity are the intercepts on the LFER, which all lie close to log 55.5, where 55.5 is the molarity of pure water. The intercept is thus very close to the size predicted for a bidentate ligand by a simple model of the chelate effect[45,46] discussed in section 3.3. However, as seen in Figure 2.19, the intercepts are frequently lower than predicted by considerations of ideas on the chelate effect.[46] Thus, for BAMTPH with six donor atoms, one would have expected an intercept of 5 log 55.5, or 8.72 log units, whereas the observed intercept in Figure 2.19 is only about one log unit. As discussed in Section 3.1.1, this lower intercept probably reflects the large amount of entropy present in the BAMTPH free ligand because of the long non-rigid arms required to hold the chelating hydroxamate groups together. In order to observe the theoretical intercept, a rigid ligand is required with a high degree of preorganization, which is true for ligands such as catecholate or kojate in Figure 2.18. With very long connecting bridges, the intercepts in the correlations tend to be much lower.

The coordinating tendencies of alkoxide groups, as far as can be judged, also appear to be ruled by the affinity of the metal ion for hydroxide ion. Thus, if one makes the reasonable assumption that the log K for deprotonation of the HIDA complexes (Chart 2.1) of metal ions refers to deprotonation of the coordinated hydroxyethyl group, then one would expect log K for protonation of the coordinated alkoxide group to become smaller as log $K_1(OH^-)$ for the metal ion increased, as seen in Figure 2.20. A similar relationship is obtained for the analogous log K for HEDTA complexes.

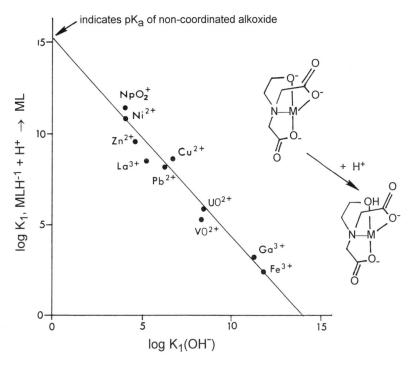

Figure 2.20. Relationship between the protonation constant of the complex of metal ions with HIDA, and the log $K_1(OH^-)$ value for each metal ion. The intercept at log $K_1(OH^-) =$ 0 should correspond to the protonation constant of the uncoordinated alkoxide arm of the ligand. The relationship also shows that for a metal ion with log $K_1(OH^-) = 14.0$, the K_a of the coordinated alkoxide should be zero. Formation constants from Ref. 11.

The affinities of metal ions for ligands containing negatively charged oxygen donor atoms are strongly related to their affinity for the archetypal RO^- donor ligand, the hydroxide ion. The value of log $K_1(OH^-)$ for a particular metal ion is therefore fundamental to understanding its complexing behavior with ligands containing negative oxygen donors. We have reproduced in Table 2.5 for ready reference the log $K_1(OH^-)$ values for the elements in their more important oxidation states.

Table 2.5. Ionic radii, and formation constants for hydroxide and ammonia complexes in aqueous solution, for a selection of metal ions.[a]

Metal ion	Li^+	Na^+	K^+	Cu^+	Ag^+	Au^+
ionic radius	0.7	1.02	1.36	0.77	1.15	1.37
log $K_1(OH^-)$	0.36	-0.2	-0.5	(6.0)	2.0	(2.7)
log $K_1(NH_3)$	-0.3	(-1.1)	(-2.8)	5.9	3.3	(5.6)

	Be^{2+}	Mg^{2+}	Ca^{2+}	Sr^{2+}	Ba^{2+}
ionic radius	0.27	0.72	1.00	1.18	1.35
log $K_1(OH^-)$	8.6	2.58	1.3	0.8	0.6
log $K_1(NH_3)$	(1.7)	0.23	-0.2	(-0.2)	(-0.2)

	Zn^{2+}	Cd^{2+}	Hg^{2+}	Sn^{2+}	Pb^{2+}
ionic radius	0.74	0.95	1.02	(0.96)	1.19
log $K_1(OH^-)$	5.0	3.9	10.6	10.4	6.3
log $K_1(NH_3)$	2.21	2.55	8.8	(2.6)	1.6

	Al^{3+}	Ga^{3+}	In^{3+}	Tl^{3+}	Bi^{3+}
ionic radius	0.58	0.62	0.80	0.89	1.03
log $K_1(OH^-)$	9.01	11.4	10.0	13.4	12.9
log $K_1(NH_3)$	(0.8)	(4.1)	(4.0)	(9.1)	(5.1)

	Sc^{3+}	Y^{3+}	La^{3+}	Gd^{3+}	Lu^{3+}
ionic radius	0.75	0.90	1.03	0.94	0.86
log $K_1(OH^-)$	9.7	6.3	5.5	6.1	6.5
log $K_1(NH_3)$	(0.7)	(0.4)	(0.2)	(0.45)	(0.7)

	Cr^{2+}	Mn^{2+}	Fe^{2+}	Co^{2+}	Ni^{2+}	Cu^{2+}
ionic radius	(0.80)	0.83	0.78	0.74	0.69	0.57
log $K_1(OH^-)$		3.4	(3.6)	(3.9)	4.1	6.3
log $K_1(NH_3)$	(1.7)	1.0	(1.4)	2.1	2.7	4.04

	V^{3+}	Cr^{3+}	Mn^{3+}	Fe^{3+}	Co^{3+}	VO^{2+}
ionic radius	0.64	0.62	0.65	0.65	0.54	
log $K_1(OH^-)$	11.7(?)	10.07	14.4(?)	11.81	13.52	8.3
log $K_1(NH_3)$	(3.8)	(3.4)	6.6(?)	(3.8)	(7.3)	3.0

	Th^{4+}	U^{4+}	Zr^{4+}	Hf^{4+}	Pu^{4+}
ionic radius	0.94	0.89	0.71	0.71	0.86
log $K_1(OH^-)$	10.8	13.3	14.3	13.7	13.2
log $K_1(NH_3)$	(0.4)	(4.2)	(2.0)	(2.4)	(4.0)

	UO_2^{2+}	Am^{3+}	Pd^{2+}
ionic radius		0.98	0.64
log $K_1(OH^-)$	8.2	8.2	13.0
log $K_1(NH_3)$	(2.0)	(2.7)	9.6

[a] Ionic radii are for the octahedral ions, expect for Cu(II), Pd(II) (square planar) and Be(II) (tetrahedral), and are from Ref. 34. The experimental log $K_1(OH^-)$ and log $K_1(NH_3)$ values are at ionic strength zero, and are from Ref. 11. Estimated values of log K_1 are given in parentheses, and have been estimated largely as described in Refs. 28, 47, 48 and 60. Some values considered doubtful by the present authors are indicated by a question mark in parentheses after the given value.

2.5.3 The neutral saturated nitrogen donor

Water and hydroxide are the archetypes for ligands with neutral oxygen donors and negative oxygen donors respectively, and similarly ammonia is the archetypal saturated nitrogen donor. As discussed in section 2.1, gas-phase results indicate strong increases of basicity of the saturated nitrogen with alkyl substitution, along series of the type $NH_3 <$ $MeNH_2 < Me_2NH < Me_3N$, and $MeNH_2 < EtNH_2 < i\text{-}prNH_2 < t\text{-}BuNH_2$. This has important consequences for the chemistry of ligands containing these groups. A particular problem that arises with ammonia as a model for all neutral nitrogen donor groups is that most metal ions do not form stable ammonia complexes in aqueous solution because the reaction with hydroxide is more favorable according to:

$$M(NH_3)^{n+} + H_2O \longrightarrow M(OH)^{(n-1)+} + NH_4^+ \qquad 2.9$$

Thus when ammonia is added to metal ions such as Fe^{3+} or Th^{4+} in aqueous solution, only metal hydroxides are obtained. This presents a limitation to building up a more complete picture of how the affinity of metal ions for saturated nitrogen donor ligands varies across the periodic table.

A method has been devised for estimating $\log K_1(NH_3)$ for metal ions such as Fe(III).[60] It is observed (section 3.3) that the effect of donor groups on the formation constants of simple (i.e. without high levels of preorganization) polydentate ligands relates to the stability constants of the same donor groups in unidentate ligands. The equilibrium where an IDA ligand displaces an ODA ligand below is thus not very different from that where an ammonia displaces a coordinated water to form a complex:

ODA complex   +   IDA   IDA complex   +   ODA

It is found (Figure 2.21) that a plot of $\log K_1(IDA)$ - $\log K_1$ (ODA) versus $\log K_1(NH_3)$ gives a reasonably good relationship. Such LFER can be drawn up for several pairs of ligands[61] where one ligand has a saturated nitrogen donor where the other has a saturated oxygen. The $\log K_1$ values for many such ligand pairs are known for metal ions such as

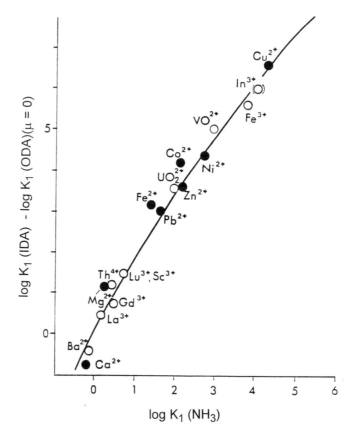

Figure 2.21. Estimation of log $K_1(NH_3)$ for metal ions for which the amines are unstable to hydrolysis. Log K_1 (IDA) - log K_1 (ODA), is the change in complex stability that occurs on changing the ethereal oxygen donor atom in ODA to a saturated nitrogen in IDA (ODA is oxydiacetate, IDA is iminodiacetate, see Chart 2.1). Log K_1 (IDA) - log K_1 (ODA) is plotted against experimentally known[11] (●) log $K_1(NH_3)$ values, and those estimated[60] (O) using several correlations of the type shown in this diagram, and also estimated from equation 2.8.

Fe(III), and consistent values of log K_1 (NH_3) can be estimated for these metal ions. These estimated values of log K_1 (NH_3) for a wide variety of metal ions are seen in Table 2.5, and are shown on Figure 2.21 as open circles. An encouraging aspect of these estimated log K_1 values is that, according to equation 2.9, all of the ammonia complexes of these metal ions for which log K_1 (NH_3) was estimated should be hydrolyzed in water, in accord with observation. An exception is Pb(II), where the log K_1 (NH_3) of 1.6 predicted by LFER such as that in Figure 2.21 suggests that Pb(NH_3)$^{2+}$ would be stable in the presence of higher concentrations of NH_4^+, and the experimentally determined log K_1 (NH_3) is 1.55 in 5M NH_4NO_3. The log K_1 (NH_3) values estimated from LFER of the type in Figure 2.21 accord well with the predictions of equation 2.9.

Two aspects of correlations of the type seen in Figure 2.21 are worth noting. One is that the slopes are generally well over unity, being in the vicinity of 1.5 to 2.0. These higher slopes seem likely to be due to the greater basicity of the secondary nitrogens in IDA, or primary nitrogens in AEIDA (Figure 2.22), as compared to the zeroth nitrogen of

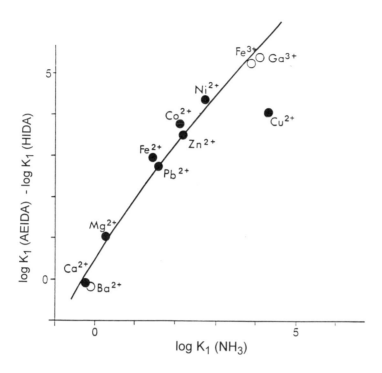

Figure 2.22. Diagram analogous to that in Figure 2.21. The pair of ligands AEIDA and HIDA are seen in Chart 2.1. The point for Cu(II) is thought to deviate downwards because, as discussed in the text, the structures of the two Cu(II) complexes are different. Log K_1 (AEIDA) - log K_1 (HIDA) is plotted against experimentally known[11] ($\bullet$) log $K_1(NH_3)$ values, and those estimated[60] (O) using several correlations of the type shown in this diagram, and also estimated from equation 2.8.

ammonia. The second point about the correlations is that they demonstrate that the effect on complex stability of replacing a donor atom of one type with another in simple polydentate ligands that are not highly preorganized is directly related to the affinity which the metal ion has for unidentate analogues of these groups, in this case ammonia. An assumption of the LFER in Figure 2.21 is that linearity requires structural similarity between the two complexes. Thus, the AEIDA complex of Cu(II) probably has a different structure to the HIDA complex, and so the point for Cu(II) in Figure 2.22 is displaced downwards. Because of Jahn-Teller distortion, coordination to the axial site on the Cu(II) will be less favored. In both complexes this less favored axial coordination site on the Cu(II) is probably occupied by the weakest donor group, which would be an acetate in the AEIDA complex, but an ethanolic oxygen in the HIDA complex, making the complexes structurally different:

acetates occupy favored
equatorial sites

acetate forced into
less favored axial site

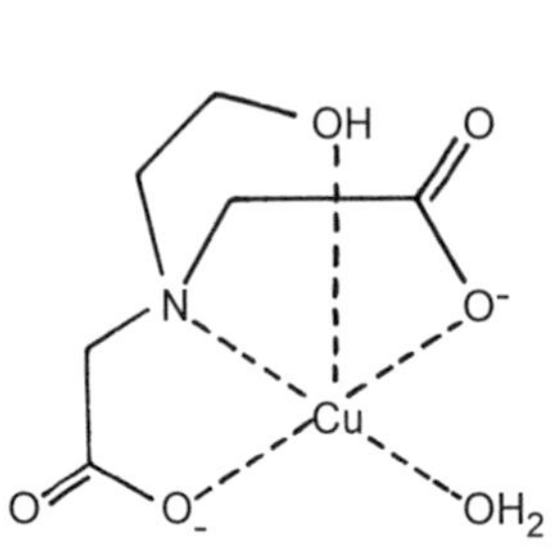

HIDA complex

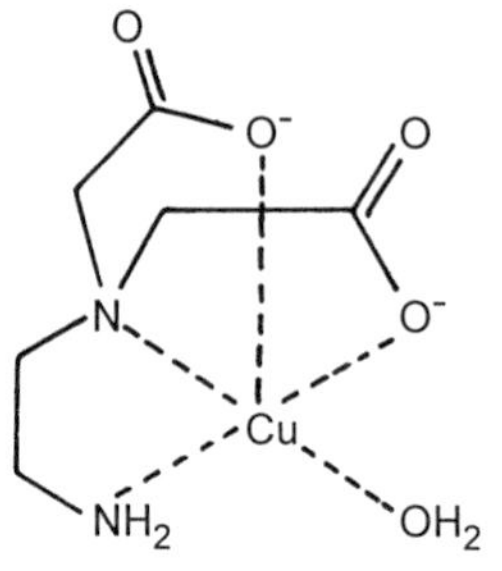

AEIDA complex

Donor atom basicity, as discussed in section 2.1.1, increases with increasing alkyl substitution for nitrogen donors along the series zeroth < primary < secondary < tertiary. This has effects on the thermodynamics of complex formation and the electronic spectra of complexes that can be separated from steric effects, which often counter the effects of increasing donor atom basicity. This is seen for the series of Cu(II) complexes below, where the increasing inductive effect on the nitrogens as they change from zeroth to primary and then secondary causes the enthalpy of complex formation (ΔH) and the energy of the d-d band to increase in parallel:

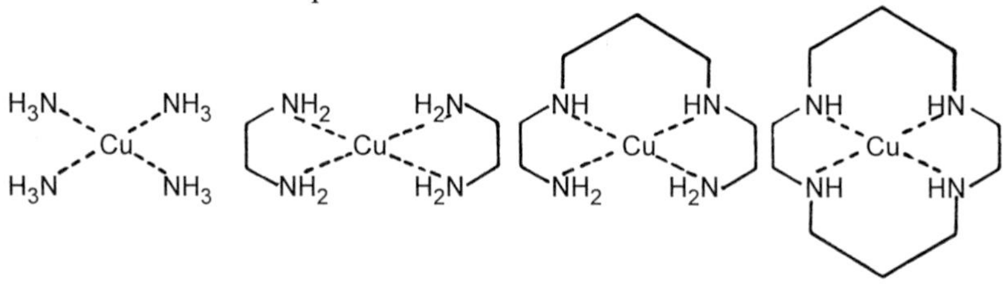

Nitrogens	all zeroth	all primary	2 primary 2 secondary	all secondary
ΔH,[a] kcal mol-1	-22.0	-25.5	-27.7	-32.4
v(d-d), cm-1	17000	18300	19000	19900

[a] Ionic strength 0.5, Ref. 11.

As discussed in section 2.3 increasing LF strength as exemplified by increasing energy of the d-d bands indicates increased overlap in the M-N bond. The idea that these effects are due to inductive effects is often masked by steric effects, so that they have been termed "hidden" inductive effects.[61] Thus, when alkyl groups are added in a sterically inefficient way as N-methyl or N-ethyl groups to EN to give DI-NMEEN or DI-NETEN (Chart 2.1) instead of the expected increase in complex stability, there is a marked drop for Cu(II) from log K_1 = 10.5 for the EN complex to log K_1 = 8.14 in the DI-NETEN complex. This drop in complex stability is accompanied by a drop in LF strength as indicated by the energy of the d-d band in these complexes seen below:

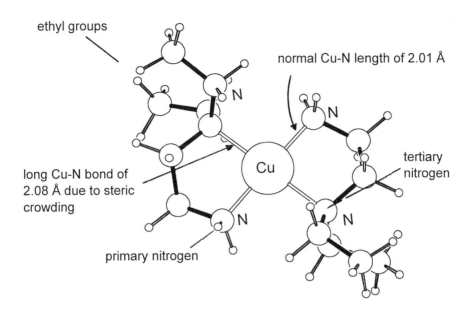

ethyl groups

normal Cu-N length of 2.01 Å

N

Cu

tertiary nitrogen

long Cu-N bond of 2.08 Å due to steric crowding

N

primary nitrogen

N

N

Bis(N,N-diethylethylenediamine)complex of copper(II)[62]

	EN	DI-NETEN
	H_2N⌒NH_2	H_2N⌒$N(CH_2CH_3)_2$
log K_1 Cu(II)	10.5	8.1
log β_2 Cu(II)	19.6	13.7
$\nu_{(d\text{-}d)}$(cm-1)	18300	17000

The drop in complex stability in the DI-NETEN complex of Cu(II) is reflected in a lengthening[62] of the Cu-N bond to the tertiary nitrogen to the rather long value of 2.08 Å in [Cu(DI-NETEN)$_2$]$^{2+}$ as compared with the normal Cu-N bond length of 2.01 Å to the primary nitrogen in this complex. Clearly, the steric interference of the N-ethyl groups on this complex with the rest of the ligand atoms leads to the long Cu-N bond, accompanied by low complex stability and a weak in-plane LF strength.

The steric problems caused by added alkyl groups thus mean that bases such as trimethylamine are very poor ligands. However, the process of C-methylation with some metal ions of low coordination number, where the steric crowding produced by introduction of the C-methyl groups is not too high, may lead to increases of complex stability which parallel the increases in donor atom basicity, discernible for protonation reactions in the gas phase. This has been discussed in section 2.1.2 for the Ag(I) ion with the series of primary amines RNH$_2$ which increase[17] in basicity as R is varied methyl < ethyl < iso-propyl < t-butyl. The Ag(I) ion is of low coordination number, probably two, in aqueous solution, and steric hindrance is also lessened by the very long Ag-N bonds. One therefore finds an increase in log K$_1$ with more negative σ* of the R substituent on

RNH_2, where a more negative R signifies a stronger electron releasing effect (see section 2.1.1 and reference 19). Figure 2.23 shows how log β_2 for RNH_2 complexes of Ag(I) increases with more negative σ^* along the series methyl < ethyl < *iso*-propyl < *t*-butyl. For CH_3Hg^+ the steric effects cause a turn-around with a fall in log K_1 as the alkyl group becomes too bulky, suggesting that for CH_3Hg^+ stronger solvation may provide more serious steric hindrance. Also shown in Figure 2.23 are the log K_1 values for Pb(II) with the series of N-alkyl substituted bis(2-hydroxypropyl)amines where the alkyl group varies from methyl to *t*-butyl. One sees here that the log K_1 values increase for Pb(II) with increasing electron donating ability of the alkyl substituent, and that steric effects do not predominate. This suggests that the very long Pb-N bonds hold the N-alkyl substituents well away from the rest of the complex, so minimizing adverse steric effects.

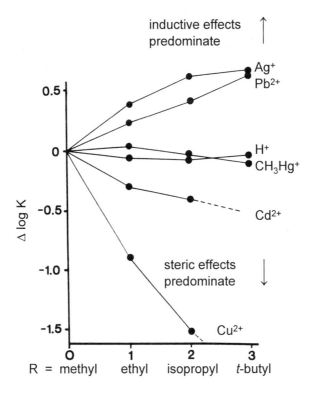

Figure 2.23. Effect of bulk and electron donating ability of R, the N-substituent, on formation constants of complexes of metal ions with the primary amines $(CH_3)_xH_{(3-x)}CNH_2$. The change in complex stability, Δ log K, on change of the N-alkyl substituent R is calculated in each case relative to R = methyl. The bulk and electron donating ability increase with X, the number of methyls on the o carbon atom, along the series of alkyl substituents methyl, ethyl, *iso*-propyl, *t*-butyl. The electron donating ability of the N-alkyl group predominates with larger metal ions Ag(I) with the series $R-NH_2$, Pb(II) with the series R- $N(CH CH_3 \cdot CH_2OH)_2$, are evenly balanced with CH_3Hg^+ and H^+ and the ligand series $R-NH_2$, while steric effects predominate with the smaller metal ions Cd(II) and Cu(II) with the series of ligand R- $NHCH_2CH_2NH_2$. Formation constant data from Refs. 11, 17, 61, and R. D. Hancock et al., to be published.

2.5.4 Unsaturated nitrogen donors

Figure 2.4 shows that in the gas phase the unsaturated nitrogen donor pyridine is a stronger base than any of the saturated N-donor ligands. However, in aqueous solution pyridine is a weaker base than the saturated nitrogen donors. A contribution to the weaker basicity of pyridine than saturated N-donors in aqueous solution must come from the fact that pyridine is a tertiary amine, and so incapable of dispersing charge to the solvent by hydrogen bonding. Figure 2.4 also shows that other unsaturated bases such as the nitriles are also strong bases in the gas phase, but are rather poor ligands in water. This too must arise because of inability to disperse charge to the solvent by hydrogen bonding. The order of basicity of some representative N-donors in water appears to be:

imidazole	pyridine	thiazole	pyrazine	azo compound	Schiff base	nitrile

| pK 9.2 | 7.0 | 5.25 | 2.44 | 0.65 | - | |

The nitrogen is sp^2 or sp hybridized in these ligands, which leads to greater s character in the orbitals used for bonding to the metal ion, and hence more covalent bonding. These ligands can thus exert very high ligand field strengths, even though their proton basicity may be significantly less than that of sp^3 hybridized nitrogens.

An important aspect of the unsaturated nitrogen donors is the possibility of π-bonding between the ligand and the metal ion. Thus, with the pyrazine ligand it has been found that[63] the Ru(II) and Os(II) complexes are greatly stabilized relative to the complexes of other nitrogen donors by what can best be rationalized as π-bonding between the nitrogen and the metal:

Strong evidence for this is the fact that the protonation constant of the pyrazine, once bound to the ruthenium(II), is much higher than that of the free ligand, which accords with the idea of shifting of negative charge to the nitrogen in the canonical structure above.[62]

A particular problem associated with heterocyclic aromatic bases such as pyridines and imidazoles is the steric hindrance to coordination of metal ions caused by the hydrogens *ortho* to the nitrogen donors. As discussed in chapter 3, these problems can be largely overcome by joining pyridyl, and other groups, together, to produce more sterically efficient multidentate ligands.

2.5.5 Ligands with heavier donor atoms S, Se, P, As

Unidentate ligands containing these groups have been studied in solution mainly with "soft" metal ions such as Ag(I) or Hg(II). An interesting aspect of the coordination of these groups with soft metal ions is that the soft-soft interactions lead to complex-formation reactions with very negative ΔH contributions to the free energy of complex-formation, in contrast to hard-hard reactions which are entropy controlled. Thus, we see for the group II metal ions Zn(II), Cd(II), and Hg(II), that the complex-formation reaction between the hardest metal ion, Zn(II), and the hardest ligand, F⁻, is favored by entropy, and disfavored by the enthalpy contribution, while for the softest metal ion, Hg(II), and the softest ligand, I⁻, the complex-formation reaction is favored by a very large enthalpy contribution:

Ligand	F^-	Cl^-	Br^-	I^-
Zn(II) ΔH	+3.8	+1.3	+0.4	-
$-T\Delta S$	-5.4	-1.2	+0.3	-
Cd(II) ΔH	+1.2	+0.3	-0.8	-2.3
$-T\Delta S$	-1.8	-2.1	-1.5	-0.9
Hg(II) ΔH	+1.0	-4.8	-10.6	-18.0
$-T\Delta S$	-2.4	-3.6	-2.1	+0.6

Data refer to enthalpy (ΔH) and entropy ($-T\Delta S$) of complex formation in kcal.mol⁻¹, from Ref. 11.

This type of observation was remarked upon by Ahrland, Chatt, and Davies[64] in their classic paper on "a and b type" (hard and soft) acids and bases, namely that the formation of complexes involving soft-soft interactions was due to favorable enthalpy contributions, while complexes involving hard-hard interactions were favored by entropy contributions.

The observation of Linear Free Energy Relation (LFER) diagrams, noted in section 2.4, indicates that for metal ions classified as "soft", there is a linear free energy relationship between the formation constants with one metal ion such as Ag(I), and another such as Hg(II). That this is controlled by enthalpy contributions is seen in Figure 2.24, where the free energies and enthalpies of complex formation[65] for Ag(I) complexes are plotted against the free energies and enthalpies, respectively, of formation of the corresponding Hg(II) complexes. Figure 2.24 shows that the relationship involving the enthalpies of complex-formation parallels the relationship involving the free energies of complex formation. The existence of the LFER thus depends on the variation in enthalpy. The variation in entropies by contrast is much smaller, and does not control the LFER.[65]

For neutral soft donors such as sulfur or phosphorus highly stable complexes are formed really only with very soft metal ions such as Ag(I), Hg(II), or Pd(II). This is exemplified by formation constants[11] for a selection of metal ions with thiourea and the thioether $S(CH_2CH_2OH)_2$ below. The ionicity parameter H_A for each metal ion[47] is included as a measure of hardness:

Lewis Acid	Ag(I)	Hg(II)	H^+	Cd(II)	Zn(II)	Pb(II)
Increasing softness					Increasing hardness	
ionicity[a]	-10.6	1.6	3.2	3.3	4.3	6.7
log K_1 for[b]						
$(NH_2)_2C=S$	7.1	11.4	0.5	1.5	0.5	0.1
$(HOCH_2CH_2)_2S$	3.5	6.4	-	-0.3	-0.1	-
$Ph_2PC_6H_4SO_3^-$	8.1	14.5	0.2	0.9	-	-
$HOCH_2CH_2S^-$	13.4	~25	9.8	8.0	~5	6.0

[a] As defined in section 2.4. Note that in this classification increasing hardness implies decreasing softness. [b] Formation constant data from Ref. 11.

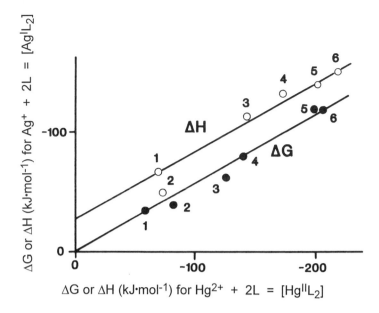

Figure 2.24. Linear Free Energy (●) and linear Enthalpy (O) relationship between the free energies and enthalpies of formation of [Hg(II)L$_2$] and of [Ag(I)L$_2$] complexes. Ligands are thiodiglycol (1), tris(3-sulphophenyl)arsine (2), thiourea (3), 4-(diphenyl-phosphino)benzenesulphonate (4), cyanide (5), and 2-hydroxyethyldiethylphosphine (6). Redrawn after Ref. 65.

Ligands such as thiourea or the phosphine above are extremely soft, and the stability of their complexes falls off rapidly with increasing hardness of the metal ion. On the other hand, the mercaptan ligand is somewhat harder, and it forms fairly stable complexes with metal ions of intermediate hardness such as Zn(II) or In(III). It is thus perhaps better classified as "intermediate" in the HSAB classification. It is, however, an extremely strong ligand. Therefore it is important in ligands such as dimercaptosuccinic acid used for treating metal intoxication with metals such as mercury.[66]

We have seen in Section 2.4 that much of "hard and soft acid and base"[44] behavior is[47,48] related to donor atom size. Thus hard behavior in metal ions such as Co(III) may be related to the difficulty of accommodating large donor atoms in tightly packed coordination spheres of small metal ions, rather than a high level of ionicity in the M-L bond. This is an important consideration in understanding much of the chemistry of ligands having large soft donor atoms. Thus, as seen in Figure 2.25, it is difficult to understand why the two sulfur donors of the ligand DME-DAC are not in the plane of the nitrogen donors, but lie one above and one below this plane.[67] This large bulk of donor atoms such as S, Cl, or P may thus be important in limiting their ability to form stable complexes in aqueous solution with small metal ions with tightly packed coordination spheres. Further effects here may be the formation of complexes of lower than usual coordination number so as to accommodate the larger donor atoms, as seen in complexes such as [NiCl$_4$]$^{2-}$ or [FeCl$_4$]$^-$, as compared to the more usual six coordination found for these metal ions.

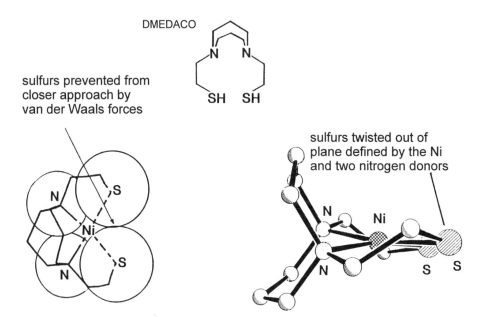

Figure 2.25. The structure[67] of [Ni(DMEDACO)]$^{2+}$, showing how the large sulfur donors are prevented from lying in the coordination plane of the square-planar Ni(II) complex by overlap of the van der Waals radii of the two sulfurs. Drawing using the ALCHEMY program (TRIPOS ASSOCIATES, St. Louis, MO) shows the van der Waals radii of the sulfurs.

Chart 2.1

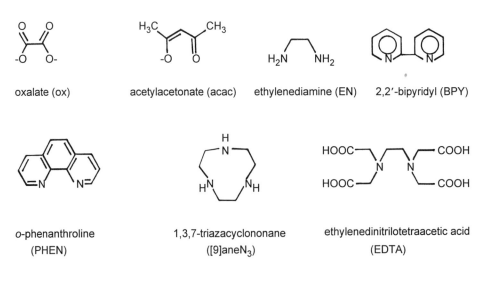

oxalate (ox)

acetylacetonate (acac)

ethylenediamine (EN)

2,2′-bipyridyl (BPY)

o-phenanthroline
(PHEN)

1,3,7-triazacyclononane
([9]aneN$_3$)

ethylenedinitrilotetraacetic acid
(EDTA)

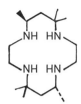

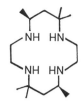

Chart 2.1 continued

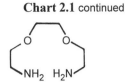

meso-5,5,7,12,12,14-hexa-
methyl-1,4,8,11-tetraaza-
cyclotetradecane (teta)

rac-5,5,7,12,12,14-hexa-
methyl-1,4,8,11-tetraaza-
cyclotetradecane (tetb)

ethylenebis(oxyethyleneamine)
(EDODA)

catecholate

5-nitrosalicylate

5-hydroxy-2-hydroxy-
methyl-4-pyrone (kojate)

malonate

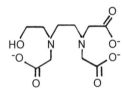

1-hydroxycyclohepta-3,5,7-trien-2-one (tropolonate)

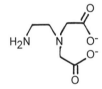

BAMTPH

3-hydroxy-2-methyl-4-pyrone (maltolate)

2-hydroxyethyliminodiacetic acid
(HIDA)

N-(2-hydroxyethyl)-N,N′,N′-
triacetic acid (HEDTA)

2-aminoethylimino-
diacetic acid (AEIDA)

oxydiacetic acid (ODA)

iminodiacetic acid (IDA)

1,4,8,11-tetraazacyclo-
tetradecane (cyclam)

N,N′,N′′,N′′′-tetramethyl-
1,4,8,11-tetraazacyclo-
tetradecane (TMC)

59

References

1. M. S. B. Munson, *J. Am. Chem. Soc.*, **1965**, *87*, 2332.
2. J. I. Brauman, J. M. Riveros, and L. K. Blair, *J. Am. Chem. Soc.*, **1971**, *93*, 3914.
3. D. K. Bohme, E. Lee-Ruff, L. B. Young, *J. Am. Chem. Soc.*, **1971**, *93*, 4608.
4. M. Taagepera, W. G. Henderson, R. T. C. Brownlee, J. L. Beauchamp, D. Holtz, and R. W. Taft, *J. Am. Chem. Soc.*, **1972**, *94*, 1369.
5. R. W. Taft, J. F. Wolf, J. L. Beauchamp, G. Scorrano, and E. M. Arnett, *J. Am. Chem. Soc.*, **1978**, *100*, 1240.
6. R. W. Taft, M. Taagepera, J. L. M. Abboud, J. F. Wolf, D. J. DeFrees, W. J. Hehre, J. E. Bartmess, and R. T. McIver, Jr. *J. Am. Chem. Soc.*, **1978**, *100*, 7765.
7. M. Taagepera, D. DeFrees, W. J. Hehre, and R. W. Taft, *J. Am. Chem. Soc.*, **1980**, *102*, 424.
8. J. E. Bartmess, J. A. Scott, and R. T. McIver, Jr. *J. Am. Chem. Soc.*, **1979**, *101*, 6046.
9. R. W. Taft, in "Kinetics of Ion-Molecule Interactions", P. Ausloss, Ed., Plenum, New York, 1979, p. 271.
10. R. T. McIver, Jr. *Scientific American*, **1980** (Nov), 148.
11. A. E. Martell and R. M. Smith, "Critical Stability Constants", Plenum, New York, 1974-1988, Vols. 1-6.
12. R. D. Hancock and A. E. Martell, *Comm. Inorg. Chem.*, **1988**, *6*, 237.
13. A. F. Trotman-Dickenson, *J. Chem. Soc.*, **1949**, 1293.
14. R. L. Woodin and J. L. Beauchamp, *J. Am. Chem. Soc.*, **1978**, *100*, 501.
15. W. A. Pavelich and R. W. Taft, *J. Am. Chem. Soc.*, **1957**, *79*, 4935.
16. L. P. Hammett, "Physical Organic Chemistry", McGraw-Hill, New York, 1940.
17. N. B. Chapman and J. Shorter, "Advances in Linear Free Energy Relationships"; Plenum Press, London, 1972.
18. J. Shorter, *Q. Rev. Chem. Soc.*, **1970**, *24*, 433.
19. R. D. Hancock, *J. Chem. Soc., Dalton Trans.*, **1980**, 416.
20. R. H. Staley and J. L. Beauchamp, *J. Am. Chem. Soc.*, **1975**, <u>97</u>, 5920.
21. W. R. Davidson and P. Kebarle *J. Am. Chem. Soc.*, **1976**, *98*, 6133.
22. J. S. Uppal and R. H. Staley *J. Am. Chem. Soc.*, **1982**, *104*, 1235.
23. J. S. Uppal and R. H. Staley *J. Am. Chem. Soc.*, **1982**, *104*, 1238.
24. M. M. Kappes and R. H. Staley, *J. Am. Chem. Soc.*, **1982**, *104*, 1813.
25. R. W. Jones and R. H. Staley, *J. Phys. Chem.*, **1982**, *86*, 1387.
26. R. R. Corderman and J. L. Beauchamp, *J. Am. Chem. Soc.*, **98**, 3998.
27. M. M. Kappes and R. H. Staley, *J. Am. Chem. Soc.*, **1982**, *104*,1819.
28. R. D. Shannon, *Acta Crystallogr., Sect. A*, **1976**, *A32*, 751.
29. R. G. Pearson and R. J. Mawby, *Halogen Chem.*, **1967**, *3*, 55.
30. F. J. C. Rossotti, in "Modern Coordination Chemistry", J. Lewis and R. G. Wilkins, Eds., Interscience, New York, 1960, p. 1.
31. a) A. C. Knipe, *J. Chem. Ed.*, **1976**, *53*, 618. b) W. P. Weber and G. W. Gokel, "Phase Transfer Catalysts in Organic Synthesis", Springer-Verlag, Berlin, 1977. c) M. Hiraoka, "Crown Compounds: their Characteristics and Applications", Elsevier, Amsterdam, 1982.
32. E. Larsson, *Z. Phys. Chem.*, A, **1934**, *169*, 215.
33. G. Hefter, *Coord. Chem. Rev.*, **1974**, *12*, 221.
34. B. P. Hay, J. R. Rustad, and C. Hostetler, *J. Am. Chem. Soc.*, **1993** ,*115*, 158.
35. A. B. P. Lever "Inorganic Electronic Spectroscopy", 2nd Ed., Elsevier, Amsterdam, 1984.
36. M. Gerloch and R. C. Slade, "Ligand Field Parameters", Cambridge University Press, 1973.
37. B. N. Figgis, "Introduction to Ligand Fields", Wiley, New York, 1966.
38. J. K. Burdett, *J. Chem. Soc., Dalton Trans.*, **1976**, 1725.
39. H. Irving and R. J. P. Williams, *Nature, Lond.*, **1948**, *162*, 746.
40. R. D. Hancock, F. Marsicano, and E. Rudolph, *J. Coord. Chem.*, **1980**, *10*, 23.
41 H. Irving and D. H. Mellor, *J. Chem. Soc.*, **1962**, 5222.
42. L. Fabbrizzi, P. Paoletti, and A. B. P. Lever, *Inorg. Chem.*, **1976**, *15*, 1502.
43. D. K. Cabbiness and D. W. Margerum, *J. Am. Chem. Soc.*, **1969**, *91*, 6540.
44. R. G. Pearson, *J. Am. Chem. Soc.*, **1963**, *85*, 3533.
45. A. W. Adamson, *J. Am. Chem. Soc.*, **1954**, *76*, 1578.
46. R. D. Hancock and F. Marsicano, *J. Chem. Soc., Dalton Trans.*, **1976**, 1096.
47. R. D. Hancock and F. Marsicano, *Inorg. Chem.*, **1978**, *17*, 560.
48. R. D. Hancock and F. Marsicano, *Inorg. Chem.*, **1980**, *19*, 2709.

49. J. O. Edwards, *J. Am. Chem. Soc.*, **1954**, *76*, 1540.
50. S. Yamada and M. Tanaka, *J. Inorg. Nucl. Chem.*, **1975**, *37*, 587.
51. A. Yingst and D. H. McDaniel, *J. Am. Chem. Soc.*, **1967**, *89*, 1067.
52. R. S. Drago, G. C. Vogel, and T. E. Needham, *J. Am. Chem. Soc.*, **1971**, *93*, 6014.
53. a) S. H. Wu, D. S. Lee, C. S. Chung, *Inorg. Chem.*, **1984**, *23*, 2548. b) C. S. Chung, *Inorg. Chem.*, **1991**, *30*, 1685.
54. G. Klopman, *J. Am. Chem. Soc.*, **1968**, *90*, 223.
55. C. J. Pedersen, *J. Am. Chem. Soc.*, **1967**, *89*, 2459.
56. a) R. D. Hancock, *Pure Appl. Chem.*, **1986**, *58*, 1445. b) R. D. Hancock and A. E. Martell, *Chem. Rev.*, **1989**, *89*, 1875.
57. H. Irving and J. J. R. Fausto da Silva, *J. Chem. Soc.*, **1963**, 1144.
58. K. V. Damu, R. D. Hancock, P. W. Wade, J. C. A. Boeyens, D. G. Billing, and S. M. Dobson, *J. Chem. Soc.*, **1991**, 293.
59. a) R. D. Hancock and B. S. Nakani, *S. Afr. J. Chem.*, **1982**, *35*, 153. b) A. Evers, R. D. Hancock, A. E. Martell, and R. J. Motekaitis, *Inorg. Chem.*, **1989**, *28*, 2189.
60. F. Mulla, F. Marsicano, B. S. Nakani, and R. D. Hancock, *Inorg. Chem.*, **1985**, *24*, 3076.
61. R. D. Hancock, B. S. Nakani, and F. Marsicano, *Inorg. Chem.*, **1983**, *22*, 2531.
62. A. Walsh and B. J. Hathaway, *J. Chem. Soc., Dalton Trans.*, **1984**, 15.
63. H. Taube, *Coord. Chem. Rev.*, **1976**, *26*, 33.
64. S. Ahrland, J. Chatt, and N. R. Davies, *Q. Rev. Chem. Soc.*, **1958**, *12*, 265.
65. R. D. Hancock and F. Marsicano, *J. Chem. Soc., Dalton Trans.*, **1976**, 1832.
66. P. M. May and R. A. Bulman, *Prog. Med. Chem.*, **1983**, *20*, 226.
67. D. K. Mills, J. H Reibenspiess, and M. Y. Darensbourg, *Inorg. Chem.*, **1990**, *29*, 4364.

CHAPTER 3

CHELATING LIGANDS

3.1 The Chelate Effect

The Chelate Effect[1] produces increased stability for the complexes of chelating ligands as compared with those of open-chain analogues. This is seen for the formation constants of complexes of Ni(II) with n-dentate polyamines, as compared with the analogous complexes with ammonia.[2]

Table 3.1. The chelate effect for complexes of Ni(II) (s = 1) with polyamines. Polyamines are of formula $NH_2(CH_2CH_2NH)_{(n-1)}H$.[a]

polyamine	EN	DIEN	TRIEN	TETREN	PENTEN[b]
denticity, n	2	3	4	5	6
log β_n (NH$_3$)	5.08	6.85	8.12	8.93	9.08
log K_1 (polyamine)	7.47	10.7	14.4	17.4	19.1

[a] Log K data from Ref. 2, ionic strength = 0.5 M. [b] Ligand is non-linear, with formula corresponding to $(NH_2CH_2CH_2)_2NCH_2CH_2N(CH_2CH_2NH_2)_2$

The origin of the chelate effect has caused much controversy.[3] Schwarzenbach originally explained the chelate effect in terms of the idea that once the first donor atom had attached itself to the metal ion, the second and subsequent donor atoms could move in only a restricted volume about the metal ion. This idea is illustrated in Figure 3.1. This in effect meant that the entropy of these subsequent donor atoms was greatly reduced as compared with an equal number of unidentate ligands. Schwarzenbach's model predicted that the chelate effect would manifest itself as a more favorable entropy of complex formation than would be the case for the analogous complex containing unidentate ligands, which seems to agree with observation. This model also predicted that the stability of complexes with larger chelate rings would be of lower complex stability than those with five membered chelate rings, because of the larger volume to which the chelate ring would be restricted when coordinated to the metal ion by only one donor atom. This appears to be true in general, but the model does not accord with observation in that it predicts that the decrease in formation constants that occurs as chelate ring size is

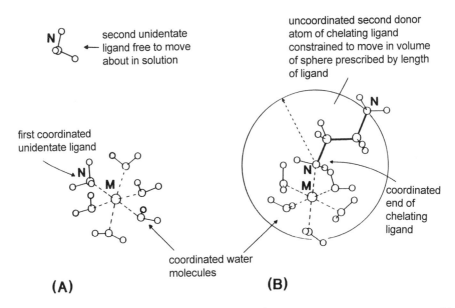

first coordinated
unidentate ligand

second unidentate
ligand free to move
about in solution

uncoordinated second donor
atom of chelating ligand
constrained to move in volume
of sphere prescribed by length
of ligand

coordinated
end of
chelating
ligand

coordinated water
molecules

(A)

(B)

Figure 3.1. Diagram illustrating the Schwarzenbach[1] model of the Chelate Effect. In (a) the second monodentate ligand (ammonia) is free to translate in solution, while in (b) the second donor atom of the chelating ligand (ethylenediamine) is constrained to move in a sphere, whose radius is prescribed by the length of the bridge connecting the two donor atoms.

increased should be an entropy effect, whereas, at least for chelate ring sizes less than seven,[4] it is predominately an enthalpy effect.

3.2 The Standard Reference State and the Chelate Effect

An alternative explanation for the Chelate Effect was put forward by Adamson.[5] He noted that the units used for the concentration of the species involved in the formation constants were such that the formation constant β_n would have units of l^n mol^{-n}. This is caused by the "asymmetry of the standard reference state" which arises because the solvent is given an activity of unity, whereas the concentrations of all other species in the equilibrium are expressed as mol.l^{-1}. One finds therefore that for unidentate and chelate ligands with the same number of donor atoms coordinated to the metal ion, the β_n values have different values of n, with different units, and so one should not really compare the constants as we have done in Table 3.1. The chelate effect even becomes reversed[5] when concentrations are expressed in mol.ml^{-1} instead of mol.l^{-1}, with complexes of unidentate ligands now appearing to be the more stable. Adamson suggested that concentrations should be expressed as mole fractions, which makes the formation constants dimensionless, so that the comparisons in Table 3.1 can be made. The effect of Adamson's proposal is that, at infinite dilution or low concentration, the total number of moles present in the solution is effectively the molarity of pure water, i.e. 55.5 M at 25 °C. Thus, the mole fraction of each species in the equilibrium is obtained by dividing its molarity by 55.5. The result of this is that n.log 55.5 must be added to all log β_n values to make them comparable, and when this is done, the chelate effect largely disappears.[5] This statement has been the cause of much controversy, but, as discussed below, the problem is quite easily resolved.

3.3   Equations for predicting the stability of Complexes of Chelating Ligands

Adamson's proposal leads[6] to equation 3.1 for relating the formation constant of an n-dentate chelating ligand to that of the analogous complex containing unidentate ligands:

$$\log K_1 \text{ (polydentate)} = \log \beta_n \text{ (unidentate)} + (n - 1) \log 55.5 \qquad 3.1$$

The predictions of equation 3.1 for polyamine complexes are usually rather lower than observed, as seen in Table 3.2. The reason for this is that the primary and secondary nitrogen donors ($pK_a = 10.6$) of the polyamines are more basic than the zero order nitrogen of ammonia ($pK_a = 9.2$). One can correct for this by adding an inductive effect factor to equation 3.1 of 1.152 ($= 10.6/9.2$), to give equation 3.2 for polyamines:

$$\log K_1 \text{ (polyamine)} = 1.152 \cdot \log \beta_n \text{ (ammonia)} + (n - 1) \log 55.5 \qquad 3.2$$

The predicted values of $\log K_1$ for polyamines are then in good agreement with the observed values, as seen in Table 3.2.

Table 3.2. Values of formation constant, $\log K_1$, predicted for polyamines considering asymmetry of standard state only (equation 3.1), and correcting for inductive effects (equation 3.2), from $\log \beta_n$ values of analogous ammonia complexes, for Ni(II) ($s = 1$). The observed values are shown for comparison. [a]

polyamine	EN	DIEN	TRIEN	TETREN	PENTEN
denticity, n	2	3	4	5	6
$\log K_1$ (polyamine) for Ni(II)					
calcd by eq. 3.1	6.82	10.33	13.34	15.89	17.78
calcd by eq. 3.2	7.58	11.37	14.67	17.25	19.16
observed[b]	7.47	10.96	14.4	17.4	19.1

[a] Polyamine abbreviations as for Table 1. [b] Observed values from Ref. 2, ionic strength = 1.0 M.

Table 3.3 shows a listing of $\log K_1$ values for polyamines calculated for a variety of metal ions using equation 3.2. These results mean that the formation constants of complexes of polyamines can be accounted for in terms of (1) the entropy contribution from the asymmetry of the standard reference state, and (2) an inductive effect contribution from the ethylene bridges connecting the donor atoms of the polyamine together. This model suggests that the chelate effect should be primarily an entropy effect from contribution (1), with an enthalpy contribution from the inductive effect contribution in (2). Examination of the thermodynamics of complex formation of EN complexes of Cu(II) and Ni(II) supports this idea, as seen in Table 3.4.

Table 3.3. Values of log K_1 for n-dentate polyamines of the formula $NH_2(CH_2CH_2NH)_{(n-1)}H$, as calculated by equation 3.2, and observed[a]

polyamine		EN	DIEN	TRIEN	TETREN	PENTEN
denticity, n		2	3	4	5	6
log K_1 [Cu(II)]	calc	10.76	15.92	20.20	21.28	
	obsd	10.54	15.9	20.1	22.8	
log K_1 [Ni(II)]	calc	7.58	11.37	14.67	17.25	19.16
	obsd	7.47	10.96	14.4	17.4	19.1
log K_1 [Fe(II)]	calc	4.38	6.82	8.67	10.02	10.87
	obsd	4.34	6.23	7.76	9.85	11.1
log K_1 [Pb(II)]	calc	4.92	7.51	9.95	11.18	12.26
	obsd	5.04	7.56	10.35	10.5	

[a] Observed values at ionic strength = 0.10 M and 25 °C from Ref. 2.

Table 3.4. The thermodynamic origin of the chelate effect shown by a comparison of the thermodynamics of formation of complexes of Ni(II) and Cu(II) with ammonia and EN (ethylenediamine).[a]

Complex[b]	log β_n	ΔH[c]	ΔS[c]	log (CH)[d]	ΔH (CH)[d]	ΔS(CH)[d]
$[Cu(NH_3)_2]^{2+}$	7.83	-11.1	-1	2.71	-2.0	7
$[Cu(EN)]^{2+}$	10.54	-13.1	6			
$[Cu(NH_3)_4]^{2+}$	13.00	-22.0	-14	6.6	-3.5	21
$[Cu(EN)_2]^{2+}$	19.6	-25.5	7			
$Ni(NH_3)_2]^{2+}$	5.08	-7.8	-3	2.27	-1.2	7
$[Ni(EN)]^{2+}$	7.35	-9.0	4			
$[Ni(NH_3)_4]^{2+}$	8.12	-15.6	-15	5.38	-2.7	18
$[Ni(EN)_2]^{2+}$	13.54	-18.3	3			
$[Ni(NH_3)_6]^{2+}$	9.08	-24	-39	8.63	-4	29
$[Ni(EN)_3]^{2+}$	17.71	-28.0	-10			

[a] Data from Ref. 2, 25 °C, ionic strength = 1.0 M. [b] Coordinated waters omitted for simplicity. [c] ΔH in kcal·mol[-1], ΔS in cal·mol[-1]·deg[-1]. [d] log K(CH), ΔH(CH) and ΔS(CH) are the formation constant, enthalpy, and entropy of the chelate effect in polyamine ligands, and correspond to the equilibrium $[M(NH_3)_{2y}]^{2+} + y \cdot EN \rightleftharpoons [M(EN)_y]^{2+} + 2y(NH_3)$.

The resolution of the problem over whether the Adamson or Schwarzenbach model of the chelate effect is correct, and whether indeed the chelate effect exists at all is quite simple. The proposals of Adamson[5] and Schwarzenbach[1] on the origin of the chelate effect are essentially the same. In the Schwarzenbach model the translational entropy of the second unidentate ligand is set close to zero by making it move in a restricted volume, while in the Adamson approach the translational entropy of the second unidentate ligand is set at zero by making each reactant fill completely the space of the standard reference state once the constants are expressed as mole fractions. However, in practical terms the Adamson approach seems to be the simpler, since it makes no assumptions about ligand geometry or the length of the bridge connecting the two or more donor atoms. Therefore, equation 3.1 will be used as the basis for analyzing the chelate effect.

Equation 3.2 works well, as seen for polyamines in Table 3.3. It should be noted that for Pb(II) and Fe(II) only log K_1 values are known for ammonia. The log $\beta_n(NH_3)$ values for n greater than 1 are estimated for these metal ions as in equation 3.3.

$$\log K_1 \text{ (polyamine)} = 1.152 \cdot n \cdot \log K_1 \text{ (ammonia)} - \sum_{i=1}^{n-1} (i)\lambda_N + (n-1) \log 55.5 \qquad 3.3$$

In equation 3.3 the term in λ_N is included, where λ_N is the stepwise decrease between log K_y and log $K_{(y+1)}$ values for ammonia complexes, and has[6] a mean value of 0.5. Equation 3.3 is thus equivalent to equation 3.2, except that the log β_n (ammonia) values of equation 3.2 are generated using a fixed value of λ_N of 0.5 plus the log K_1 (ammonia) values.

Equation 3.3 has been extended[6] to include acetate groups. This has been achieved by adding a second set of terms analogous to those for the nitrogen donors, with a separate λ_O term that handles the stepwise decrease in log K_n for acetate complexes as more acetates are added. Initially,[6] λ_O was set equal to 0.19 log K_1(acetate) for each metal ion, but was later found to be better obtained by empirically adjusting it to a best fit for each metal ion.

$$\log K_1 \text{ (amino acid)} = 1.152 n \cdot \log K_1 \text{ (ammonia)} - \sum_{i=1}^{n-1} (i)\lambda_N + m \cdot \log K_1 \text{ (acetate)} - \sum_{i=1}^{m} (i)\lambda_O \qquad 3.4$$

$$+ \ (m + n - 1) \log 55.5$$

In equation 3.4, m is the number of acetate groups on the amino acid (e.g., m is four for EDTA). The empirically adjusted values of λ_O are much smaller than those found as log $K_1(CH_3COO^-)$ - log $K_2(CH_3COO^-)$ for metal ions. This was rationalized[6] as the large experimental values of λ_O reflecting the effect on log K_n of mutual electrostatic repulsion between the negatively charged acetate groups as the value of n is increased. This effect is largely removed if the acetate groups are bound together in a single chelating ligand. Table 3.5 shows the usefulness of equation 3.4 in predicting log K_1 values for amino acid type ligands. Equations similar to 3.4 have been developed[7] with some success to include other types of group, such as pyridyl, imidazolate, and phenolate groups.

For ligands containing pyridyl groups, paralleling the results for polyamines, equation 3.1 predicts log K_1 values for ligands such as bipyridyl and terpyridyl which are much too low, if log β_n values for pyridine (PY) are used in place of log $\beta_n(NH_3)$. Similarly, the log K_1 values for ligands containing both saturated amine and pyridyl groups, such as AMPY, are much too low. This can be remedied by using empirical values for log K_1 (PY) calculated from the known log K_1 (BPY) values, using equation 3.1. This parallels the use of a factor of 1.152 in equation 3.2 to correct for the greater basicity of primary nitrogens compared with ammonia. However, the higher values for the empirically derived log K_1(PY) values probably do not reflect greater basicity, but rather a diminution of the steric interference produced by the *ortho* hydrogens on pyridine

Table 3.5. Values of log K_1 for various polyaminocarboxylate ligands, calculated with equation 3.4, and observed.[a]

Metal Ion		Ca(II)	La(III)	Ni(II)	Pb(II)	Al(III)	Fe(III)
GLY[b]	calc	2.19	4.06	6.26	5.65	5.43	9.57
	obsd	1.39	4.00	6.18	5.23	5.47	10.66
IDA	calc	4.85	8.04	9.37	9.39	9.61	13.99
	obsd	3.47	7.20	9.14	8.29	9.42	14.08
NTA	calc	7.75	12.17	12.45	13.06	13.46	18.14
	obsd	7.71	12.45	12.83	12.66	13.38	17.88
EDMA	calc	3.70	6.03	10.62	8.74	7.59	15.18
	obsd			10.88	8.67		
EDDA	calc	5.85	9.51	13.73	12.84	11.77	19.78
	obsd	(5.51)[c]	(8.36)	14.53	11.58		(18.24)
EDTA	calc	11.90	17.92	19.86	19.75	19.14	27.65
	obsd	12.37	18.14	20.36	19.80	19.20	27.64
DTMA	calc	3.71	6.50	14.49	11.32	9.25	20.30
	obsd			14.81			

[a] At ionic strength of zero. Where values are reported only at other ionic strengths, these have been corrected to zero by comparison with metal ions and ligands of the same charge reported at both ionic strengths. Observed values are from Ref. 2. The log K_1 (NH_3) values used are from Table 2.4. The log K_1 (CH_3COO^-) values used are calculated as 0.34 [log K_1 (OH^-)], where 0.34 = pK_a (acetic acid)/pK_w. This overcomes the problem that many of the acetate complexes of larger metal ions are themselves chelates. The value of λ_0 used are as follows: Ca(II), -0.24; La(III), -0.15; Ni(II), 0.03; Pb(II), 0.07; Al(III), 0.33; Fe(III), 0.60. [b] Ligand abbreviations: GLY = glycine; IDA = iminodiacetic acid; NTA = nitrilotriacetic acid; EDMA = ethylenediaminemonoacetic acid; EDDA = ethylenediamine-N,N'-diacetic acid; EDTA = ethylenedinitrilotetraacetic acid; DTMA = diethylenetriaminemonoacetic acid. [c] Figures in parentheses are for ethylenediamine-N,N-diacetic acid rather than ethylenediamine- N,N'-diacetic acid.

rings, as seen in Figure 3.2. The *ortho* hydrogens on pyridine rings must produce considerable steric hindrance once coordinated to a metal ion, and this will be greatly reduced when the pyridines are bonded to other coordinating groups at the *ortho* position. Some calculated and observed values for ligands containing pyridyl groups coordinated to Ni(II) and Mn(II) are listed in Table 3.6.

As has been pointed out in section 2.5.2, plots of log K_1 for chelating ligands containing negatively charged oxygen donors *versus* log $K_1(OH^-)$ for the metal ions concerned are linear, with intercepts close to log 55.5. Plots of this type are seen for acetylacetonate (ACAC) and tropolonate (TROP) in Figure 3.3. Both plots have intercepts close to the expected value of log 55.5. The points on the relationships show larger than usual deviations from linearity, which is discussed below in terms of the effect of chelate ring size on complex stability.

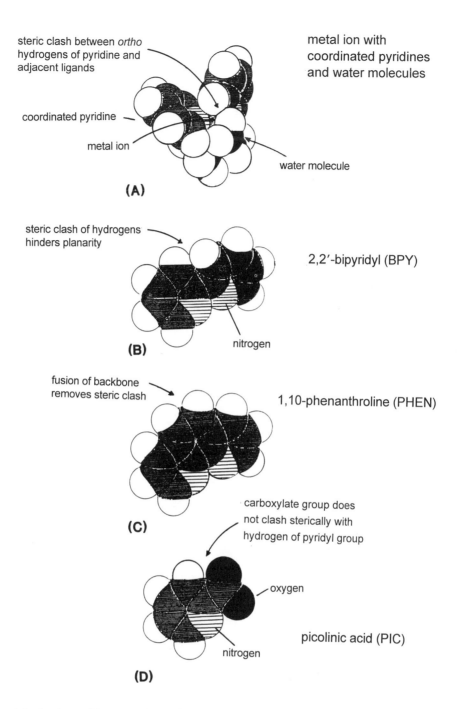

steric clash between *ortho* hydrogens of pyridine and adjacent ligands

metal ion with coordinated pyridines and water molecules

coordinated pyridine

metal ion

water molecule

(A)

steric clash of hydrogens hinders planarity

2,2′-bipyridyl (BPY)

nitrogen

(B)

fusion of backbone removes steric clash

1,10-phenanthroline (PHEN)

(C)

carboxylate group does not clash sterically with hydrogen of pyridyl group

oxygen

picolinic acid (PIC)

nitrogen

(D)

Figure 3.2. Steric problems produced by the *ortho* hydrogens of the pyridyl group in pyridyl containing ligands. Space filling drawings of (A) pyridine groups coordinated to a metal aquo ion, showing how the *ortho* hydrogens create van der Waals repulsions between adjacent coordinated ligands, (B) 2,2′-bipyridyl (BPY), showing how clashes between the *ortho* hydrogens hinder ligand planarity, (C) 1,10-phenanthroline (PHEN), where the clash produced by adjacent *ortho* hydrogens on BPY is removed by forming an extra ring, and (D) picolinic acid, showing that the carboxylate group of the ligand does not clash sterically with the *ortho* hydrogens of its pyridyl group.

Table 3.6. Values of log K_1 observed for ligands containing pyridyl and saturated amino groups, and calculated using equation 3.1.[a]

Ligand[b]		AMPY	IDPY	NTPY	EDDPY	EDTPY	AMPY-DA	BPY	TERPY
log K_1[c] Ni(II)	calc	7.26	0.84	14.04	14.26	19.04	12.77	(7.04)	10.64
	obsd	7.11	8.70	14.45	14.40	18.0	12.1	7.04	0.7
log K_1[c] Mn(II)	calc	2.85	4.24	5.21	5.44	5.8	(2.62)	4.01	
	obsd	2.66	4.16	5.6	5.90	0.3[d]	2.62	4.4	

[a] log K_1 (PY) = 2.9 for Ni(II) and 0.69 for Mn(II), derived as described in the text from log K_1 (BPY) and equation 3.1. [b] Ligand abbreviations: AMPY = 2-aminomethylpyridine; IDPY = imino-bis(methyl-2-pyridine); NTPY = nitrilotris(methyl-2-pyridine); EDDPY = ethylenebis(iminomethyl-2-pyridine); EDTPY = ethylenedinitrilotetrakis(methyl-2-pyridine); AMPY-DA = N-(2-pyridylmethyl)-iminodiacetic acid; BPY = 2,2'-bipyridyl; TERPY = 2,2',2''-terpyridyl. [c] Observed log K_1 values from Reference 2. [d] This anomalously high value may reflect a change in spin state or coordination number for the Mn(II) ion. An anomalously high log K_1 is also observed[2] for the Mn(II) complex of the hexadentate nitrogen donor ligand PENTEN (tetrakis-2-aminoethyl)-ethylenediamine.

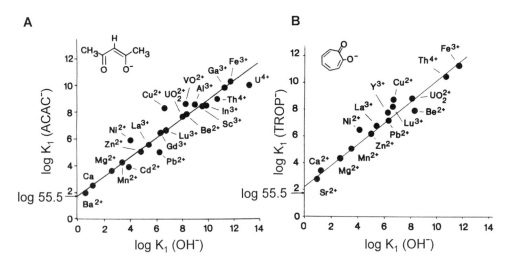

Figure 3.3. Plot of (A) log K_1 for acetylacetonate and (B) tropolonate for a variety of metal ions, *versus* log K_1(OH⁻) for each metal ion. The intercept is the value of log 55.5 expected as the entropy contribution to the chelate effect.[8] The deviation of the points for Pb(II) and Be(II) on each relationship is discussed in the text in terms of the relationship between metal ion size, chelate ring size, and complex stability. This effect is further analyzed in Figure 3.12.

3.4 Rule of Average Environment for Chelating Ligands

Equation 3.4 works because the observed formation constants for amino acid ligands are an additive function of the formation constants found for the unidentate ligands, once corrections for small differences in inductive effects, for example, have been taken into account, as well as the entropy contributions from the asymmetry of the standard reference state. The observed additivity suggests that a "rule of average environment"[9] would apply to sets of ligands such as EN, OX and GLY, or PHEN, catecholate, and OXINE, below:

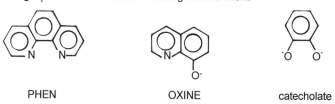

	EN	GLY	OX
Log K_1 Cu(II)	10.48	8.15 (8.36)[a]	6.23
Log K_1 Ni(II)	7.35	6.18 (6.26)[a]	5.16
Log K_1 Cr(II)	5.48	4.74	(4.00)[a]
Log K_1 Pb(II)	5.04	4.87 (4.98)[a]	4.91

[a] log K_1 calculated from rule of average environment.

PHEN   OXINE   catecholate

In Figure 3.4 is shown the relationship between log K_1 (OXINE) and the mean of log K_1 (PHEN) and log K_1 (catecholate) for a variety of metal ions. No values of log K_1 are currently available for log K_1 (catecholate) for Ca^{2+}, In^{3+}, and Pb^{2+}, and these were estimated with the help of Figure 2.18 as 3.5, 19.5, and 12.5 respectively. Figure 3.4 can now be used to make predictions that would be hard to arrive at by any other means. For example, no data are available that would even allow a guess at what log K_1 (PHEN) might be for Th(IV). However, log K_1 (OXINE) is 10.5 for Th(IV), and log K_1 (catecholate) is[2] 19.8 (Figure 2.18 would give an estimate of 20.6). This leads to a prediction of log K_1 (PHEN) for Th(IV) of 10.5 x 2 - 19.8 = 1.2. As would be expected, the very hard Th(IV) ion has only a limited affinity for PHEN.

For some systems, the rule of average environment does not hold without some modification. Thus, for the set of ligands PIC (picolonic acid), BPY, and OX, the formation constants estimated for PIC as {log K_1 (BPY) + log K_1(OX)}/2 are consistently lower than observed values by about 1.4 log units. This is shown in Figure 3.5, where the relationship between log K_1 (PIC) and {log K_1 (BPY) + log K_1 (OX)}/2 has a slope of unity, but an intercept of 1.4 log units. This can be understood in terms of the role of the *ortho* hydrogens of BPY in producing difficulty in coordinating to metal ions, as summarized in Figure 3.2. By contrast, the carboxylate group of picolinic acid, as seen in Figure 3.2, does not interfere sterically with the adjacent *ortho* hydrogens of the pyridine ring. It seems reasonable that this lack of steric interference on the part of the carboxylate group would account for the higher than expected (in terms of the rule of average environment) stability of the PIC complexes. Figure 3.5 can be used to make a variety of predictions, such as log K_1 (BPY) for UO_2^{2+} = 2.2, or PuO_2^{2+} = 2.9. It was the correlation in Figure 3.5 that suggested that log K_1 (BPY) for La^{3+} should be 1.2. Experimentally, log K_1 (BPY) was found[10] to be 1.1 for La^{3+}.

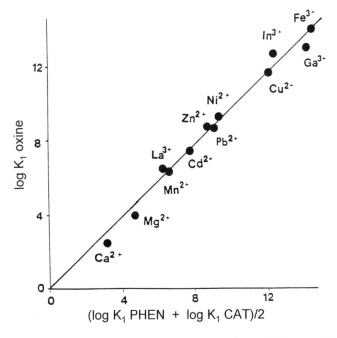

Figure 3.4. The Rule of Average Environment in complex stability, as illustrated by the relationship between log K_1 for 8-hydroxyquinoline (oxine) complexes, and the mean of log K_1 (PHEN) (PHEN = 1,10-phenanthroline) and log K_1 (CAT) (CAT = catecholate, 1,2-dihydroxybenzene) for the complex of each metal ion. Data from Ref. 2, ionic strength = 0.10 M, and 25°C.

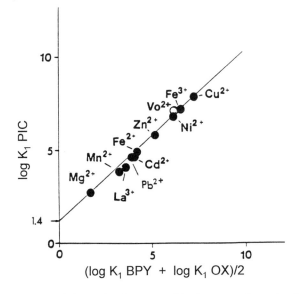

Figure 3.5. Modification of the Rule of Average Environment brought about by the steric clashes between the *ortho* hydrogens on 2,2'-bipyridyl (BPY) on becoming planar, shown in Figure 3.2. The relationship is for log K_1 (PIC) *versus* the mean of log K_1 (BPY) and log K_1 (OX) (PIC = picolinic acid, BPY = 2,2'-bipyridyl, OX = oxalate). The intercept is thought to be due to the fact that there are no steric clashes involving the *ortho* hydrogens of PIC, whereas these are quite marked for BPY (Figure 3.2). Data at 25°C and ionic strength = 0.10 M, from Ref. 2.

Other approaches to estimating the formation constants of chelating ligands have been detailed[11] elsewhere. These include plots of log K_1 for one metal ion such as Cu(II) against log K_1 for a similar metal ion, in this case Ni(II).[12] This is shown in Figure 2.6. What has become clear at this point is that, in general, the formation constants of polydentate ligands can be derived from the formation constants of unidentate analogues, following quite simple approaches based on additivity. However, the complexes of nearly all of the ligands discussed so far involve five membered chelate rings only. Equations such as 3.3 or 3.4 can be modified to take into account the observation that complexes with six membered chelate rings usually form complexes which are lower in stability than their five membered ring analogues. Thus, empirically adjusting the values of λ_N and λ_O in equation 3.4 can produce satisfactory prediction of log K_1 for many ligands containing six membered chelate rings. A similar approach has been adopted by Harris.[7] However, such empirical approaches fail to predict correctly those examples where a combination of five and six membered chelate rings leads to an increase in log K_1. The effect of chelate ring size is of fundamental importance in controlling complex stability and selectivity. The shifts that occur in formation constant on change of chelate ring size are steric in origin, and to comprehend them more fully, the steric aspects of complex formation must be considered.

3.5   The Size of the Chelate Ring and Complex Stability

It is usually observed that an increase of chelate ring size leads to a decrease in complex stability. This observation was originally interpreted by Schwarzenbach[1] in terms of his model in which the longer connecting bridges of ligands that form larger chelate rings allow the second donor atom to translate in a larger volume after the first donor atom had become attached. This lowered the probability of attachment of the second donor atom, which meant that the drop in complex stability as chelate ring size increased should be an entropy effect. As seen in Table 3.7, the drops in complex stability for TN complexes relative to EN complexes, or TMDTA relative to EDTA complexes, are almost entirely due to enthalpy effects. Increase in chelate ring size from five membered to six membered in these complexes leads to drops in complex stability that are wholly caused by less favorable enthalpy contributions. Where decreasing complex stability with increase in chelate ring size has not been interpreted in entropy terms, these enthalpy based changes have[4] been attributed to steric strain, or repulsion between lone pairs on donor atoms. The steric strain interpretation can be checked by molecular mechanics (MM) calculation.

The importance of steric strain to the enthalpy of complex formation, and hence to the stability of complexes, lies in the change in steric strain, ΔU, that occurs on complex formation:

$$
\begin{array}{llll}
M & + & n{\cdot}L & \longrightarrow & ML_n \\
U_M & & n{\cdot}U_L & & U_{ML_n} \\
\Delta U & = & U_{ML_n} & - U_M & - n{\cdot}U_L
\end{array}
\qquad 3.5
$$

In equation 3.5, U_M, U_L, and U_{ML} are the strain energies of the free metal ion, the ligand L, and the complex formed between the metal ion and n ligands, respectively. The complexes of Ni(II) with polyamines with differing sizes of chelate rings are ideal to test the idea that decrease of complex stability, which is almost entirely an enthalpy effect, is largely due to steric strain. There is available[2] a large body of data on the formation constants and enthalpies of complex formation of these complexes, as well as crystallographic studies, which allow[13] for the development of appropriate force field

Table 3.7 (a). Thermodynamics of complex formation of EDTA (five membered chelate ring involving both N-donors) compared with TMDTA (six membered ring involving both N-donors).[a]

Metal Ion	Ionic Radius[c]	EDTA[b]			TMDTA		
		log K_1	ΔH	ΔS	log K_1	ΔH	ΔS
Cu^{2+}	0.57	18.70	-8.2	58	18.82	-7.7	60
Ni^{2+}	0.69	18.52	-7.6	59	18.07	-6.7	60
Zn^{2+}	0.74	16.44	-4.9	59	15.23	-2.3	62
Cd^{2+}	0.95	16.36	-9.1	44	13.83	-5.4	45
Ca^{2+}	1.00	10.61	-6.6	26	7.26	-1.7	27
La^{3+}	1.03	15.46	-2.9	61	11.28	+3.8	64
Pb^{2+}	1.18	17.88	-13.2	38	13.70	-6.4	41

[a] Units for ΔH are kcal.mol[-1], for ΔS are cal·deg[-1]·mol[-1]. Data from Ref. 2. [b] Ligand abbreviations: EDTA = ethylenedinitrilotetraacetic acid; TMDTA = trimethylenediaminetetraacetic acid. [c] Units are Å, from Ref. 14.

Table 3.7(b). Thermodynamics of complex formation for EN (five membered chelate ring) and TN (six membered chelate ring).[a]

Metal ion	Complex[c]	EN[b]			TN		
		log K	ΔH	ΔS	log K	ΔH	ΔS
Cu(II)	ML	10.48	-12.6	6	9.68	-11.4	6
Cu(II)	ML_2	19.57	-25.2	5	16.79	-22.4	2
Ni(II)	ML	7.33	-9.0	3	6.30	-7.8	3
Ni(II)	ML_2	13.41	-18.3	0	10.48	-15.0	-2
Cd(II)	ML	5.42	-6	5	4.47	-5	4
Cd(II)	ML_2	9.60	-13.3	-1	7.18	-10.0	-1

[a] Units for ΔH are kcal.mol[-1], for ΔS are cal·deg[-1]·mol[-1]. Data from Ref. 2, at 25°C, ionic strength 0.10 M. [b] Ligand abbreviations: EN = ethylenediamine; TN = 1,3-diaminopropane. [c] For the complex indicated as ML, log K refers to the equilibrium M + L $\rightleftharpoons$ ML, and for the complexes indicated as ML_2, log K refers to the equilibrium M + 2L $\rightleftharpoons$ ML_2.

parameters for the bonds involving the high-spin Ni(II) ion. Of specific interest here is the pair of complexes [Ni(2,2,2-tet)(H$_2$O)$_2$]$^{2+}$ and [Ni(2,3,2-tet)(H$_2$O)$_2$]$^{2+}$, because, unlike the usual situation, the complex with 2,3,2-tet, which forms one six membered ring and two five membered chelate rings, is more stable[2] than that with 2,2,2-tet, which forms all five membered rings. Therefore, values of U_L were calculated for the free ligands EN and TN, and U_L was estimated to differ for DIEN and DPTN by 0.37 kcal.mol[-1] per extra methylene group,[15] which was the calculated difference in U_L for EN and TN. Values of U_{ML} were calculated for the pairs of complexes shown in Table 3.8. The differences in ΔU in Table 3.8 correspond quite well with the differences in ΔH. Of importance is the

Table 3.8. Changes in enthalpy of complex formation of polyamine complexes of Ni(II) on increasing the chelate ring size from five to six membered, compared with the difference in strain energy calculated by molecular mechanics calculation.[a]

complex[b]	U[c]	-ΔU[d]	ΔH[e]	-Δ(ΔH)
Ni(EN)	1.14		-9.0	
		1.53		1.2
Ni(TN)	4.04		-7.8	
Ni(EN)$_2$	3.35		-18.3	
		3.07		3.3
Ni(TN)$_2$	7.16		-15.0	
Ni(EN)$_3$	4.57		-28.0	
		7.44		6.7
Ni(TN)$_3$	13.12		-21.3	
Ni(DIEN)	6.08		-11.9	
		1.46		1.3
Ni(DPTN)	8.28		-10.6	
Ni(DIEN)$_2$	11.87		-25.3	
		7.97		7.7
Ni(DPTN)$_2$	21.32		-17.6	
Ni(2,2,2-tet)	9.44		-14.0	
		-2.49		-3.9
Ni(2,3,2-tet)	7.32		-17.9	

[a] The difference in strain energy, -ΔU, should be compared with the difference in enthalpy of complex formation, -Δ(ΔH); units are kcal.mol-1. [b] EN = ethylenediamine, TN = 1,3-diaminopropane, DIEN = 1,4,7-triazaheptane, DPTN = 1,5,9-triazanonane, 2,2,2-tet = 1,4,7,10-tetraazadecane, 2,3,2-tet = 1,4,8,11-tetraazaundecane. All complexes are for octahedral high-spin Ni(II), and charges and coordinated waters on the complexes required to make the coordination number up to six, are omitted for simplicity. [c] Ref. 12 and 13. [d] Corrected[12,13] for differences in strain energy of free ligands of 0.37 kcal.mol-1 per extra methylene group. [e] Ref. 2, ionic strength 0.10 M, 25°C.

fact that the complex of the ligand 2,3,2-tet, with a six membered chelate ring, is more stable than that of 2,2,2-tet, which forms five membered chelate rings only. In agreement with general opinion, the complex of 2,3,2-tet is more stable than that of 2,2,2-tet because of steric effects. One might take the view that the bite in the six membered ring in 2,3,2-tet enables the ligand to span the coordination sites on Ni(II) better than 2,2,2-tet, which is too short. Table 3.8 demonstrates the potential of MM calculations for rationalizing and predicting the thermodynamics of complex formation.[14,15]

Examination of the crystal structures of the complexes of Ni(II) with polyamines with five membered or six membered chelate rings suggests that the N-N distance across the chelate ring of TN is greater than that of EN. One might conclude from this that one way in which to increase selectivity of a ligand for large metal ions is to increase the size of the chelate ring. However, a large amount of stability constant data[2] reveals that exactly the opposite is true; an increase in chelate ring size leads to large drops in complex stability for large metal ions, and may even lead to an increase in complex stability for small metal ions. So closely related to metal ion size is this effect that it can be quantified as seen in Figure 3.6. In Figure 3.6(a) Δ log K, the change in complex stability for divalent metal ions in passing from EDTA to TMDTA, is plotted[13] as a function of ionic radius.[14] The ionic radii used are for octahedral coordination, except for the Jahn-Teller distorted Cu(II), where the square-planar radius is used. Plotted in Figure 3.6(b) is the change in ΔH in passing from EDTA to TMDTA complexes. Figure 3.6 shows that the effect on complex stability of increase of chelate ring size from five membered in EDTA complexes to six membered in TMDTA complexes is strongly dependent on metal ion radius, and further, that the effect is entirely due to enthalpy changes.

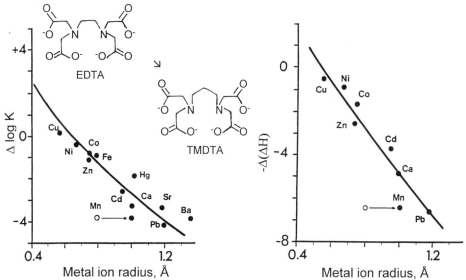

Figure 3.6. (a) Effect on complex stability of increase of chelate ring size, from five membered in EDTA to six membered in TMDTA. The change in formation constant in passing from EDTA to TMDTA, Δ log K, is plotted as a function of octahedral ionic radius[14] of the metal ions. The large deviation for the point for Mn(II) may be corrected, as discussed in the text, by using the effective ionic radius calculated from the crystal structure[16] of [Mn(EDTA)H₂O]²⁻, where the Mn(II) is actually seven coordinate rather than octahedral. Formation constants from Ref. 2. **(b)** The relationship between chelate ring size and the enthalpy of complex formation of EDTA and TMDTA complexes. the values of Δ(ΔH) for each metal ion, M, are the ΔH values for the reaction [M(EDTA)] + TMDTA = [M(TMDTA)] + EDTA. The fact that the changes in complex stability on increase of chelate ring size are entirely an enthalpy effect supports the interpretation of the effect as due to steric strain rather than entropic effects. Enthalpy values from Ref. 2.

In Figure 3.7 is shown a plot of Δ log K for passing from 2,2,2-tet to 2,3,2-tet, as a function of ionic radius. Examination of available formation constants[2] suggests that relationships such as those seen in Figures 3.6 and 3.7 are quite general for ligand pairs where a five membered chelate ring is increased in size to six membered. This observation leads to the formulation of a rule of ligand design: "Increase of chelate ring size from five membered to six membered leads to an increase of selectivity of a ligand for smaller compared to larger metal ions". This rule holds in most situations. An important exception is where the increase in chelate ring size has been carried out for

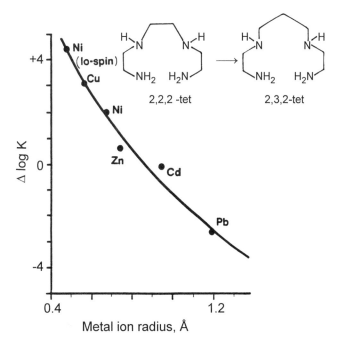

Figure 3.7. Effect on complex stability of increase of chelate ring size from five membered in 2,2,2-tet to six membered in 2,3,2-tet. The change in formation constant, log K_1, in passing from 2,2,2-tet to 2,3,2-tet, Δ log K_1, is plotted as a function of octahedral ionic radius[14] of the metal ions. The ligands 2,2,2-tet and 2,3,2-tet are shown on the Figure. Formation constants from Ref. 2.

sufficient chelate rings that steric difficulties arise in connecting adjacent six membered rings together, which is discussed in section 3.9. As discussed in Chapter 4, this rule also holds for most macrocyclic ligands, and appears to dominate macrocyclic hole size considerations in determining metal ion selectivity. The following discussion details the origin of the metal ion size selectivity exhibited by five and six membered rings.

3.6   The Geometry of the Chelate Ring, and Preferred Metal Ion Sizes

One can readily understand why change of chelate ring size from five membered to six membered results in an increase in complex stability for small metal ions relative to large metal ions, by reference to the low-strain form of cyclohexane.[17,18] Cyclohexane, seen in Figure 3.8, has in its chair conformer the minimum strain energy possible for a cycloalkane. All torsional angles are 60°, and the C-C-C bond angles are all the ideal value of 109.5°. The six membered chelate ring of TN involving two nitrogens and a metal ion in place of three of the carbon atoms of cyclohexane will also be of very low strain energy (Figure 3.8) as long as the metal ion is of about the same size and geometry as an sp^3 hybridized carbon atom. The ideal metal ion for coordination to TN thus has an N-M-N bond angle of 109.5°, and a short M-N bond length of about 1.6 Å. This is shown in Figure 3.9, where the strain energy of the TN ring has been calculated[19] as a function of both N-M-N angle and M-N bond length.

The best size metal ion for the five membered chelate ring can also be derived by considering the cyclohexane ring. Removal of two adjacent carbon atoms, and conversion of the next two carbon atoms into nitrogens (Figure 3.8) in cyclohexane yields a minimum strain ethylenediamine (EN) ring. Extrapolation of the lone pairs on the two

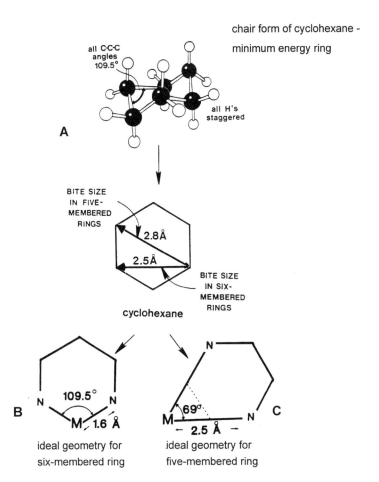

Figure 3.8. The relationship between the chair form of cyclohexane and the geometric requirements for a metal ion to form a minimum strain chelate ring involving the five membered chelate ring of EN (ethylenediamine) or the six membered ring of TN (1,3-diaminopropane). For an alkane to have the minimum strain energy, all torsion angles must be 60°, and the C-C-C angles should be 109.5°, as shown for the chair form of cyclohexane at (A). To maintain minimum strain energy, these torsion and bond angles should also be maintained in chelate rings, which can be derived from cyclohexane as shown. The resulting minimum strain requirements for the six membered chelate ring of TN are shown at (B), and those for the five membered chelate ring of EN are shown at (C). Redrawn after Ref.18.

nitrogens focuses them on a point some 2.5 Å away, after slight adjustment to align the two lone pairs in the same plane. This simple exercise shows that the minimum strain energy should be observed for the five membered EN chelate ring with metal ions having M-N bond lengths of 2.5 Å, and N-M-N angles of about 70°. This is confirmed by MM calculations[19] of strain energy of the EN chelate ring as a function of M-N length and N-M-N bond angle, seen in Figure 3.10.

The relationship of complex stability to chelate ring size, seen in Figures 3.6 and 3.7, is thus readily understood in terms of the best-fit size of metal ion for complexing with five and six membered rings. In particular, smaller metal ions tend to have lower coordination numbers, and so, as the M-N bond length becomes smaller, so the N-M-N

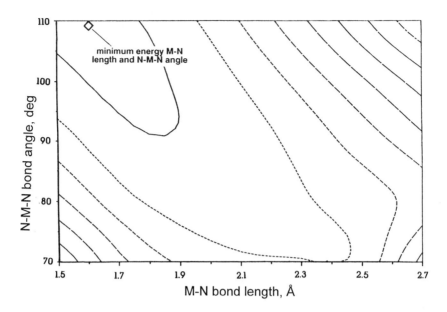

Figure 3.9. Strain energy of the six membered chelate ring of TN as a function of initial strain free M-N bond length and N-M-N angle, calculated as described in the text. The contour lines indicate energies increasing in steps of 1 kcal.mol^{-1} from the lowest energy at 0.5 (———), 1.5 (-----), 2.5 (– – – –), kcal.mol^{-1}. Also indicated (◇) is the lowest energy M-N length and N-M-N angle for a six membered TN chelate ring. Redrawn after Ref. 19.

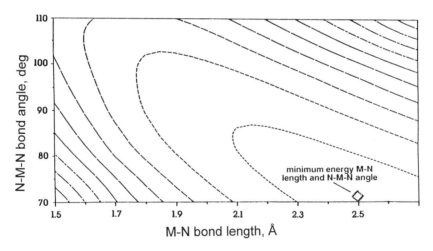

Figure 3.10. Strain energy of the five membered chelate ring of EN as a function of initial strain free M-N bond length and N-M-N angle, calculated as described in the text. The contour lines indicate energies increasing in steps of 1 kcal.mol^{-1} from the lowest energy point at 0.5 (———), 1.5 (-----), 2.5 (– – – –), kcal.mol^{-1}. Also indicated (◇) is the lowest energy M-N length and N-M-N angle for a five membered EN chelate ring. Redrawn after reference 19.

angles should become larger. Conversely, as metal ions become larger, their coordination numbers tend to become higher, and they will therefore have smaller N-M-N angles as the M-N bond becomes longer. Some metal ions may tend to be exceptions to this rule, such as Ag(I) and Hg(II), which, in spite of being large, will tend to have low coordination numbers, preferring in many instances coordination numbers as low as two. Thus, the rather less negative Δ log K value seen for Hg(II) in Figure 3.6, than might have been expected from its large ionic radius,[14] may reflect the idea that the longer bridge in the six membered ring might allow the Hg(II) to approach more closely the placement of the two nitrogens at an angle of 180° to each other. Figure 3.6 for simplicity uses octahedral ionic radius as a measure of metal ion size and therefore of how well the metal ion should fit into a particular size of chelate ring. This is of course only a rough approximation, since many of the metal ions, particularly the large metal ions such as Pb(II), are known[20] to have higher coordination numbers, and therefore larger ionic radii, in their EDTA complexes. It would be more accurate to use crystallographically determined M-N bond lengths and N-M-N angles for each metal ion, and actually calculate the strain in each EN type chelate ring in the EDTA complex, and compare it with the strain in the six membered ring in the TMDTA complex. Such a cumbersome procedure would not generally be useful. However, such ideas go a long way to explain the deviation for the point for Mn(II) in Figure 3.6. Crystallographic studies[16] show that Mn(II) in its EDTA complex is seven coordinate (Figure 3.11), with a M-N length of 2.38 Å and a N-M-N angle of 75.3°. This corresponds to an ionic radius of 1.0 Å for Mn(II),

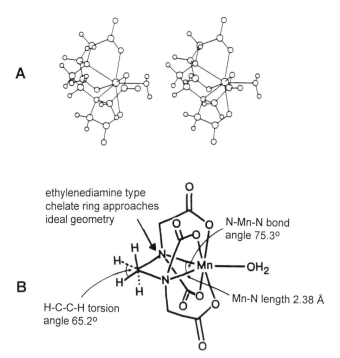

Figure 3.11. Structure of the $[Mn(EDTA)H_2O]^{2-}$ anion[16] in stereoview at (A). At (B) is shown how the seven coordinate arrangement allows a very close approach to ideal geometry in the ethylenediamine type chelate ring of the ligand. Drawn using the ALCHEMY program (TRIPOS Associates, St. Louis, MO) from coordinates in Ref. 16.

which is shown to improve the position of the point on Figure 3.6. Other factors to be taken into account might be that the M-N bonds in the Ba(II) or Sr(II) complexes of EDTA would be much less covalent than those in, say, the Ni(II) complex, and the M-N bond length deformation and N-M-N angle deformation constants would be much smaller for Ba(II) and Sr(II). This might account for the slight upward deviation for the Sr(II) and Ba(II) points in Figure 3.6. Here again, however, one would like to keep the simplicity of using octahedral radii as a measure of the M-N bond lengths for all metal ions. The inaccuracies introduced by this approximation, both with regard to true M-N bond lengths, and the differences between the force constants associated with the M-N bond in different metal ions, do not generally justify a more complex approach. Only for the purposes of understanding deviations such as that for the Mn(II) EDTA complex in Figure 3.6 is a more complex approach justified.

One might ask at this stage whether the analysis in Figure 3.8 requires that complexes of small metal ions with six membered chelate rings be more stable than those with five membered rings. A quick scan of the literature[2] does not suggest that this is the case. However, most metal ions commonly studied are not nearly small enough to satisfy the requirements of the six membered ring. The Cu(II) ion may be regarded as small, but the Cu-N bond length for square planar Cu(II) is[17] about 2.03 Å, considerably longer than the 1.6 Å required for strain free coordination to a six membered chelate ring. However, the small Be(II) and B(III) give very short M-L bond lengths, and here one finds a ready explanation for the greater stability of the complexes of these Lewis acids with a variety of six membered chelate rings as compared with their five membered chelate ring analogues:

	Ionic radius(Å)[a]	OX	MAL	TROP	ACAC	TIRON	CTA
chelate ring size		5	6	5	6	5	6
log K_1 for[b]							
Be(II)	0.27	4.96	6.18	7.40	7.90	14.6	16.3
Cu(II)	0.57	6.23	5.80	9.23	8.25	15.6	15.9
Pb(II)	1.18	4.20	3.98	7.54	4.71	14.8	13.0

[a] Ionic radii (Å) from Ref. 18. [b] Formation constants from Ref. 2, at 25 °C and ionic strength = 0.10 M.

For the very small B(III) the complex with CTA is[2] considerably more stable than that with TIRON. In organic chemistry[21] one also has the observation that five membered rings are generally less stable than six membered rings in, for example, cycloalkanes or lactams, which is readily understood from the present discussion of chelate rings.

In Figure 3.3, as noted in section 3.3, the correlation of log K_1 for acetylacetonate, and to a lesser extent tropolonate, with log K_1(OH⁻) is rather noisy, which is to say that many of the points show quite considerable deviations. This "noise" can be readily understood from ideas developed here on the interplay of chelate ring size, metal ion size, and complex stability. One notes that acetylacetonates have six membered chelate rings, while the similar tropolonates have five membered chelate rings. In Figure 3.12 is plotted $\Delta \log K$, which is $\log K_1$ (tropolonate) - $\log K_1$ (acetylacetonate), *versus* metal ionic radius. One sees the usual response of complex stability to chelate ring size in Figure 3.12. Thus, the large Pb(II) ion has relatively low stability with the six membered chelate ring of acetylacetonate, while the very small Be(II) ion shows little destabilization caused by the six membered ring of acetylacetone.

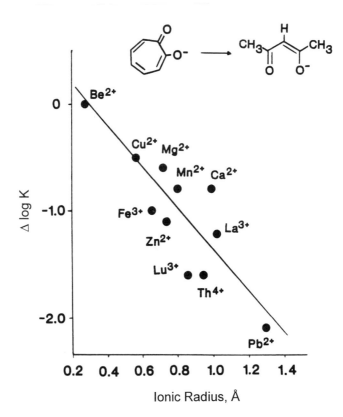

Figure 3.12. How the size of the chelate ring in tropolone complexes (five membered chelate ring) affects the stability of its complexes relative to those of acetylacetonate (six membered chelate ring). The plot shows the change in complex stability, $\Delta \log K$, on increase of chelate ring size [$\Delta \log K = \log K_1$ (acetylacetonate) - $\log K_1$ (tropolonate)], as a function of metal ion radius.[14] Formation constants at 25 °C and ionic strength 0.1, from Ref. 2.

3.7 Chelate Rings of Other Sizes

3.7.1 Chelate rings larger than six membered

For chelate rings larger than six membered, it does appear[2] that there is a considerable contribution from entropy to the further decrease in complex stability that occurs. The thermodynamics[2] of complex formation of Ni(II) complexes of EDTA type ligands of the structure $(^-OOCCH_2)_2N(CH_2)_nN(CH_2COO^-)_2$ as n increases are as follows:

n, the number of methylene groups in the bridge	2	3	4	5	6
log K_1 [Ni(II)]	18.52	18.07	17.27	13.8	13.71
ΔH (kcal.mol^{-1})	-7.6	-6.7	-7.0	-6.7	-8.5
ΔS (cal.deg^{-1}.mol^{-1})	59	60	56	41	34

(data at 25 °C and ionic strength 0.10 M., from Ref. 2.)

The effect on log K_1 for EDTA type ligands as the bridge connecting the two nitrogen donors together becomes longer is seen for a selection of metal ions in Figure 3.13. Figure 3.13 shows that change in complex stability relative to the EDTA complex, Δ log K, for a variety of metal ions, as a function of n, the number of methylene groups in the bridge of the chelate ring involving the two nitrogen donors, as n increases from two to six. Figure 3.13 shows an initial change in complex stability as n increases from two to three which is related to metal ion size, as expected for increase in chelate ring size from five to six membered. Thereafter, there is a steady decrease in complex stability as n

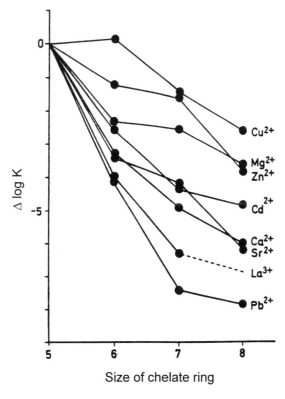

Figure 3.13. Variation of log K_1 with chelate ring size in EDTA type analogues of the formula ($^-$OOCCH$_2$)$_2$N(CH$_2$)$_n$N(CH$_2$COO$^-$)$_2$. Chelate ring size varies from 5 ($n = 2$) to 8 ($n = 5$). The variation is plotted as the change in stability, Δ log K, relative to the EDTA complex ($n = 2$) for each metal ion. Formation constants at 25°C and ionic strength 0.1, from Ref. 2. Redrawn after Ref. 22.

increases up to six, which is not strongly metal ion dependent. This suggests that increase in chelate ring size beyond six membered is not particularly useful as a ligand design tool for producing discrimination between different metal ions.

There is currently[23] great interest in designing complexing agents for Fe(III) for treating iron overload conditions accompanying treatment of the hereditary disease thalassemia. One line of research has been to mimic naturally occurring catecholate and hydroxamate type ligands[23] such as enterobactin or β-desferriferrioxamine-B (DFO) seen in Figure 3.14. Ligands such as MECAM reported by Raymond and coworkers,[24] or BAMTPH reported by Martell and coworkers,[25] are typical examples of synthetic mimics of the naturally occurring iron transport ligands, or siderophores.

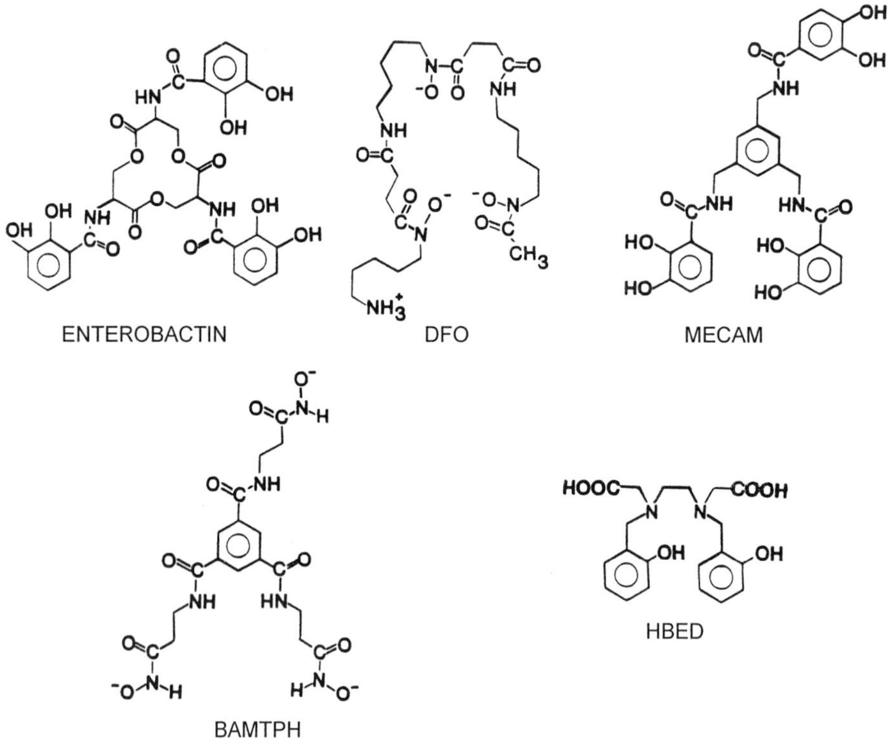

Figure 3.14. Some naturally occurring and synthetic ligands proposed as iron(III) complexing agents.

The catecholate or hydroxamate functional groups contain only negatively charged oxygen donors, or carbonyl groups (in hydroxamates), which oxygens cannot serve as points of attachment for further bridges to other chelating groups, as is the case with saturated nitrogens or ethereal oxygens. This is shown in Figure 3.15. The very long connecting chains required to connect together more than one catecholate or hydroxamate group can therefore be regarded as part of very large chelate rings. For example, the large chelate rings in DFO are effectively fourteen membered chelate rings. The relationship of log K_1 for DFO *versus* log K_1 (OH⁻) for a variety of metal ions is seen in Figure 3.16. What is important in Figure 3.16 is that the intercept is far below the value of 5·log 55.5 = 8.7 expected (see equation 3.2, section 3.3) for a hexadentate ligand. The most reasonable interpretation for this observation would seem to be that the very long bridges in the DFO

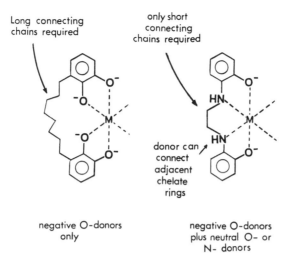

Figure 3.15. How the exclusive use of negatively charged oxygen donors leads to the need for very large chelate rings in ligands which are more than bidentate. This problem is overcome with the incorporation of some neutral nitrogens as donor atoms.

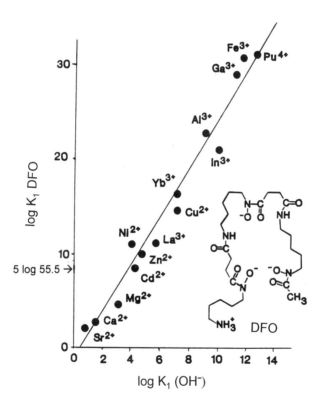

Figure 3.16. The relationship of log K_1 for DFO (desferriferrioxamine-B) complexes, against log K_1 (OH⁻) for the same metal ions. The diagram shows that the intercept is far lower than the 5·log 55.5 expected from equation 3.2. This type of effect is attributed to the high entropy (lack of preorganization)of the free DFO ligand. All data from reference 2, at 25 °C and ionic strength 0.10.

ligand produce a high level of entropy in the free ligand. This is an unfavorable contribution to the free energy of complex formation, which is subtracted from the 5·log 55.5 entropy of the chelate effect, leading to a lower than expected intercept in a diagram such as Figure 3.16. The intercepts in diagrams such as Figure 3.16 should be useful in understanding entropy effects in ligands with negative oxygen donors. Figure 3.16 suggests that the long connecting bridges in ligands such as enterobactin and DFO are serious drawbacks to achieving maximum complex stability, and more compact ligand designs such as the ligand HBED in Figure 3.14, where nitrogens are used to connect adjacent chelate rings together, may ultimately be more successful.

Another example of thermodynamic effects in ligands with chelate rings that are large is seen in dicarboxylic acid ligands such as adipic acid (hexanedioic acid). In Figure 3.17 is plotted log K_1 for succinate against log K_1 (OH⁻) for a variety of metal ions. It is seen that there is a break in the relationship, and that at low stabilities the values of log K_1 (adipate) become independent of log K_1 (OH⁻). The broken line in Figure 3.17 corresponds to log K_1 for an outer-sphere complex involving a doubly charged cation and a doubly charged anion, as calculated using the Fuoss[8] equation. One suggests that in Figure 3.17 the linear portion of the relationship would extrapolate back to a very low intercept, which would correspond to the very large entropy induced into the free adipate

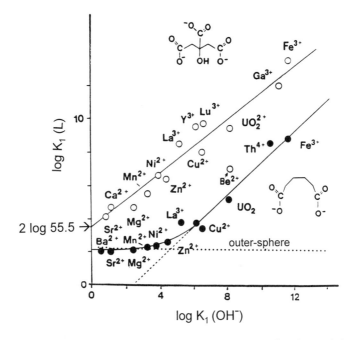

Figure 3.17. Relationship between the formation constants for citrate (O), and for adipate (●),with log K_1 (OH⁻) for a variety of metal ions. The relationship for citrate shows the intercept of 2·log 55.5 expected[27] for a tridentate ligand, but the value for Be(II) is rather low, probably due to steric difficulties of all three donor atoms coordinating effectively to the donor sites on the Be(II) ion. The relationship for adipate shows a negative intercept because of the entropy associated with immobilizing the long bridge between the donor atoms. The dotted line is log K_1 calculated from the Fuoss[8] equation for an outer-sphere complex between a dipositive cation and a dinegative anion. It is suggested[27] that the complexes falling on the dotted line are outer-sphere in nature. Formation constants at 25 °C and ionic strength zero, from Ref. 2.

ligand by its long connecting bridge. The broken line indicates the formation of outer-sphere complexes with adipate by metal ions at lower log K_1 (OH⁻) values rather than inner-sphere complexes, which would be of lower stability judging from extrapolation of the linear portion of the relationship back to lower log K_1 (OH⁻) values. In contrast to adipic acid, citrate in Figure 3.17 has the intercept of 3.4 log units expected for a tridentate ligand.

3.8   More Highly Preorganized Chelating Ligands

More highly preorganized[27] ligands have as free ligands their donor atoms held fixed more nearly in the arrangement required in the final complex. This offers[28] potential entropic advantages in complex formation, as well as enthalpic advantages. The enthalpic advantages arise because the increase in strain energy on complex formation (equation 3.5) will be lower than if the ligand had adopted a rather different conformation of lower energy than that finally adopted in the complex. There may also be enforced proximity of the lone pairs on the donor atoms of the free ligand, leading to electrostatic repulsion, that is relieved on complex formation. Other factors in preorganization, more relevant possibly to macrocycles (Chapter 4), are that there may be desolvation[29] of the donor atoms of the ligand in the confined space of the ligand cavity.

An interesting early[30] example of preorganization is to be found in the ligand CDTA, seen in Figure 3.18(a). Here the cyclohexanyl bridge leads to a very considerable increase in complex stability relative to the EDTA complexes for many metal ions, which is largely entropy controlled. A simple interpretation of this effect is that in EDTA the free ligand adopts the *skew* conformation to minimize steric and electrostatic repulsion between acetate groups. In the CDTA free ligand (Figure 3.18(b)) the nitrogens are held in the *trans* position required for complex formation, and so increases in complex stability of as much as five log units relative to EDTA are observed. These increases in complex stability are size related such that the increases are larger for smaller metal ions, as seen in Table 3.9.

Table 3.9. Thermodynamics of complex formation of EDTA, DMEDTA, and CDTA complexes of some metal ions, showing the effect of the *trans* methyl groups on the ridge of DMEDTA, and the cyclohexenyl bridge of CDTA, on complex formation.[a]

		EDTA	DMEDTA	CDTA
Ca(II)	log K_1	10.6	12.3	13.2
	ΔH	-6.5	-3.5	-3.7
Cu(II)	log K_1	18.7	21.6	21.9
	ΔH	-8.2		-6.1
Fe(III)	log K_1	25.0	28.2	30.0
	ΔH	-2.7		
La(III)	log K_1	15.46	16.98	
	ΔH	-18.9	-16.6	

[a] EDTA = ethylenedinitrilotetraacetate acid, DMEDTA = *dl*-2,3-diaminobutane-N,N,N',N'-tetra-acetic acid, CDTA = *trans*-1,2- diaminocyclohexane-N,N,N',N'-tetraacetic acid. Data from Ref. 2 at 25 °C, ionic strength = 0.10 M. ΔH in kcal·mole⁻¹.

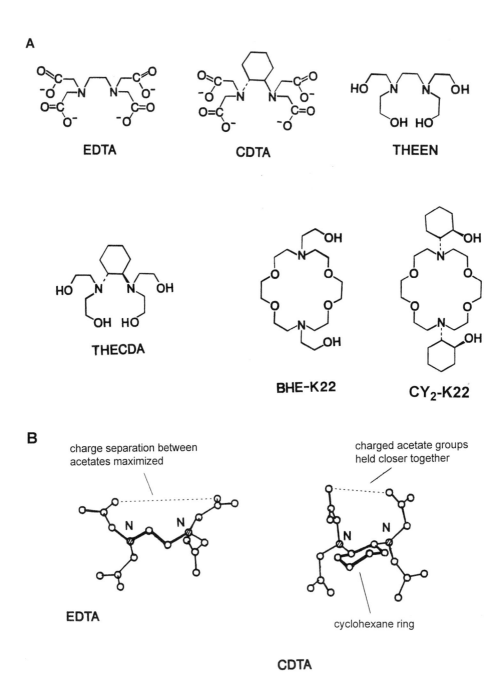

Figure 3.18. **(a)** Some ligands with cyclohexenyl bridges, and their analogues with simple ethylene bridges. **(b)** The EDTA ligand in its minimum energy skew form, with the charged acetate groups far apart, and CDTA with the nitrogen donors held in the *trans* form, so that the acetate groups cannot move so far apart. It is thought that this latter fact contributes to the extra stability of complexes of CDTA as compared to EDTA. Hydrogens have been omitted from the two ligands for simplicity.

It is surprising that this effect has not been further exploited in ligand design. Recently,[31] ligands such as THECDA and Cy_2-K22 (Figure 3.18(a)) have been reported. The

It is surprising that this effect has not been further exploited in ligand design. Recently,[31] ligands such as THECDA and CY$_2$-K22 (Figure 3.18(a)) have been reported. The increases in complex stability relative to THEEN for the ligands studied were quite modest, but also size related in a manner similar to CDTA, as seen in Figure 3.19. It was suggested[30] that the smaller increases in log K$_1$ for THECDA complexes relative to the THEEN analogues, than was observed for CDTA relative to EDTA, was due to the lack of charge on the hydroxyethyl groups of THECDA, so that the gain in complex stability in preorganizing them was quite small as compared to preorganizing charged acetate groups in CDTA.

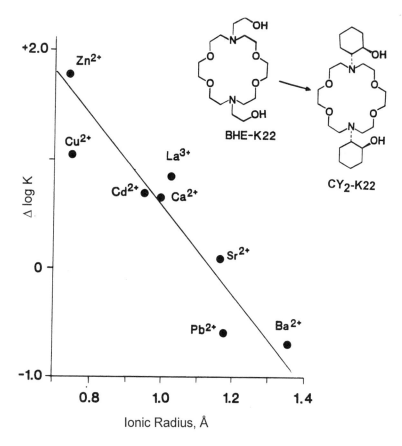

Figure 3.19. The dependency on metal ion size of the effect on complex stability produced by the cyclohexenyl bridge. The value of $\Delta \log K$ is $\log K_1$ for the CY$_2$-K22 complex minus $\log K_1$ for the BHE-K22 complex. The ionic radii refer to the octahedral radii, and are from Ref. 14.

Other types of preorganization[32] of chelating ligands are found in doubly bridged polyamines, as seen in the ligands BAE-PIP, BAE-HPIP, and BAE-DACO, shown in Figure 3.20(a). Here the double bridges between a pair of nitrogen donor atoms lead to greater rigidity. The ligand BAE-PIP is actually of a very low level of preorganization, since the free ligand will undoubtedly have the piperazine type ring formed by the double bridge in the chair conformer (Figure 3.20), whereas the boat conformer is required for complex formation. Molecular mechanics calculation suggests[32] that the energy required

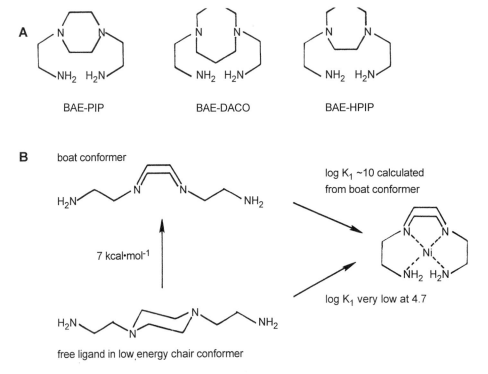

A

BAE-PIP

BAE-DACO

BAE-HPIP

B

boat conformer

log K$_1$ ~10 calculated from boat conformer

7 kcal·mol^{-1}

log K$_1$ very low at 4.7

free ligand in low energy chair conformer

Figure 3.20(a). Some reinforced open-chain polyamines (A), and the effect of the low energy boat conformation in the free ligand BAE-PIP on the stability of its complexes (B).

to change BAE-PIP from the chair to the boat conformation is about 7 kcal.mol^{-1}, and it is this unfavorable contribution to complex formation that undoubtedly accounts, at least in part, for the low stability of complexes of BAE-PIP. A further contributing factor must be the long best fit M-N length of 3.8 Å, and small best fit N-M-N angle of 38°, for chelating with a piperazine ring.[32] These requirements are far from the M-N lengths and N-M-N angles of any metal ion. The ligand BAE-HPIP forms complexes of much greater stability than BAE-PIP (Figure 3.20), due to a better fit of the metal ion. The HPIP (homopiperazine) ring coordinates with least strain[32] to metal ions with M-N lengths of 2.10 Å, and N-M-N angles of 74°, which is much closer to the requirements of metal ions such as Ni(II). The free HPIP ring also has no other conformations of greatly lowered energy as compared with that required for complex formation. The ligand BAE-DACO shows no marked increase in complex stability as compared with BAE-HPIP. This probably reflects the fact that (see Figure 3.20(b)) the coordinated DACO ligand must be of high energy. Either both rings involving the metal ion and the coordinated DACO have the chair conformation, but there is serious H--H van der Waals repulsion, or one of these rings has the boat conformation, which relieves H--H repulsion, but leads to higher strain energy because of the boat conformation. As seen in Figure 3.20(b), a boat conformation chelate ring of the PN ligand is about 4.0 kcal.mol^{-1} higher in energy than a chair conformation. The way to overcome this problem is to follow a simple rule. If the hydrogen atoms on two adjacent carbon atoms, not bonded to each other, are involved in strong van der Waals repulsion, leading to destabilization of the complex, this may be removed by creating a C-C bond between the two adjacent carbon atoms. This results in the ligand bispidine (Figure 3.20(b)) which has no problems of H--H repulsion, and which should lead to complexes of considerable stability due to the very high levels of preorganization of bispidine.

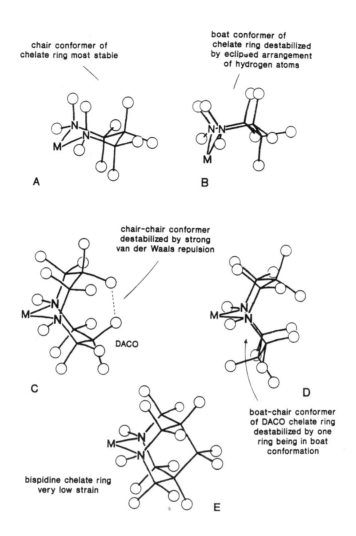

chair conformer of chelate ring most stable

boat conformer of chelate ring destabilized by eclipsed arrangement of hydrogen atoms

A

B

chair-chair conformer destabilized by strong van der Waals repulsion

DACO

C

D

boat-chair conformer of DACO chelate ring destabilized by one ring being in boat conformation

bispidine chelate ring very low strain

E

Figure 3.20(b). The chair (A) and boat (B) conformers of the six membered 1,3-propanediamine (PN) chelate ring, showing how the eclipsed arrangement of the hydrogens in (B) leads to a strain energy about 4 kcal.mol^{-1} higher than in (A). At (C) is shown the chair-chair conformer of the DACO chelate ring, showing the strong H--H van der Waals repulsion that destabilizes the complex. At (D) is shown the chair-boat conformer of DACO chelate ring, where destabilization is produced by the presence of a boat conformation ring. At (E) is shown the bispidine chelate ring, showing how the presence of a methylene group in bispidine in place of the two sterically interfering hydrogens in (C) should lead to a very much more strongly complexing ligand.

3.9 The Effect of Mixtures of Chelate Rings of Different Sizes on Complex

Stability

The discussion so far has considered only the effect of individual chelate rings of different sizes on complex stability. If one looks at the series of ligands from 2,2,2-tet through 3,3,3-tet, it is apparent that for a metal ion such as Cu(II), there is a peak in complex stability at the 2,3,2-tet ligand. The stability of the complex with 3,3,3-tet is very much lower.

ligand	2,2,2-tet	2,3,2-tet	3,2,3-tet	3,3,3-tet
log K_1 [Cu(II)]	20.1	23.2	21.7	17.1
ΔH [Cu(II)]	-21.5	-27.7	-24.8	-19.5

Data from Ref. 2, ionic strength 0.10 M, and 25 °C.

The reason for the low stability of the 3,3,3-tet complex lies in the ease with which chelate rings of different sizes can be joined together. In Figure 3.21 is shown a drawing of an energy minimized six membered chelate ring involving PN (1,3-diaminopropane), coordinated to a Cu(II) ion. A second six-membered chelate ring of the same type, in its minimum energy conformation, is then formed so as to share a nitrogen donor of the first chelate ring, so that the two rings are joined by the shared nitrogen donor atom. It is seen that the sp³ orbitals of the shared nitrogen are not oriented so as to overlap with the sp³ orbitals of the carbon atom from the second formed chelate ring. The reason for the low stability of polyamine complexes having only six-membered chelate rings is generally that they cannot be joined together about a square planar or octahedral metal ion without causing a high level of steric strain. In contrast, in Figure 3.21 is also shown the fusion of an adjacent six membered and five membered chelate ring about a square-planar metal ion. In their minimum strain conformations, the fusion of the five and six membered rings can be carried out with almost no steric strain. This accounts for the high stability of the 2,3,2-tet complex. Although not shown in Figure 3.21, the same approach shows that adjacent five membered rings do not fuse in the low-strain fashion observed for a five and a six membered chelate ring. Figure 3.21 suggests that, in general, ligands which form alternating five and six membered chelate rings, will form more stable complexes than those which form only five membered chelate rings, or only six membered chelate rings. This is, however, also dependent on metal ion size. For larger metal ions, the low affinity for six membered chelate rings will be the more important, and the 2,2,2-tet complex will be more stable than the 2,3,2-tet complex. Also of great importance, as discussed in 4.6.1, is metal ion geometry. For a tetrahedral metal ion, a series of six-membered rings will be preferred.

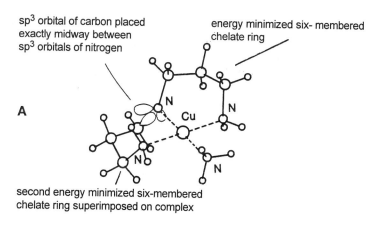

sp³ orbital of carbon placed exactly midway between sp³ orbitals of nitrogen

energy minimized six-membered chelate ring

A

second energy minimized six-membered chelate ring superimposed on complex

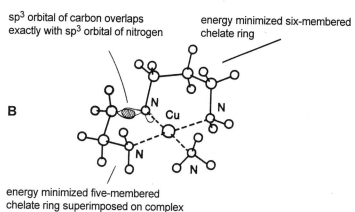

sp³ orbital of carbon overlaps exactly with sp³ orbital of nitrogen

energy minimized six-membered chelate ring

B

energy minimized five-membered chelate ring superimposed on complex

Figure 3.21. Why two adjacent six membered chelate rings lead to low complex stability, while a six membered chelate ring coupled to a five membered chelate ring leads to high complex stability with a metal ion the size of Cu(II). At A are shown two six membered chelate rings, energy minimized using molecular mechanics. It is seen that formation of the C-N bond required to fuse them together will lead to considerable steric strain, since the sp³ orbitals required to form the bond do not overlap. On the other hand, at (B) is shown that adjacent five membered and six membered chelate rings overlap with virtually no steric strain.

3.10   Steric and Inductive Effects in Chelating Ligands

It has been pointed out (section 2.1.2) that both donor strength and the tendency to cause steric hindrance to complex formation increase along the series $NH_3 < NH_2R < NHR_2 < NR_3$ for R = CH$_3$, for example, and also as the nature of R changes along the series $CH_3 < CH_3CH_2 < (CH_3)_2CH < (CH_3)_3C$ in ligands such as the RNH_2 series. The steric hindrance to complex formation so often outweighs the inductive effects that the latter tend to be overlooked, and when they do manifest themselves, are misinterpreted and ascribed to other effects. These inductive effects have been termed[33] "hidden" inductive effects.

An example of steric effects outweighing inductive effects is seen for N-methyl substitution of ethylenediamines, as in the following:

	EN	N-MEEN	N,N'-DIMEEN	N,N,N',N'-TMEEN
	NH$_2$ NH$_2$	CH$_3$NH NH$_2$	CH$_3$NH NHCH$_3$	(CH$_3$)$_2$N N(CH$_3$)$_2$
log K$_1$ Cu(II)	10.48	10.33	10.02	7.20
log K$_1$ Ni(II)	7.35	7.17	6.89	3.57
log K$_1$ Cd(II)	5.4	5.47	5.20	3.87

Data from Ref. 2, 25°C, ionic strength 0.10 M.

From the literature[2] it would appear that N-alkyl substitution usually leads to a drop in complex stability. Several examples of inductive effects predominating are known, however. One such example is seen for ligands derived from ethylenediamine by C-methyl substitution:

	EN	DIMEEN	TMEEN	DICHEN
log K$_1$ [Cu(II)]	10.48	11.27	11.63	12.20
log β_2 [Ni(II)]	13.54	14.01	14.56	14.90

Data from Ref. 2, 25°C, ionic strength 0.10 M.

This general pattern of behavior follows the observations for monodentate amines, where C-methyl substitution is found to produce less serious effects than N-methyl substitution, so that inductive effects may be observed[22] with metal ions less seriously affected by steric effects, such as Ag(I) discussed in section 2.5.3 (page 54). A remarkable example of lack of response to steric effects is found in the very large Pb(II) ion, which forms a complex with NH$_3$ with log K$_1$ of only 1.6, but with t-butylamine log K$_1$ is about 5.0.[34]

In this section we have seen that chelating ligands tend to behave additively as far as complex stability goes, relative to their unidentate analogues. Interesting effects of preorganization can be introduced, however, with more rigid bridges, and with increased inductive effects, provided that the increased inductive effects are not outweighed by accompanying steric effects. The most efficient way to increase inductive effects without increasing adverse steric effects is to create an extra chelate ring. The logical end of this process is the connection of the donor atoms of the ligand to form a complete ring, the macrocyclic ring. With the macrocyclic ligands, discussed in the next chapter, we pass to ligands on a much higher level of preorganization than is true for most chelating ligands.

References

1. G. Schwarzenbach, *Helv. Chim. Acta*, **1952**, *35*, 2344.
2. A. E. Martell and R. M. Smith, *Critical Stability Constants*, Plenum, New York, Vols. 1-6, 1974, 1975, 1977, 1977, 1982, and 1986.
3. A. E. Martell, in *Essays in Coordination Chemistry*, Eds. W. Schneider, G. Anderegg, and R. Gut, Berkhauser Verlag, Basel, 1964, pp 52-54.
4. R. D. Hancock and A. E. Martell, *Comments Inorg. Chem.*, **1988**, *6*, 237.
5. A. W. Adamson. *J. Am. Chem. Soc.*, **1954**, *76* 1578.
6. R. D. Hancock and F. Marsicano, *J. Chem. Soc., Dalton Trans.*, **1976**,1096.
7. W. R. Harris, *J. Coord. Chem.*, **1983**,13, 16.
8. R. M. Fuoss, *J. Am. Chem. Soc.*, **1958**, *80*, 5059.
9. R. D. Hancock and G. J. McDougall, *J. Chem. Soc., Dalton Trans.*, **1977**, 67.
10. R. D. Hancock, G. J. Jackson, and A. Evers, *J. Chem. Soc. Dalton Trans.*, **1978**, 1384.
11. A. E. Martell, R. J. Motekaitis, and R. M. Smith, in *Environmental Inorganic Chemistry*, A. E. Martell and K. J. Irgolic, Eds., VCH Publishers, Deerfield Beach, FL, 1985, p.89.
12. R. D. Hancock and G. J. McDougall, *J. Chem. Soc., Dalton Trans.*, **1978**, 1438.
13. R. D. Hancock, G. J. McDougall, and F. Marsicano, *Inorg. Chem.*, **1979**, *18*, 284.
14. R. D. Shannon, *Acta. Cryst, Sect. A*, **1976**, *A32*, 751.
15. R. D. Hancock, *Pure Appld. Chem.*, **1986**, *58*, 1445.
16. S. Richards, B. Pederson, J. V. Silverton, and J. L. Hoard, *Inorg. Chem.*, **1984**, *3*, 27.
17. R. D. Hancock, Progr. *Inorg. Chem.*, **1989**, *36*, 187.
18. R. D. Hancock, *Acc. Chem. Res.*, **1990**, *23*, 253.
19. R. D. Hancock, P. W. Wade, M. P. Ngwenya, A. S. de Sousa, and K. V. Damu, *Inorg. Chem.*, **1990**, *29*, 1968.
20. P. G. Harrison, M. A. Healy, and A. T. Steel, *Inorg. Chim. Acta*, **1982**, *67*, L15.
21. See for example, R. T. Morrison and R. N. Boyd, *Organic Chemistry*, 5th Edition, Allyn and Bacon, Boston, 1987, p 442.
22. R. D. Hancock, *J. Chem. Soc., Dalton Trans.*, **1980**, 460.
23. K. N. Raymond, G. Mueller, and B. F. Matzanke, *Top. Curr. Chem.*, **1984**, *123*, 49.
24. W. R. Harris, K. N. Raymond, and F. L. Weitl, *J. Am. Chem. Soc.*, **1981**, *103*, 2667.
25. a) I. Yoshida, I. Murase, R. J. Motekaitis, and A. E. Martell, *Can. J. Chem.*, **1983**, *61*, 2740. b) Y. Sun, A. E. Martell, and R. J. Motekaitis, *Inorg. Chem.*, **1985**, *24*, 4343. c) M. J. Miller, *Chem. Rev.*, **1989**, *89*, 1563.
26. K. G. Ashurst and R. D. Hancock, *J. Chem. Soc., Dalton Trans.*, **1977**,1701.
27. D. J. Cram, T.Kaneda, R. C. Helgeson, S. B. Brown, C. B. Knobler, E. Maverick and K. N. Trueblood, J. Am. Chem. Soc., **1985**, *107*, 3645.
28. R. D. Hancock and A. E. Martell, *Chem. Rev.*, **1989**, *89*, 1875.
29. a) D. Hinz and D. W. Margerum, *Inorg. Chem.*, **1974**, *13*, 2941. b) D. K. Cabbiness and D. W. Margerum, *J. Am. Chem. Soc.*, **1969**, *91*, 6540.
30. G. Schwarzenbach, H. Senn, and A. Anderegg, *Helv. Chim. Acta*, **1957**, *40*, 1889.
31. A. S. de Sousa, G. J. B. Croft, C. A. Wagner, J. P. Michael, and R. D. Hancock, *Inorg. Chem.*, **1991**, *30*, 3525.
32. R. D. Hancock, M. P. Ngwenya, A. Evers, P. W. Wade, J. C. A. Boeyens, and S. M. Dobson, *Inorg. Chem.*, **1990**, *29*, 264.
33. R. D. Hancock, B. S. Nakani, and F. Marsicano, *Inorg. Chem.*, **1983**, *22*, 2531.
34. R. D. Hancock, F. Marsicano, T. M. Mali, and H . Maumela, to be published.

CHAPTER 4

COMPLEXES OF MACROCYCLES AND OTHER MORE HIGHLY
PREORGANIZED LIGANDS

Macrocyclic ligands have traditionally been divided into two classes, those with oxygen donors, such as the crown ethers and cryptands, discovered by Pedersen[1] and Lehn,[2] and the nitrogen donor macrocycles, first investigated by workers such as Curtis[3] and Busch.[4] The sulfur-donor macrocycles[5-7] resemble the nitrogen donor macrocycles more closely in their coordinating properties, and so tend to be grouped with them. The division into mainly oxygen-donor macrocycles on the one hand, and nitrogen donor macrocycles on the other, derives from a tendency on the part of investigators to study one or the other type of ligand, which in turn rests on the very different types of metal ions which are strongly complexed by the two classes of ligand. The oxygen donor macrocycles tend to complex well with metal ions such as the larger alkali and alkaline earth metal ions, and the larger post-transition metal ions such as Pb(II), Tl(I), or Hg(II). The nitrogen donor macrocycles complex well with transition metal ions, as well as the post-transition metal ions. The sulfur-donor macrocycles are weakly complexing with all metal ions, but complex best with the same group as the nitrogen donor macrocycles. This chapter is not an attempt to cover all aspects of macrocyclic chemistry, but rather an attempt to provide insight into the factors that govern the complexation of metal ions by macrocyclic ligands. For other aspects of macrocyclic chemistry, a selection of excellent books and reviews is available.[8-17]

Of particular importance here is the origin of the macrocyclic effect,[18] which is the extra thermodynamic stability shown by complexes of macrocycles as compared to those of their open-chain analogues, and the selectivity displayed by macrocycles for metal ions. Two aspects of metal ion selectivity of macrocycles are of importance. First, the preference of some metal ion types to bind with oxygen donor macrocycles, or with nitrogen donor macrocycles, must be accounted for, and one should say at once that the origin of this preference is not as simple as it might at first appear. The second important aspect of macrocyclic chemistry is that of selectivity for metal ions on the basis of their size. A popular concept in macrocyclic chemistry has been that of *size match selectivity*, which is that a macrocyclic ligand will complex best with metal ions that fit most closely into the cavity of the macrocycle. That this idea should be right at first sight seems fairly obvious, and yet, as will be shown, the situation here is also much more complex. Other factors that must be dealt with are the higher ligand field strengths observed[19] in complexes of many nitrogen donor ligands than of their open chain analogues, and the accompanying ability to stabilize unusual oxidation states.[20,21] Finally, the synthesis of all the above ideas on factors that control complexation of metal ions by macrocyclic ligands should allow for a discussion of ligand design for a variety of applications of macrocyclic ligands.

4.1   The Thermodynamics of the Macrocyclic and Cryptate Effects

The thermodynamic macrocyclic effect is the extra stability observed in complexes of macrocycles as compared with their open-chain analogues, and corresponds to the equilibrium:

$$[M(\text{open-chain})]^{n+} \quad + \quad \text{macrocycle} \quad \rightleftharpoons \quad [M(\text{macrocycle})]^{n+} \quad + \quad \text{open-chain}$$

where "open-chain" refers to the open chain analogue of the macrocyclic ligand. The macrocyclic effect was first noted by Cabbiness and Margerum[18] in complexes of tetraaza macrocycles as compared to their open-chain analogues. It was concluded[22] that the macrocyclic effect arose entirely from a favorable enthalpy contribution. Whether one finds that the macrocyclic effect arises from enthalpy or entropy, or a mixture of both, depends on how one chooses the open-chain analogue for the macrocycle. For example, one could consider either 2,2,2-tet or 2,3,2-tet (see Figure 4.1 for key to ligand abbreviations) as a suitable open-chain analogue for 13-aneN$_4$.

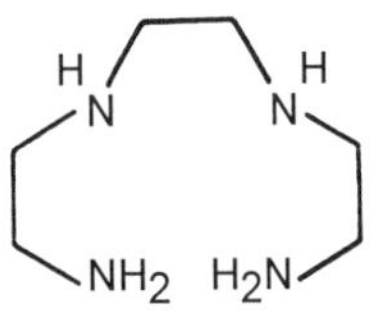

2,2,2-tet

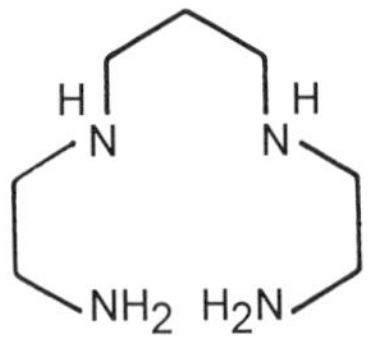

2,3,2-tet

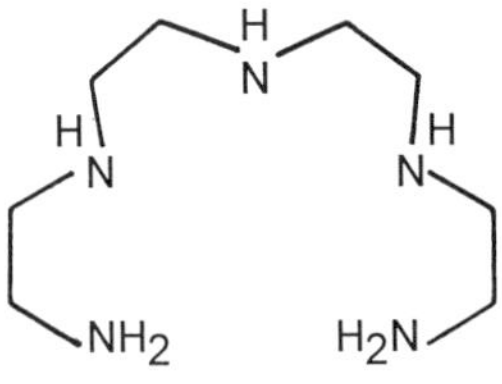

TETREN

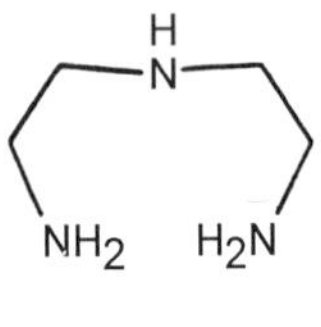

DIEN

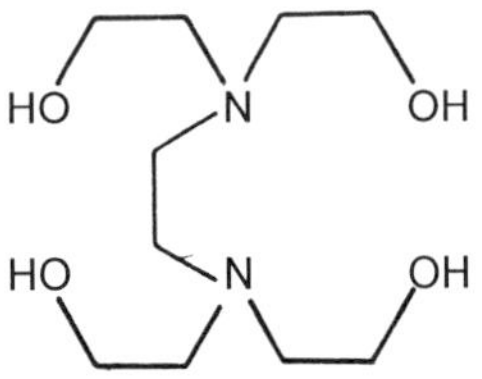

THEEN

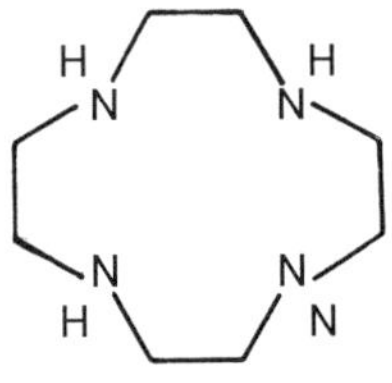

12-aneN$_4$

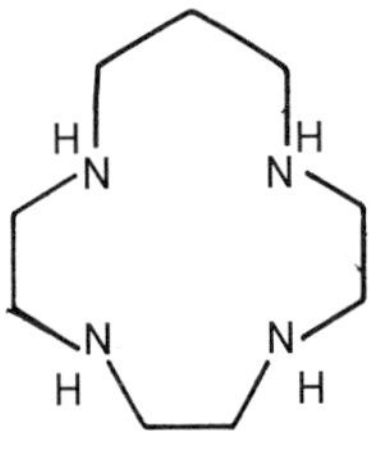

13-aneN$_4$

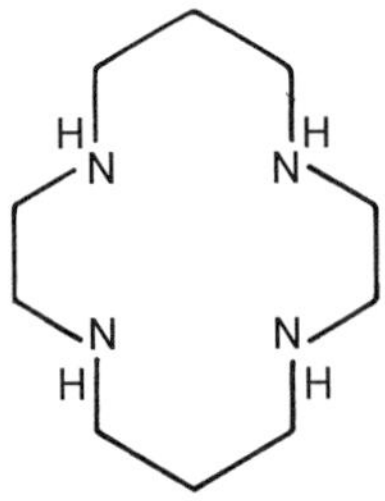

14-aneN$_4$

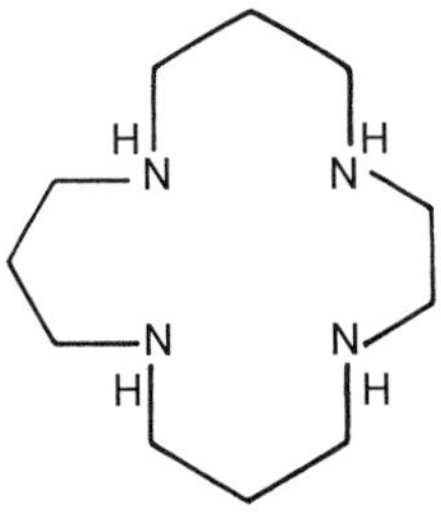

15-aneN$_4$

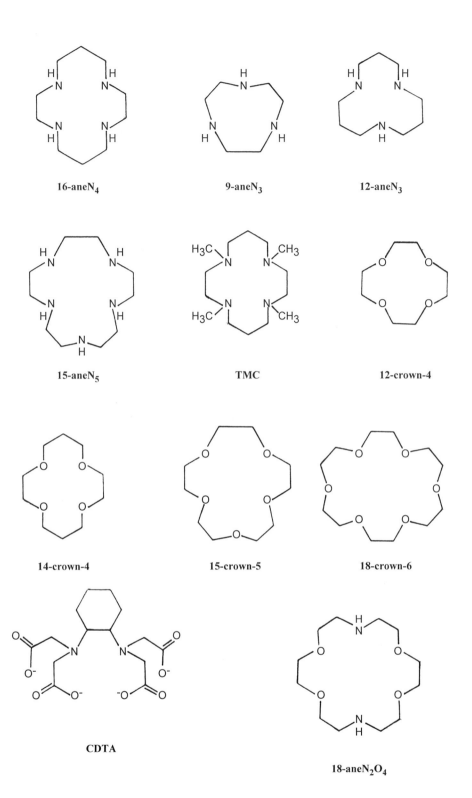

16-aneN$_4$

9-aneN$_3$

12-aneN$_3$

15-aneN$_5$

TMC

12-crown-4

14-crown-4

15-crown-5

18-crown-6

CDTA

18-aneN$_2$O$_4$

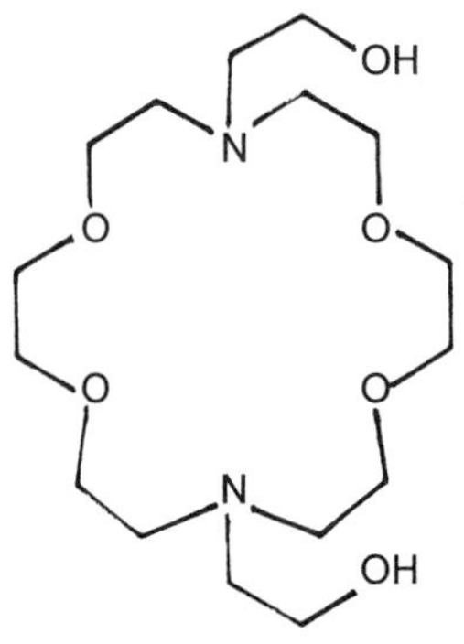

BHE-K22

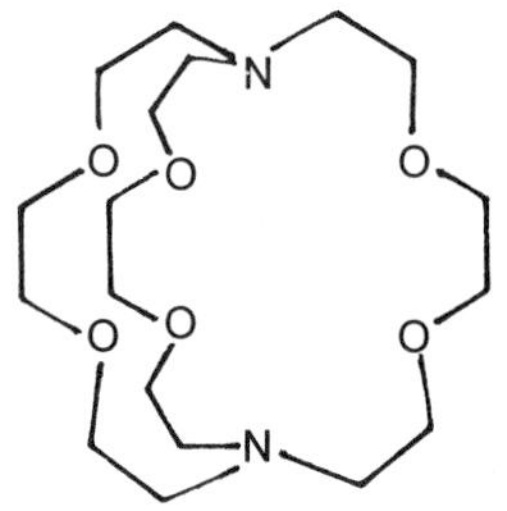

Cryptand-222

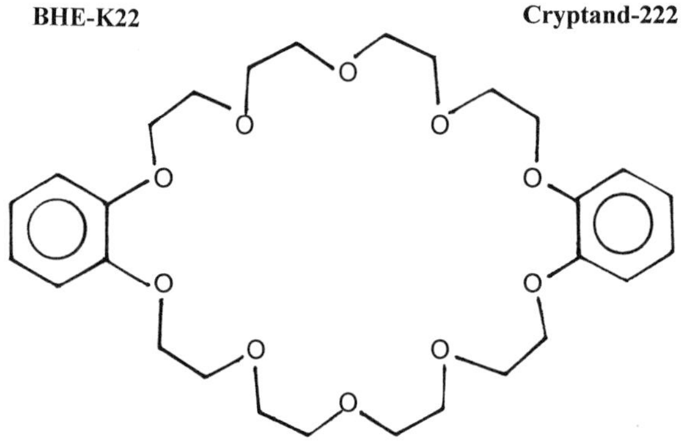

Dibenzo-30-crown-10

Figure 4.1. Ligands discussed in this chapter.

An important aspect of the choice of open-chain analogue[23] is the amount of steric strain present in the complex of the macrocycle and its open-chain analogue. Thus, one of the first measurements of the enthalpy[24] of complex formation of a tetraaza macrocycle was of Cu(II) with 12-aneN$_4$. When comparison was made with the open-chain 2,2,2-tet, the macrocyclic effect was found to be entirely an entropy effect. The Cu(II) coordinates with very considerable steric strain[25] in a ligand such as 12-aneN$_4$, and cannot, for example, assume the preferred square planar coordination geometry, which is still possible for the complex of the less sterically constrained 2,2,2-tet ligand. *One cannot compare a complex of low steric strain, here the 2,2,2-tet complex of Cu(II), with one of high steric strain, the 12-aneN$_4$ complex, as a basis for analyzing the macrocyclic or any other effect,* unless such strains are construed as being within the scope of the definition of the macrocyclic effect. Rather,[26] one should attempt to find complexes of low steric strain, and for the tetraazamacrocycles, the best comparison appears to be of the complexes of Cu(II) and Ni(II) of 2,3,2-tet with those of 14-aneN$_4$. This comparison in Table 4.1 shows the macrocyclic effect to be entirely an enthalpy effect, when complexes of low steric strain are compared with each other. For Zn(II) in Table 4.1, there is an entropic contribution to the macrocyclic effect, which may reflect the fact that 14-aneN$_4$ is not a best fit for Zn(II), since log K$_1$ for the Zn(II) complexes of smaller macrocycles such as 12-aneN$_4$ is higher.

Table 4.1. Thermodynamic contributions to the macrocyclic effect in complexes of tetraaza macrocycles.[a]

		Metal Ion		
		Cu(II)	Ni(II)	Zn(II)
$\log K_1$	14-aneN$_4$	26.5	19.4	15.5
	2,3,2-tet	23.2	15.9	12.6
	log K (MAC)	3.3	3.5	2.9
ΔH	14-aneN$_4$	-32.4	-24.1	-14.8
	2,3,2-tet	-27.7	-18.6	-11.9
	ΔH (MAC)	-4.7	-5.5	-2.9
ΔS	14-aneN$_4$	13	8	21
	2,3,2-tet	13	10	18
	ΔS (MAC)	0	-2	3

[a] Data for log K, ΔH and ΔS from Ref. 24 and 27-32. Units for ΔH are kcal·mol^{-1}, and for ΔS cal.deg^{-1}·mol^{-1}.

For the crown ethers, possibly the best comparison for the macrocyclic effect comes from the thermodynamics of complex formation of 18-crown-6 compared to pentaglyme, as seen in Table 4.2. For the cryptate effect, the best comparison that can be made[34] from the standpoint of the complexes of the ligands having similar coordination numbers and geometry, and not too different levels of steric strain, is probably that of the ligand BHE-K22 with cryptand-222, the thermodynamics of which are seen in Table 4.3. Tables 4.2 and 4.3 indicate that the macrocyclic effect in complexes of crown ethers, and the cryptate effect, are from a thermodynamic standpoint predominantly enthalpic effects, and for the cryptates the entropy contributions to the cryptate effect are strongly unfavorable. In subsequent sections we will attempt to rationalize these facts.

4.2 The Preferred Geometry of Chelate Rings Containing Neutral Oxygen and Nitrogen Donors

In section 3.6 the preferred geometry of chelate rings of size five and six was discussed, based on the chelate rings of ethylenediamine and 1,3-diaminopropane, which have nitrogen donors. In these chelate rings the nitrogen, with a coordinated metal ion, has close to tetrahedral coordination geometry. One might expect from simple approaches such as the VSEPR (valence shell electron pair repulsion) approach[40] that neutral oxygen would be sp^3 hybridized and coordinate to metal atoms so as to produce tetrahedral coordination geometry, with the unshared pair of electrons on the oxygen occupying one of the coordination sites. As pointed out by Hay *et al.*,[41] the coordination of simple ethers to metal ions is such that the two carbon atoms and the metal ion attached to the oxygen donor all lie in the same plane to give the oxygen approximately trigonal planar coordination. A very large number of crystal structures for THF (tetrahydrofuran) coordinated to metal ions, available in the Cambridge Crystallographic data base,

Table 4.2. Thermodynamics of the macrocyclic effect in complexes of crown ethers, as evidenced by the thermodynamics of complex formation of 18-crown-6 complexes and the open chain analogue, pentaglyme.[a]

		Metal Ion		
		Na$^+$	K$^+$	Ba^{2+}
log K$_1$	18-crown-6	4.36	6.06	7.04
	pentaglyme	1.44	2.1	2.3
	log K (MAC)	2.92	3.96	4.74
ΔH	18-crown-6	-8.4	-13.4	-10.4
	pentaglyme	-4.0	-8.7	-5.69
	ΔH (MAC)	-4.4	-4.7	-4.8
ΔS	18-crown-6	-8	-17	-3
	pentaglyme	-7	-20	-8
	ΔS (MAC)	-1	3	5

[a] In 100% methanol, from Refs. 15 and 33, pentaglyme = $CH_3(OCH_2CH_2)_5OCH_3$. Units for ΔH are kcal·mol^{-1}, and for ΔS cal·deg^{-1}·mol^{-1}.

Table 4.3. Thermodynamics of the cryptate effect as illustrated by comparison of the ligands BHE-K22 and cryptand-222.[a]

		Metal Ion			
		Sr(II)	Ba(II)	Ag(I)	Pb(II)
log K$_1$	cryptand-222	8.0	9.5	9.6	12.04
	BHE-K22	4.0	5.3	7.27	9.2
	log K (CRYPT)	4.0	4.2	2.3	2.8
ΔH	cryptand-222	-10.6	-14.2	-12.8	-13.8
	BHE-K22	-2.4	-4.3	-8.6	-8.1
	ΔH (CRYPT)	-8.2	-9.9	-4.2	-5.7
ΔS	cryptand-222	1	-4	1	10
	BHE-K22	10	10	4	15
	ΔS (CRYPT)	-9	-14	-3	-5

[a] ΔH is in kcal·mol^{-1}, ΔS is in cal·mol^{-1}·deg^{-1}. The thermodynamic quantities associated with the cryptate effect refer to the equilibrium [M(BHE-K22)]$^{n+}$ + cryptand-222 $\rightleftharpoons$ [M(cryptand)]$^{n+}$ + BHE-K22. Thermodynamic data for cryptand-222 from Refs. 27 and 35-39, and for BHE-K22 from Ref. 34. For key to ligand abbreviations see Figure 4.1.

supports the idea that coordination of THF gives trigonal planar coordination around the oxygen donor. An example of this is seen for the [Li(THF)$_4$]$^+$ cation[42] in Figure 4.2(a).

Inspection of several of these complexes suggests that the ideal M-O-C angle for coordination of metal ions to ethereal oxygens, and also by extension, alcoholic oxygens, will be about 126°. The observation of Hay et al.[41] has the important consequences that the geometry of minimum-strain chelate ring that involves neutral oxygen donors will be rather different from that involving neutral nitrogen donors. The minimum-strain geometry for chelate rings of size five and size six, involving neutral oxygen and neutral nitrogen donors, is seen in Figure 4.3. One sees that chelate rings involving neutral oxygen donors coordinate with minimum strain with larger metal ions than analogous

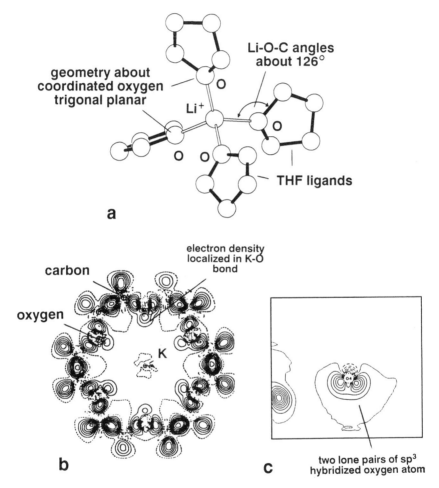

Figure 4.2. (a) Structure of the [Li(THF)$_4$]$^+$ cation (THF = tetrahydrofuran), showing the planar coordination geometry around the oxygen donor atoms, with the oxygens lying in the plane of the Li$^+$ ion and the two adjacent carbon atoms. This structure is typical of the complexes of THF and other simple ethers with metal ions such as the alkali and alkaline earth cations. Redrawn with coordinates from Ref. 42. **(b)** Electron density difference map[46] for the [K(18-crown-6)]$^+$ complex, and **(c)** for a single oxygen atom from the 18-crown-6 ligand. Note that in (b) the electron density of the lone pairs of oxygen is concentrated in the K-O bond, while in the free ligand two distinct lone pairs are present. Figures (b) and (c) redrawn after Ref. 46.

N-donors

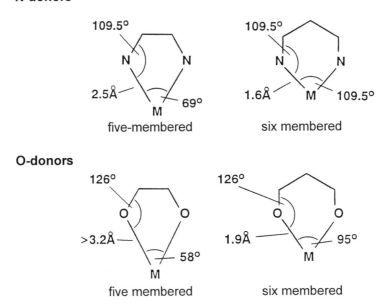

five-membered   six membered

O-donors

five membered   six membered

Figure 4.3. The metal to ligand bond lengths, and bite angles, that produce the minimum strain energy structures for chelate rings of size five and size six with saturated chelate rings containing nitrogen donors and oxygen donors.

chelate rings involving the neutral nitrogen donor. An important aspect of the coordination of chelate rings containing neutral oxygen donors is that low strain will occur when the required trigonal planar geometry around the coordinated oxygen can be achieved. For saturated nitrogen donors, achievement of close to tetrahedral coordination geometry will be important. Figure 4.4 shows how a metal ion such as Li^+ achieves[43] the required planarity about the coordinated oxygens with crown ethers such as 12-crown-3 and 16-crown-4.

The fact of the preferred planar coordination around the coordinated neutral oxygen suggests that the hybridization of the oxygen atom might be sp^2. One would expect some structural consequences in the immediate vicinity of the neutral oxygen in complexes of crown ethers. One unusual feature that has been noted[44] in the structures of [K(18-crown-6)]$^+$, and observed in the structures of free ligands and complexes of other crown ethers,[45] is a shortening of the C-C bond. Thus, in a study[44] of [K(18-crown-6)](ClO$_4$) at 20 K, the mean C-C bond lengths were found to be 1.49 Å, compared with usual C-C bond lengths of 1.54 Å. This appears to be caused by the electronegativity of the oxygens since electronegative elements attached to carbon cause a shortening of the C-C bond. The C-C bonds in simple ethers are longer because the C-C bonds are attached to only one oxygen. Knochel and coworkers[46] have carried out an X-ray crystallographic study of the electron density distribution in crown ethers as free ligands and also complexed to alkali metal ions. Of interest here is that (Figure 4.2) in free 18-crown-6 the electron density of the lone pairs on the oxygens is located in two distinct orbitals as expected from sp^3 hybridization. In the K^+ complex of 18-crown-6 the electron density is located in the K-O bond as though the hybridization of the oxygen donor was indeed sp^2.

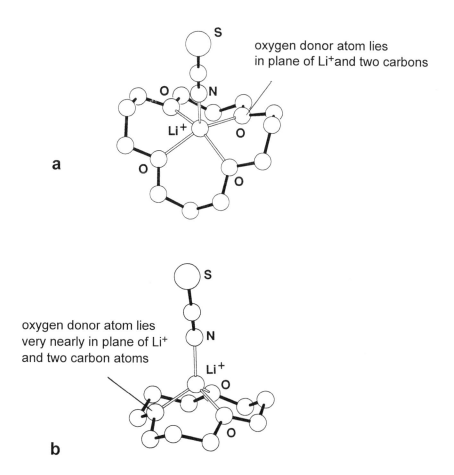

Figure 4.4. Structures of the complex cations with Li^+ of (a) 16-crown-4 and (b) 12-crown-3, showing how the required planar coordination around the oxygen donor atoms is achieved. The anion coordinated to Li in both cases is thiocyanate. Redrawn with coordinates available in Ref. 43.

4.3   The Origin of the Macrocyclic Effect

There have been[47] several suggestions on the origin of the macrocyclic effect. Factors that have been considered to play a role are:

1) preorganization of the ligand,[17,18,22,24,48] in the sense of the free ligand having a limited number of conformers, some of which have structures that are similar to the conformation required to complex the metal ion;

2) desolvation of the donor atoms in the confined space of the macrocyclic cavity;[18,22]

3) intrinsic basicity effects[49-51] due to the electron releasing (inductive) effects of ethylene bridges between donor atoms; and

4) enforced repulsion between the lone pairs of the donor atoms in the cavity of the macrocycle, which is relieved on complex formation.[48,50-52]

Under effect 1), suggestions such as "prestraining",[50] "preorientation",[24] and "multijuxta-positional fixedness"[4] have been grouped under the term "preorganization" suggested by Cram,[17] which is used to cover all these effects. The ideas suggested under effect 1) all carry the intuitively appealing idea that the entropy of the free macrocyclic ligand is lower than that of its open-chain analogue. In addition, there is less steric strain involved in taking the macrocycle from its minimum energy conformations in the free ligand to those conformations required in the complex, than is found to be the case for the open-chain analogue.

The simplest cases of preorganization, such as the CDTA ligand, discussed in 3.8, involve the ligand already being in the right conformation for complex formation. In the case of 18-crown-6, and also cryptand-222, the structures of the free ligands in the crystalline state,[45,53] and in solvents of low dielectric constant, are those seen in Figure 4.5.

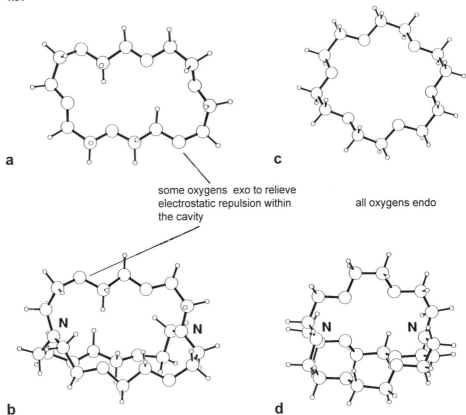

a

some oxygens exo to relieve electrostatic repulsion within the cavity

c

all oxygens endo

b

d

Figure 4.5. The structures of 18-crown-6 and cryptand-222 observed[46,53] in the solid state at (a) and (b) respectively, and at (c) and (d), the conformers required for forming complexes with large metal ions.

These are rather different from the conformations required to form the complex, also shown in Figure 4.5. In both 18-crown-6 and cryptand-222, the conformations of the free ligands have methylene groups folded into the cavity of the ligand, with oxygen donor atoms in *exo* positions. Molecular mechanics calculations suggest[54-56] that the reason for the folding of ligands such as 18-crown-6 and cryptand-222 is to reduce electrostatic repulsion between the lone pairs on the donor atoms of the free ligand. Because of this electrostatic repulsion, the energy of the D_{3d} conformer of 18-crown-6, observed in the complex, is considerably higher than that of the C_i conformer of the free ligand, and the same is true for the D_{3d} conformer of cryptand-222 required for complex formation. The free ligands 18-crown-6 and cryptand-222 are thus not preorganized in the sense that they are already in the right conformation for complex formation. When these ligands are dissolved in water, it is likely[56] that solvation lessens the repulsion between the lone pairs in the free ligand, and a D_{3d}-like conformation may result.

In relation to the macrocyclic effect, in comparing 18-crown-6 and cryptand-222 with their open chain analogues, 1) the solvation of the donor atoms of 18-crown-6 and cryptand-222 is likely to be less than that of their open chain analogues, 2) the increase in strain energy on complex formation may be less for 18-crown-6 and cryptand-222 than their open-chain analogues, and 3) lone-pair - lone-pair repulsion is likely to be higher in 18-crown-6 and cryptand-222 than in their open chain analogues, even though it can be decreased somewhat in the cyclic ligands by adopting *exo* orientations for some of the donor atoms, as seen in Figure 4.5. The important result is that in macrocyclic ligands, all three of the above effects lead to a high energy state for the free ligand, which is relieved on complex formation.

Molecular mechanics calculations on nitrogen donor macrocycles[49,50] suggest similar contributions to the macrocyclic effect to those for the oxygen donor macrocycles. What would be required to assess the role of ligand solvation better, and the other proposals listed above, in producing the macrocyclic and cryptate effects, would be complete Born-Haber cycles of the type shown below for the formation of the complexes of both the macrocyclic ligand and its open chain analogue:

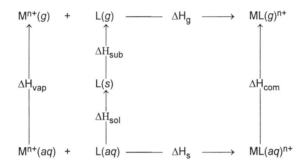

For the ligands cyclam and 2,3,2-tet, it has been shown[57] that the heat of solution (ΔH_{sol} + ΔH_{sub} in the above cycle) of cyclam from the gas phase was some 5.2 kcal.mol^{-1} smaller than for 2,3,2-tet. From this result it was tentatively concluded that the entire macrocyclic enthalpy of -4.7 kcal.mol^{-1} seen in Table 4.1 in comparing the Cu(II) complexes of these two ligands was due to this difference in solvation. However, this conclusion is based on the assumption that ΔH_{com} in the above cycle is the same for both complexes, and without this result one cannot be sure of the magnitude of the overall contribution of heats of solution of the species involved in the equilibrium to the macrocyclic effect.

Izatt *et al.*[58] measured only the heat of solution of the free ligands pentaglyme and 18-crown-6 in methanol (ΔH_{sol}), and found them to be virtually identical. This tends not to support the idea of a contribution from hindered ligand solvation in causing the macrocyclic effect. However, without knowing the heat of vaporization (ΔH_{sub}) of the free ligands, one cannot be sure of the significance of this result. These authors also determined[58] the size of the macrocyclic enthalpy for Na$^+$, K$^+$, and Ba^{2+} in 90% methanol, and in 100% methanol, and found virtually no difference. This tends also to suggest that solvation effects are not important. However, what one may be seeing here is that as the solvating power of the solvent is lowered, so the contribution to the macrocyclic effect of hindered solvation is reduced, but the contribution from lone-pair - lone-pair repulsion increases.

A significant step towards solving the origin of the macrocyclic effect, at least for crown ethers, has been taken with recent studies[59-63] of complexation of alkali metal ions by crown ethers in the gas phase. If solvation is the sole cause of the macrocyclic effect, then no macrocyclic effect should be observed in the gas phase, where no solvent is present. However, comparisons of triglyme and 12-crown-4, and pentaglyme with 18-crown-6, show a definite macrocyclic effect for all the alkali metal ions, which rules out solvation as a sole cause of the macrocyclic effect. Some theoretical studies are also available on the origin of the macrocyclic effect. Reibnegger and Rode[64] report LCAO-MO calculations on complexes of 12-aneN$_4$ compared to its open-chain analogue (CH$_3$NHCH$_2$CH$_2$NHCH$_2$)$_2$, which suggest that an important contribution to the macrocyclic effect here is the greater energy required to get the open-chain analogue into the conformation required for complex formation than is true for the macrocycle. Yamabe et al.[65] have carried out a theoretical study of the M-O bond in complexes of crown ethers with alkali metal ions which has indicated a considerable level of covalence in the M-O bonds of alkali metal-crown ether complexes, which suggests that inductive effects discussed below may be important.

All of the above discussion has focused on possible contributions to the macrocyclic effect and cryptate effects that derive from the high energy state of the free ligand. This would suggest that, apart from the macrocyclic effect itself, the properties of macrocyclic complexes, once formed, should not be very different from those of their open-chain analogues. This is not the case, and we now discuss the contribution to the macrocyclic effect that would also affect the properties of the macrocyclic complexes themselves, namely the inductive effect.

4.4 Inductive Effects in Complexes of Macrocyclic Ligands

The relationship between enthalpy of complex formation and energies of ligand field (LF) bands found[66] for polyamine complexes of Cu(II), discussed in section 2.3, suggests a relationship between covalence in the M-N bond produced by inductive effects, and the enthalpy of complex formation. An indication of the importance of this is seen in the enthalpies and energies of the LF bands for complexes of Cu(II) in Chart 4.1. Also shown in Chart 4.1 is a complex of Cu(II) and Ni(II) for which ΔH is not known, but which demonstrate how the LF strength increases with donor atom basicity along the series primary < secondary < tertiary.

One sees that the higher enthalpy of complex formation for the Cu(II) complex of cyclam, as compared with 2,3,2-tet, can be explained[49] in terms of the increasing donor power of the nitrogen donors of the ligands along the series primary < secondary < tertiary. If one extrapolates from the results of the non-macrocyclic ligands to the cyclam complex, one would expect[47] a value of ΔH for the cyclam complex of Cu(II) of about -29.9 kcal·mol^{-1}. The fact that ΔH for Cu(II) with cyclam is[24] at -32.4 kcal·mol^{-1}, somewhat higher than this, suggests that the extra 2.5 kcal·mol^{-1} is derived from non-inductive effect related contributions such as hindered solvation of the donor atoms, lone-pair - lone-pair repulsion, and preorganization. One can also roughly estimate that 2.2 kcal·mol^{-1} of the macrocyclic effect would be due to inductive effects.

Chart 4.1

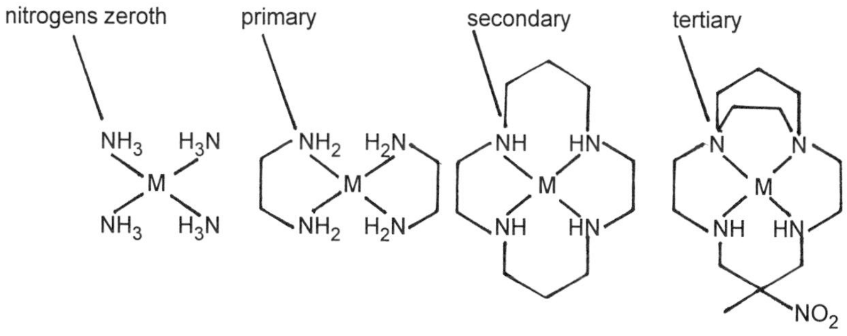

	nitrogens zeroth	primary	secondary	tertiary
ΔH [Cu(II)](kcal·mol^{-1})	-22.0	-25.5	-32.4	?
v(d-d)(cm^{-1}) Cu(II)	17000	18300	19900	21050
Ni(II)	~20000	21600	22470	23900

The strongest evidence for the role of inductive effects in macrocyclic chemistry is the chemistry of the complexes themselves. Thus, [Fe(9-aneN$_3$)$_2$]$^{2+}$ is low-spin, whereas its open-chain analogue [Fe(DIEN)$_2$]$^{2+}$ is high-spin.[67] The ability of N-donor macrocycles to stabilize high oxidation states[68] of metal ions is much greater than for open-chain polyamines. Within the group of macrocycles studied, the ability to stabilize Cu(III) and Ni(III) appears to be related[20] to high LF strength and covalence. If only free ligand effects (e.g. hindered solvation) were important in producing the macrocyclic effect, the tetraazamacrocycles would not be expected to be so very much better at stabilizing unusual oxidation states, even where the cavity of the macrocycle was much too large[68] for the metal ion. Unusual oxidation states that have been stabilized by N-donor macrocycles include Ni(III),[20,68-73] Ni(I),[71,74-78] Cu(III),[20,68,70,79,80] Co(I),[75,77,81,82] Ag(II),[70,83,84] Pd(III) and Pt(III),[6,21] and even Hg(III).[85]

A factor that has hindered the detection of the role of inductive effects in the thermodynamics and in the LF strengths of complexes of macrocyclic and open chain ligands has been steric hindrance (see section 2.5.3). Thus, in a complex such as [Ni(TMC)]$^{2+}$, one would expect the LF strength to be very high because of the presence of four tertiary nitrogen donors in the ligand. However, MM calculation shows[51,86] that the four N-methyl groups undergo very strong mutual van der Waals repulsion, as seen in Figure 4.6. The van der Waals repulsion between the N-methyl groups stretches the Ni-N bond out from the strain-free value[25] of 1.91 Å to the observed[86] value of 1.99 Å. Accompanying this distortion of the Ni-N bond length in [Ni(TMC)]$^{2+}$ is a drop in log K$_1$ from 20.1 in the 14-aneN$_4$ complex to 8.4 in[87] the TMC complex, and a drop[88] in the LF splitting parameter Dq$_{xy}$ from a value of 2043 cm^{-1} in [Ni(14-aneN$_4$)]$^{2+}$ to 1782 cm^{-1} in [Ni(TMC)]$^{2+}$. This demonstrates the important principle that in order to observe the inductive effects of added alkyl groups, the addition should be done in a sterically effective manner, or else adverse steric effects will predominate and the inductive effects will not be apparent.

Inductive effects are not usually considered in the chemistry of oxygen donor macrocycles, although from the energies of complexation of metal ions with water, alcohols, and ethers in the gas phase (sections 2.1.1 and 2.1.2), one would expect

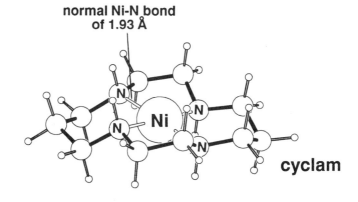

normal Ni-N bond
of 1.93 Å

cyclam

a

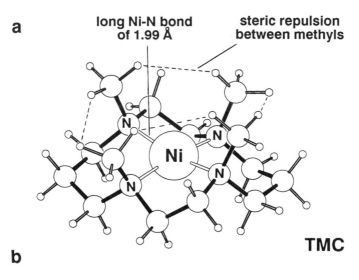

long Ni-N bond
of 1.99 Å

steric repulsion
between methyls

TMC

b

Figure 4.6. The structure of (a) the 14-aneN$_4$ complex of low-spin Ni(II) and (b) the TMC complex, showing how van der Waals repulsion between the N-methyl groups leads to a lengthening of the Ni-N bonds, a lowering of the in-plane ligand field strength, and a drop in log K$_1$, for the TMC complex.[51,86]

inductive effects to be important. There are cases that show that inductive effects must be operative. These are specifically the stabilities of the complexes of crown ethers with the large post-transition metal ions Tl(I) and Pb(II). These two ions are the same size[89] as Rb(I) and Sr(II) respectively, so that one might have expected log K$_1$ values for Tl(I) and Rb(I), and for Pb(II) and Sr(II), to be the same. However, log K$_1$ for the two post-transition metal ions is higher than for Rb(I) and Sr(II), in spite of the heat of hydration[90] being higher for Tl(I) and Pb(II), as seen in Table 4.4. One would interpret this result to mean that 1) the heats of hydration of the large post-transition metal ions is higher because of greater covalence in the M-O bond, and that 2) the stability of the 18-crown-6 complexes is also higher because Pb(II) and Tl(I) can respond to the greater donor strength of the ether oxygens of 18-crown-6 as compared to the water molecules displaced in the complex formation reaction. The fact that solvation energies of metal ions must be overcome in forming the complex with crown ethers has often been invoked

Table 4.4. The role of electronegativity in stabilizing the complexes of large post-transition metal ions with 18-crown-6 relative to those of their pre-transition analogues.

	Metal Ion			
	Sr^{2+}	Pb^{2+}	Rb^+	Tl^+
Ionic Radius[a] (Å)	1.18	1.19	1.52	1.50
Electronegativity[b]	1.0	1.8	0.8	1.8
Heat of hydration[c]	-363.5	-371.9	-81.0	-87.0
Log K_1 (18-crown-6)[d]	2.84	4.23	1.55	2.27
ΔH (18-crown-6)[e]	-3.6	-5.2	-3.8	-4.4

[a] In Å, for six-coordination, from Ref. 89. [b] From Ref. 91. [c] From Ref. 90, in kcal·mol⁻¹. [d] Mean of values reported in Ref. 15, in aqueous solution. [e] In kcal·mol⁻¹, from Ref. 27.

to explain factors such as selectivity trends. This is somewhat misleading. If the heat of hydration of the metal ion is higher, one would expect a parallel increase in the enthalpy of formation of the bonds to the crown ether. The fact that not all metal ions form complexes with crown ethers with identical log K values has to be attributed to 1) steric effects, and 2) inductive effects, as will be discussed further in section 4.6.

4.5 The Macrocyclic Effect in Mixed Donor Macrocycles

One sees in Tables 4.1 and 4.2 that there is a strong macrocyclic effect in both nitrogen donor macrocycles, and in crown ethers, of roughly three log units. For mixed donor macrocycles containing both oxygen and nitrogen donors, however, there is a decline in the macrocyclic effect as compared to the all-oxygen or all-nitrogen donor systems.[92] This is true, as seen in Table 4.5, even when the bridge added to the open-chain ligand is added across two nitrogens in both cases, as in:

BAEDOE   12-aneN₂O₂

X = NH = all nitrogen donors
X = O = mixed donor macrocycle

It was difficult to see[92] how addition of an ethylene bridge at two nitrogen donors to give the macrocycle, which nitrogens are remote from the other donors (the X groups above), could produce entirely different results for the macrocyclic effect depending on whether X was O or NH. The reason for the difference may lie in steric differences between the coordination requirements of the oxygen and nitrogen donors, indicated by Hay *et al.*[41] Thus, the trigonal planar coordination around the oxygen donors when coordinated to a metal ion may produce a very different steric response in BAEDOE complexes on cyclization to give the macrocycle 12-aneN₂O₂, as compared to the tetrahedral coordination geometry around the coordinated nitrogen donors of the analogous 2,2,2-tet and 12-aneN₄ complexes.

Table 4.5. The macrocyclic effect in complexes of macrocycles containing a mixture of nitrogen and oxygen donors, as compared to their all-nitrogen donor analogues. [a]

all-nitrogen donors			mixed donor analogues		
(structure)	*(structure)*		*(structure)*	*(structure)*	
log K_1	log K_1	log K (MAC)	log K_1	log K_1	log K (MAC)
Cu(II) 20.1	24.4	+4.3	7.89	8.7	+0.8
Cd(II) 10.6	12.71	+2.1	5.68	6.55[b]	+0.9

(structure)	*(structure)*		*(structure)*	*(structure)*	
log K_1	log K_1	log K (MAC)	log K_1	log K_1	log K (MAC)
Ni(II) 10.5	16.2	+5.7	5.62	8.49	+2.87
Pb(II) 7.5	11.0	+3.5	6.10	5.17	-0.93

[a] Data from Refs. 27 and 92. Log K (MAC) is the thermodynamic macrocyclic effect, and is equal to log K_1 for the macrocyclic ligand minus log K_1 for the open-chain analogue. [b] Actually for 1,7-dioxa-4,10-diazacyclododecane, which should not be too different from the 1,4-dioxa ligand shown.

4.6 The Selectivity of Macrocyclic Ligands for Metal Ions

The predominant thinking on the patterns of selectivity shown by macrocyclic ligands for metal ions has in the past been based on size match selectivity. By size match selectivity is meant that a metal ion will show its maximum stability with the member of a series of macrocycles where the match between the size of the metal ion and of the cavity in the macrocycle is closest. There are two ways in which the size of the macrocyclic ring can be varied. Firstly, the number of donor atoms may be varied, as in the series of crown ethers 12-crown-4, 15-crown-5, 18-crown-6, and so on, with the chelate rings usually remaining the same size. This is the typical way in which cavity size is varied in crown ethers, and the chelate rings present in the complex of the macrocycle are almost always five membered, with the exception of ligands such as 14-crown-4, where six membered chelate rings are formed. The second manner in which macrocyclic hole size can be varied is by means of varying the length of the bridges between the donor atoms, with the number of donor atoms remaining constant. This is invariably how hole size is varied for nitrogen donor macrocycles. Here, clearly, because of the strongly coordinating nature of the nitrogen donor, it is not felt to be reasonable to have differing numbers of nitrogen donor atoms when making comparisons. We now discuss the factors that appear to control the selectivity patterns of macrocyclic ligands with metal ions of different sizes.

4.6.1 The metal ion selectivities of the tetraaza macrocycles

Busch et al.[19] have used molecular mechanics (MM) to calculate the hole sizes of the tetraaza macrocycles 12-aneN$_4$ through 16-aneN$_4$. This was done by setting the M-N force constants to zero, which in the MM calculation allowed the M-N distance to relax to the value dictated by the ligand. This then gives the best-fit M-N length for the metal ion to coordinate with minimum strain in the cavity of the macrocycle. Similar calculations by other authors[50] have essentially supported these results. The calculated best-fit M-N lengths for the tetraaza macrocycles 12-aneN$_4$ through 16-aneN$_4$ are shown in Table 4.6.

Table 4.6. The best-fit M-N lengths for coordinating into the cavities of the tetraaza macrocycles when constrained to[66] square-planar coordination, or when allowed to assume whichever coordination geometry gives the lowest energy.[a] Also shown are some strain-free M-N bond lengths for metal ions[25] discussed here, placed in columns below macrocycle in which they might fit best[48] if constrained to square planar coordination geometry.

	Macrocycle				
	12-aneN$_4$	13-aneN$_4$	14-aneN$_4$	15-aneN$_4$	16-aneN$_4$
best-fit M-N length (square planar)[b]	1.83	1.92	2.07	2.22	2.38
best-fit M-N length[c] with geometry indicated	2.21 square pyramidal	-	2.07 square planar	-	1.81 tetrahedral
macrocycle conformation	++++		++--		+-+-
strain-free M-N length[d] for M^{2+}		1.91 [Ni(II), S=0]	2.03 [Cu(II)] 2.10 [Ni(II), S=1]	2.15 [Zn(II)]	2.38 [Cd(II)] 2.61 [Pb(II)]

[a] M-N lengths in Å. [b] From Ref. 48. All macrocycles have the ++-- conformation. [c] From Ref. 93. [d] From Ref. 25. Strain-free M-N lengths in Å.

In Figure 4.7 is shown the variation[25,51,94] of log K$_1$ for the series of tetraaza macrocycles 12-aneN$_4$ through 16-aneN$_4$. The variation has been shown relative to the complex with 12-aneN$_4$ as Δ log K, which is log K$_1$(X-aneN$_4$) - log K$_1$(12-aneN$_4$) for each metal ion, where X is the number of skeletal atoms forming the macrocyclic ring, which is the macrocyclic ring size. It is seen that in general the behavior does not[25] conform to the idea of size match selectivity. Thus, the very large Pb(II) ion shows a steady decrease in Δ log K$_1$ as the size of the macrocyclic ring increases, and complexes most strongly with 12-aneN$_4$, which ligand supposedly has (Table 4.6) a cavity much too small for Pb(II). The medium sized Zn(II) shows a weak preference for 12-aneN$_4$, when Table 4.6 suggests that Zn(II) should complex best with 15-aneN$_4$. Low spin Ni(II) should fit best into 13-aneN$_4$, but complexes much more strongly with 14-aneN$_4$. Only for Cu(II) and high-spin Ni(II) does the peak in complex stability with the tetraaza macrocycles occur at a ligand (14-aneN$_4$) that appears to be a best-fit for the metal ion on the basis of the match between metal ion M-N length, and best fit M-N length for coordinating in the cavity of the ligand.

The selectivities of the tetraaza macrocycles for metal ions of different sizes are not well explained by the idea of size-match selectivity. In their calculations, Busch et al.[19]

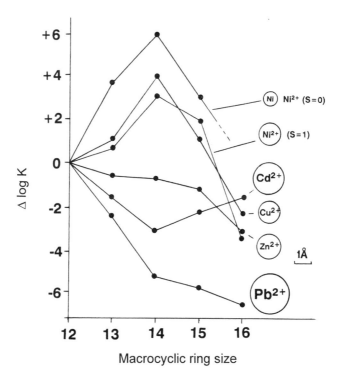

Figure 4.7. The change in complex stability ($\Delta \log K$) relative to the 12-aneN4 complex, as macrocyclic ring size is increased along the series 12-aneN$_4$ through 16-aneN$_4$, for a variety of metal ions.[25,51,94,95] The changes in complex stability, $\Delta \log K$, are log K_1 for the complex of the particular macrocycle minus log K_1 for 12-aneN$_4$ complex. The sizes of the metal ions are indicated by circles of size proportional to metal ion radius.[89]

constrained the metal ion to be square planar, and confined the tetraaza macrocycles to have the *trans*-III, or ++-- conformer shown in Figure 4.8. At least two other conformers, the *trans*-I (++++) and *cis*-V (+-+-) conformers, are commonly found in complexes of the tetraaza macrocycles (Figure 4.8). Extensive MM calculations[51] have shown that over certain ranges of bond lengths, the *trans*-I or *cis*-V conformers may be of lower energy than the *trans*-III conformer, and that metal ions that do not fit into the cavity of the *trans*-III conformer can avoid high steric strain by adopting the *trans*-I or *cis*-V conformers. This is shown in Figure 4.9 for the *trans*-I and *trans*-III conformers of 12-aneN$_4$. The MM calculations[51] show that, although the *trans*-III conformer of 12-aneN$_4$ has a small cavity, this conformer is for all M-N lengths at much higher energy than the *trans*-I conformer, and so, as is supported by the available crystal structures, is at too high an energy to be observed. In the *trans*-I conformer, the metal ion lies well out of the plane of the nitrogen donors. In this position, the factors that control selectivity for the metal ion are the same as for open-chain ligands, namely the size of the chelate ring. We may thus interpret the steady decrease in log K_1 for the Pb(II) complex of the tetraaza macrocycles with increase of macrocyclic ring size (Fig. 4.7) in terms of the fact that, as the macrocyclic cavity size increases, the number of six membered rings is increasing, and large metal ions such as Pb(II) do not coordinate well in six membered chelate rings.

Figure 4.10 shows the variation in strain energy, U, with M-N length, of the *trans*-I and *trans*-III conformers of 14-aneN$_4$ complexes. Here, the behavior is rather different from that found for 12-aneN$_4$ in Fig. 4.9. Instead of the *trans*-III conformer being always

a

trans-III + +-- conformer, metal ions of suitable size (e.g. Cu(II)) lie in plane of donor atoms

b

trans-I + + + + conformer, too large metal ions (e.g. (Hg(II), Cd(II)) lie above plane of donor atoms

c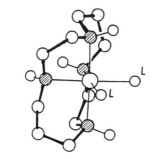

cis-V + - + - conformer allows too large (e.g. Pb(II)) octahedral metal ion to escape compression.

Figure 4.8. Important conformations of complexes of tetraaza macrocycles, as exemplified by (a) the *trans*-III (++--), (b) *trans*(I) (++++), and (c) *cis*-V (+-+-) conformers of 14-aneN$_4$. The convention followed in naming the conformers follows the work of Bosnich *et al.*[96] The "+" and "-" indicate whether the N-H hydrogen atom lies above or below the plane of the macrocycle.

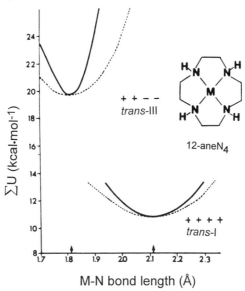

Figure 4.9. Variation in total strain energy as a function of initial (———) and final (····) energy minimized M-N bond length, as calculated[51] for the *trans*-I (++++) and *trans*-III (++--) conformers of 12-aneN$_4$ complexes. The *trans*-I conformer is at lower energy for all M-N bond lengths, so that complexes of 12-aneN$_4$ having the *trans*-III conformation are unlikely to exist. The energy minima correspond to the best-fit size of metal ion for coordinating to the macrocycle. It is apparent that the *trans*-I conformer coordinates best with larger metal ions than does the *trans*-III conformer. Redrawn after Ref. 51.

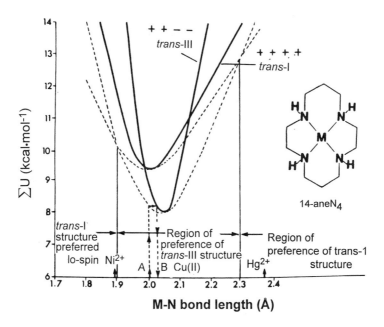

Figure 4.10. Variation in total strain energy as a function of initial (——) and final (····) energy minimized M-N bond length, as calculated[51] for the *trans*-I (++++) and *trans*-III (++--) conformers of 14-aneN$_4$ complexes. The arrows indicate the strain-free bond lengths of Cu(II) and Hg(II), with corresponding adoption of the *trans*-III conformer by Cu(II)[94] and the *trans*-I conformer by Hg(II).[97] Redrawn after Ref. 1.

at higher energy, for 14-aneN$_4$ there is a region,[51] between 1.9 and 2.3 Å, where the *trans*-III conformer is of lower energy than the *trans*-I conformer. Crystal structures of 14-aneN$_4$ complexes of metal ions of different sizes correspond quite well with the predictions of Figure 4.10. For example, Cu(II) with a M-N bond length close to 2.0 Å has, as expected, the *trans*-III conformation[95] in its 14-aneN$_4$ complex, while the Hg(II) complex, with M-N lengths of 2.37 Å, adopts[97] the *trans*-I conformation.

An alternative conformer adopted by 14-aneN$_4$ with metal ions that are too large to coordinate with low strain within the cavity of the *trans*-III conformer is the folded *cis*-V conformer. In Figure 4.11 is seen a plot of U as a function of M-N length for the *trans*-III and *cis*-V conformers of 14-aneN$_4$ octahedral complexes, with coordinated water molecules occupying the two coordination sites. There is a crossover point at a M-N length of 2.09 Å, below which the *trans*-III conformer should be more stable, and above which the *cis*-V should be more stable. In agreement with this, it is found that Cr(III), which is at the crossover point in terms of M-N length, exists as a mixture of the *cis*-V and *trans*-III forms in aqueous solution.[98] Co(III), with much shorter M-N lengths, exists in the region where the *trans*-III conformer should be more stable. The *cis*-V conformer with Co(III) can be prepared in the solid state, but rapidly rearranges to the *trans*-III form in aqueous solution.[99]

The MM calculations have shown that metal ions that are the wrong size for the cavity in a tetraaza macrocycle can escape this problem by being coordinated in positions where they lie out of the plane of the donor atoms, such as in the *trans*-I and *cis*-V conformers. In these positions, the metal ion selectivity is governed by the same factors that control selectivity for metal ions in open chain polyamines, namely the size of the chelate ring.[23,25,51] This is demonstrated in Figure 4.12, where the change in complex

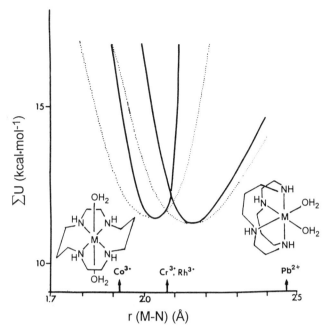

Figure 4.11. Variation in total strain energy as a function of initial (———) and final (····) energy minimized M-N bond length, as calculated[51] for the *trans*-I (++++) and *cis*-V (+-+-) conformers of 14-aneN$_4$ complexes. The arrows indicate the strain-free M-N lengths for Pb(II), Cr(III) and Co(III), in agreement with which the *cis*-V conformer is more stable for Cr(III), Rh(III) and the very large Pb(II) ion, and the *trans*-III is more stable for Co(III). Redrawn after Ref. 51.

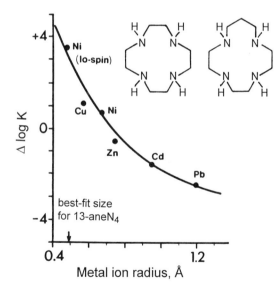

Figure 4.12. The change in complex stability, $\Delta \log K$, produced by increase of chelate ring size from all five membered rings in 12-aneN$_4$ to one of the chelate rings being six membered in the 13-aneN$_4$ complexes, plotted as a function of metal ion radius.[89] This plot suggests that the dominant factors in controlling metal ion selectivity in the tetraaza macrocycles is not macrocyclic ring size but chelate ring size. This figure should be compared with Figures 3.6 and 3.7. Redrawn after Ref. 25.

stability, Δ log K, for going from the complex with 12-aneN$_4$ to the complex with 13-aneN$_4$, is plotted against metal ion radius.[89]

Figure 4.7 suggests that Zn(II) is on the borderline of metal ion size between preferring five membered and preferring six membered chelate rings, and so shows no strong preferences. Why, then, does the small Cu(II) ion not show a steady increase in log K$_1$ with increasing number of six membered chelate rings, but rather peak at 14-aneN$_4$? In fact, Cu(II) is not small enough to coordinate well with a ligand containing only six membered chelate rings (Figure 4.7). In section 3.5 it was shown that minimum strain coordination as part of a six membered TN (1,3-propanediamine) type of ring requires a M-N length of 1.6 Å, and Cu(II) with its Cu-N bond lengths of 2.0 Å is too large. Empirically one observes that metal ions will coordinate best with ligands where their M-N bond lengths correspond to an average of the best-fit M-N lengths for the chelate rings present. Thus, 14-aneN$_4$ has two five membered chelate rings (best fit M-N length of 2.5 Å) and two six membered chelate rings (best fit M-N length of 1.6 Å), and so will coordinate best with a metal ion with a M-N bond length of (2 x 2.5 + 2 x 1.6)/4 = 2.05 Å, which is close to the 2.0 Å for Cu(II) and also high-spin Ni(II) (2.1 Å). This approach indicates that 16-aneN$_4$ will, in fact, complex best with a very small metal ion of M-N length 1.6 Å. Molecular Mechanics calculations have suggested that,[93] if the metal ions are allowed to assume any coordination geometry, then 16-aneN$_4$ will form a very low strain complex with a tetrahedral metal ion with M-N bond lengths of 1.81 Å, as shown in Figure 4.13. This best fit length is a little longer than the 1.6 Å expected above,

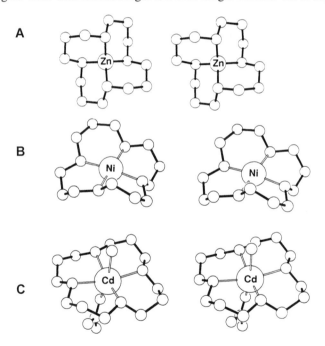

Figure 4.13. (A) The *ch,ch,ch,ch* conformer of 16-aneN$_4$ with a small tetrahedral metal ion, predicted by MM calculation,[93] and since observed[100] for the Zn(II) cation. At (B) is shown the *ch,ch,tw,tw* ++-- conformer of 16-aneN$_4$ that is observed[101] when the metal ion cannot achieve tetrahedral coordination, as is shown here for high-spin Ni(II),[101] and at (C) is shown the *ch,ch,ch,ch* ++++ conformer observed for large metal ions such as Cd(II), Hg(II), and Pb(II).[102]

because steric crowding of H-atoms raises steric strain below an M-N length of 1.81 Å. The +-+- conformation predicted for 16-aneN$_4$ was unknown at the time when the MM calculations were carried out, but was subsequently observed[100] in the complex of Zn(II) with 16-aneN$_4$ seen in Figure 4.13. Figure 4.13 shows that 16-aneN$_4$ has in fact a very small cavity, provided that the metal ion is free to assume tetrahedral coordination geometry. If the metal ion is not able to assume tetrahedral coordination geometry, then more highly strained conformers, also seen in Figure 4.13 are adopted. Thus, the Ni(II) complex adopts a +++- conformation, and the larger Cd(II), Hg(II) and Pb(II) adopt a ++++ conformation. The strain energy of the +-+-, +++-, and ++++ conformers of 16-aneN$_4$ complexes as a function of M-N length is seen in Figure 4.14. Figure 4.14 resembles Figures 4.9 and 4.10 for 12-aneN$_4$ and 14-aneN$_4$ complexes, except that for the 16-aneN$_4$ complexes the +-+- conformer at very short M-N lengths is of much lower energy than any of the other conformers. None of the metal ions so far studied are of a size small enough, and have tetrahedral coordination geometry, which accounts for the fact that log K$_1$ values for all metal ions with 16-aneN$_4$ are rather low.

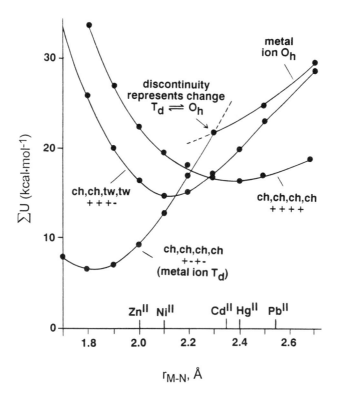

Figure 4.14. The strain energies of different conformers of 16-aneN$_4$ complex calculated as a function of initial strain free M-N length.[100] The geometry of the metal ion was optimized, as well as the structure of the ligand. At short M-N length the +-+- conformer favors the small tetrahedral Zn(II) ion, The ++++ conformer has a best-fit cavity corresponding to a M-N length of 2.4 Å. The discontinuity for the +-+- conformer at 2.3 Å corresponds to a change from tetrahedral to *cis*-octahedral coordination. At intermediate M-N lengths square planar coordination of the metal ion in the ++-- conformer, with various forms of twist-boat chelate ring present, is favored.

The studies of the tetraaza macrocycles have revealed that the metal ion selectivities of these ligands are far more complex than envisaged in the simple size match selectivity model, and in the next section we extend these ideas to the oxygen donor macrocycles, the crown ethers.

4.6.2 The metal ion selectivities of the crown ethers

It has been pointed out[12] that the idea of size match selectivity does not account well for the size match selectivities of the crown ethers. In Figure 4.15 is shown the variation in log K_1 for crown ether complexes[15] of Na^+, K^+, and Cs^+, as a function of macrocyclic ring size for crown ethers ranging in ring size from 12-crown-4 to 30-crown-10. Figure 4.15 is not really consistent with the idea of size match selectivity. The stability of the complexes of the smallest macrocycle, 12-crown-4, is hardly affected by metal ion size, since log K_1 is very nearly the same for Na^+, K^+, and Cs^+. At the other end of the size scale, K^+ forms a more stable complex with the very large dibenzo-30-crown-10 ligand than does the larger Cs^+ cation. Only for 18-crown-6 is the expected preference for K^+ observed, but one is left to wonder what the significance of this can be when one might largely summarize crown ether chemistry by saying that crown ethers tend to prefer potassium. The immediate thought that springs to mind here is that what is really controlling metal ion size selectivity is size of the chelate ring. Figure 4.3 suggests that a five membered chelate ring containing two ethereal or alcoholic oxygen donors will coordinate with minimum steric strain with a metal ion of M-O length of about 3.2 Å. This is somewhat longer than the K^+ M-O bond length of about 2.8 Å observed in, for example, the $[K(18\text{-crown-}6)]^+$ complex.[103] The reason[47] a majority of crown ethers form their most

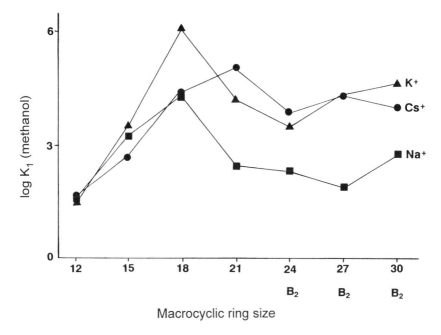

Figure 4.15. Variation in log K_1 for crown ether complexes of Na^+, K^+, and Cs^+, as a function of size of the macrocyclic ring. Data in methanol at 25 °C, from Ref. 15. The crown ethers are in simple unsubstituted forms, except for macrocyclic ring sizes 24, 27, and 30, which are dibenzo substituted.

stable complexes with K^+ is that there is a balance between opposing factors. In the gas phase, strength of interaction between the cation and the crown ether will be $Li^+ > Na^+ > K^+ > Rb^+ > Cs^+$. Opposing this is the trend that strain energy due to a mismatch between chelate ring size requirements and metal ion size increases only slowly from $Cs^+ < Rb^+ < K^+$, but rises much more rapidly thereafter $K^+ \ll Na^+ \ll Li^+$. It has already been shown in section 2.5.1 that neutral oxygen donors promote selectivity for large metal ions, and it seems reasonable to interpret this fact in terms of the idea that five membered chelate rings containing ethereal or alcoholic oxygen donors coordinate with minimum steric strain with large metal ions. The fact that the effect of neutral oxygen donors on metal ion selectivity is very similar, whether these are added as pendent groups, or so as to give a macrocyclic ring, is illustrated in Figure 4.16. In Figure 4.16 is plotted the change in

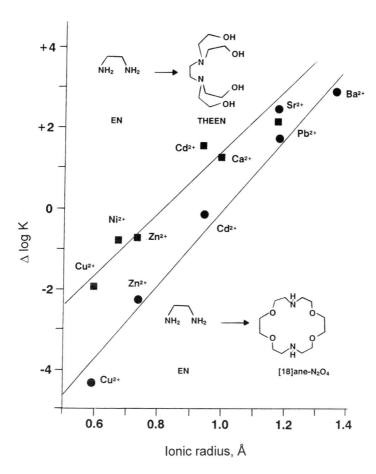

Figure 4.16. The change in complex stability, $\Delta \log K$, plotted as a function of metal ion radius,[89] (a) on passing from the EN complex to the THEEN complex, and (b) in passing from the EN complex to the 18-aneN$_2$O$_4$ complex, for a variety of metal ions. The diagram demonstrates the essential similarity on metal ion selectivity produced by adding neutral oxygen donors, whether as part of pendent groups, or as part of macrocyclic rings. Formation constant data from Ref. 27.

121

complex stability, $\Delta \log K$, on addition of a) hydroxyethyl groups to EN to give the ligand THEEN, and of b) ethereal bridges to EN to give 18-aneN$_2$O$_4$, both as a function of metal ion radius.[89] It is seen in Figure 4.16 that the response of the metal ions in terms of selectivity is very similar, whether neutral oxygen donors are being added to form pendent donors, or so as to form a macrocyclic ring.[104,105] Figure 4.16 shows that for both the macrocycle 18-aneN$_2$O$_4$ and the open-chain ligand THEEN, the effect of the neutral oxygen donors on selectivity is to increase selectivity for the larger metal ions relative to the smaller metal ions.

As with the tetraaza macrocycles, the question raised for the crown ethers is why size match selectivity does not appear to be a dominant factor in generating selectivity for metal ions. If one looks at structures of 18-crown-6 complexes, for example, as with 14-aneN$_4$, it is found that several conformations are observed which appear to depend on metal ion size (Figure 4.17). Thus Cs$^+$ is too large for the cavity in the ligand and so is coordinated above the plane of the donor atoms in the D_{3d} conformation, while K$^+$ appears to be about the right size, and so lies in the cavity exactly in the plane of the donor atoms. Metal ions such as Na$^+$ and Bi^{3+} appear to be too small for the cavity in 18-crown-6, and so conformations such as the "half-buckled" and "buckled" (c) and (d) are observed. Here the metal ions do not lie in the plane of the donor atoms, reminiscent of what happens to complexes of tetraaza macrocycles when the metal ion is the "wrong" size for the cavity in the macrocycle in its more symmetric conformations.

The strain energies of the D_{3d}, half-buckled, and buckled conformers of 18-crown-6 complex, calculated[106] as a function of M-O bond length are seen in Figure 4.18. The D_{3d} conformer is of lowest energy at M-O lengths above 2.55 Å. This differs from Figures 4.10 and 4.11 in that the D_{3d} conformer is of very much lower energy than any other

a large Cs$^+$ ion lies above plane of donor atoms in D_{3d} conformer

b K$^+$ fits cavity almost perfectly in D_{3d} conformer

c Na$^+$ too small for D_{3d} conformer, "half buckled" conformer

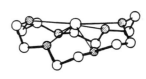

d smaller Bi^{3+} ion adopts buckled ++-++- conformer

Figure 4.17. Drawings of different conformers of the 18-crown-6 complex adopted as the metal ions become progressively smaller in (a) through (d). (a) is the D_{3d} conformer of Cs$^+$ with the too large metal ion rising up out of the plane of the donor atoms.[107] (b) shows the D_{3d} conformer of K$^+$ which fits well into the cavity.[103] (c) shows the "half-buckled" conformer of the somewhat too small Na$^+$ ion,[108] and (d) shows the buckled ++-++- conformer adopted by the much too small Bi(III) ion.[109] Redrawn after Ref. 106.

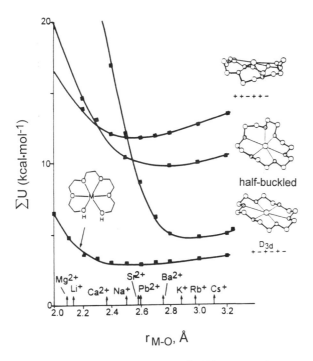

Figure 4.18. The variation in strain energy of different conformers of complexes of 18-crown-6, calculated by Molecular Mechanics, as a function of M-O bond lengths, from the coordinates for crystal structures of the different conformers shown. The D_{3d} conformer is most stable for metal ions of bond length more than 2.55 Å, while the "half-buckled" and ++-++- conformer, although of high energy, become more stable than the D_{3d} at shorter M-O lengths. Also shown is the plot of strain energy versus M-O bond length for the open-chain pentaethylene glycol complex, which indicates the metal ion size selective behavior of a non-macrocyclic analogue of 18-crown-6. The arrows show the ionic radii of the individual metal ions. Redrawn after Ref. 106.

conformer available to the complex, and exerts some kind of selectivity based on metal ion size, with the K^+ ion having a M-N bond length of 2.88 Å fitting best into the cavity of 18-crown-6. This accounts for the sharp selectivity observed in complex stability for K^+ against Na^+ and Li^+ at 18-crown-6 in Figure 4.15, and one may say that 18-crown-6 shows some measure of size match selectivity. More typical of crown ethers is the plot of strain energy for 12-crown-4 and 15-crown-5 seen in Figure 4.19. In these complexes the metal ions are placed in positions well out of the plane of the donor atoms, so that there is very little influence of the cavity on metal ion selectivity. As seen in Figure 4.19, the shapes of the strain energy versus M-O bond length curves for 12-crown-4 and 15-crown-5 complexes are very similar to the shapes of the curves for the open-chain analogues, indicating that the factors that govern selectivity in 12-crown-4 and 15-crown-5 complexes are the same as in open-chain ligands.

For very large macrocycles such as dibenzo-30-crown-10, here also potassium (Figure 4.20) forms the most stable[15] complex of the alkali metal ions. Here the macrocycle is folded[110] like the seam of a tennis ball around the metal ion, as seen in Figure 4.20. Clearly, there is no cavity in the ligand, and it exerts its selectivity[106] largely due to the constraints imposed by torsional factors in the macrocyclic ring.

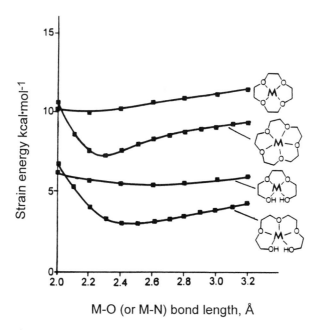

Figure 4.19. The variation in strain energy of complexes of 12-crown-4 and 15-crown-5, calculated by MM, as a function of M-O bond length, from coordinates for crystal structures.[106] Also shown is the plot of strain energy versus M-O length for the complexes of the open-chain triethylene glycol and tetraethylene glycol, which indicate the metal ion size selective behavior of non-macrocyclic analogues of 12-crown-4 and 15-crown-5. Redrawn after Ref. 106.

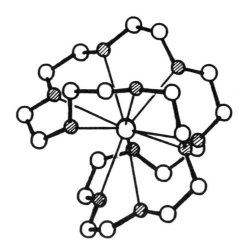

Figure 4.20. The structure of the 30-crown-10 complex of K[+] generated[106] by MM calculation from coordinates for the dibenzo complex.[110] The structure shows the "tennis ball seam" mode of coordination to the metal ion. Redrawn after ref. 106.

124

One can say that crown ethers are divided into three groups in terms of their selectivity for metal ions:

Group 1 consists of 12-crown-4 and 15-crown-5, in which the cavity is too small to allow complexation of metal ions within the cavity. Metal ions lie coordinated well out of the cavity, where the factors that govern selectivity are largely the same as in open-chain ligands.

Group 2 contains 18-crown-6, and probably other macrocycles derived from 18-crown-6 by C-substitution, as well as substitution of N-donors for O-donors. 18-crown-6 has three important conformers in its metal ion complexes, of which the D_{3d} is of considerably lower energy at longer M-O lengths. Here it is unique in that it exerts considerable size match selectivity for metal ions such as K^+ that have M-O lengths of about 2.9 Å, and that lie within the plane of the donor atoms.

Group 3 contains the larger crown ethers. These all fold around the metal ion, as seen for 30-crown-10 in Figure 4.20, and so also have no real cavity. They exert their selectivity mainly due to the torsional constraints within the macrocyclic backbone.

4.6.3   The selectivity patterns of triaza and pentaaza macrocycles

The chemistry of the triaza macrocycles[111,112] differs from the tetraaza macrocycles in that the ligands are generally too small to have a cavity. A major difference also arises from the fact that[113] the small macrocyclic ring of the triaza macrocycles is more rigid than the larger macrocyclic ring of the tetraaza macrocycles. Thus, only one type of conformer, the R,R,R (or its enantiomer, the S,S,S seen in Figure 4.21, has been observed for 9-aneN$_3$ complexes, in contrast to the several conformers known for tetraaza macrocycles. The occurrence of only one type of conformer in triaza macrocycles means that, in spite of the metal ion not lying in the plane of the donor atoms, 9-aneN$_3$ does exert selectivity based on metal ion size. In addition, there is a strong macrocyclic effect for the triaza macrocycles.[112]

Metal Ions	Ni(II)	Cu(II)	Zn(II)	Cd(II)	Pb(II)
Ionic radius:	0.69	0.75	0.74	0.95	1.18
log K$_1$ (9-aneN$_3$)	16.2	15.5	11.6	9.4	11.0
log K$_1$ (DIEN)	10.5	15.9	8.8	8.1	7.5
Δ log K(MAC)	5.7	-0.4	2.8	1.3	3.5

(log K$_1$ values from reference 27, Δ log K (MAC) = thermodynamic macrocyclic effect, Ionic radii (Å) from Ref. 89.)

Molecular Mechanics calculations show that the 9-aneN$_3$ type of ring is fairly rigid, and that in spite of the presence of five membered rings, the best fit size of metal ion is quite small at about 2.1 Å.[113] The effect of this is that the macrocyclic effect tends to be larger for smaller metal ions, as seen above. The larger than expected macrocyclic effect for the large Pb(II) ion may be related to the observation[114] that the structure of [Pb(9-aneN$_3$)(NO$_3$)$_2$] shows a stereochemically active lone pair on the Pb(II). The low value of the thermodynamic macrocyclic effect for 9-aneN$_3$ with Cu(II) must relate to the facial coordination forced by this ligand, with one of the nitrogen donors occupying the

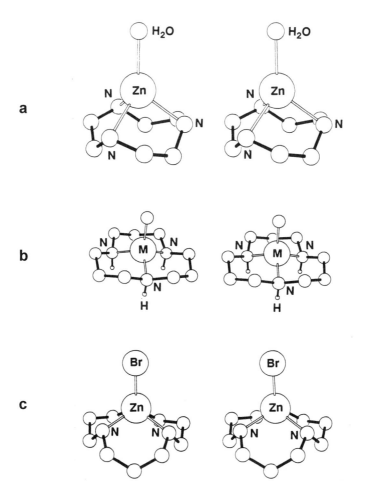

Figure 4.21. Conformers of complexes of triaza macrocycles with metal ions. a) The R,R,R type of conformer, which, together with its S,S,S enantiomer, is the only type of conformer observed for 9-aneN₃ complexes. b) The ch,ch,ch --- conformer of 12-aneN₃ complexes, which would be formed[93] with very small metal ions the size of Be(II). Note the unusual placement of the N-H hydrogens below the plane of the complex. c) The complex of Zn(II) with 12-aneN₃, which is adopted by larger metal ions.[115,116]

unfavorable axial coordination site on Cu(II), which does not happen with DIEN. The origins of this macrocyclic effect appear to lie in the same factors as discussed for the tetraaza macrocycles. Of particular importance is the fact that, for example, the Ni(II) bis-9-aneN₃ complex exhibits a very high ligand field (LF) strength.[49] Other evidence for the high ligand field strength is the fact that the Co(II) and Fe(II) complexes of 9-aneN₃ are low spin.[111,112] The fact that high LF strengths are exerted even where the metal ion does not lie in the cavity of the ligand, and the M-N bond lengths are normal argues persuasively for the idea[49] that the high LF strength often observed in the complexes of tetraza macrocycles is due to the inductive effects of the alkyl groups on the secondary nitrogens, as discussed in section 2.5.3.

Increase of size of the macrocyclic ring for triaza macrocycles so that more six membered chelate rings are produced successively in 10-aneN₃, 11-aneN₃, and 12-aneN₃ has the expected effect of lowering the complex stability of larger metal ions such as

Pb(II) more than smaller metal ions such as Ni(II). However, because these smaller more rigid macrocyclic rings themselves favor complexation of fairly small metal ions, the effect is not marked, and may be erratic. The presence of six membered chelate rings in complexes of 12-aneN$_3$ leads to preference for very small metal ions in the same way as observed[93] for 16-aneN$_4$ complexes. The conformer adopted by 12-aneN$_3$ with the smallest metal ions, seen in Figure 4.21, can only occur with very small tetrahedral metal ions of the size of Be(II) (ionic radius[89] 0.27 Å). Otherwise, with larger metal ions such as tetrahedral Zn(II) (ionic radius 0.60 Å), higher energy conformers have to be adopted,[115,116] such as the ++- conformer seen in Figure 4.21, which lead to complexes of generally low stability.

The pentaaza macrocycles with most smaller metal ions cannot encircle these metal ions so as to give five donor atoms in a plane, but must fold[112,117] onto the metal ion. Molecular Mechanics calculations by the present authors show that a metal of M-N length 2.52 Å which can accept planar coordination of the macrocycle will complex with 15-aneN$_5$ with reasonably low steric strain. The macrocyclic effect in complexes of 15-aneN$_5$ compared to the open chain analogue TETREN is large irrespective of metal ion size, so that the buckled coordination geometry does not appear to have an adverse effect on complex stability:

Metal Ion	Zn(II)	Cu(II)	Mn(II)	Cd(II)	Hg(II)	Pb(II)
ionic radius (Å)	0.74	0.75	0.80	0.95	1.02	1.18
log K$_1$15-aneN$_5$	19.1	28.3	10.6	19.2	28.5	17.3
log K$_1$ TETREN	15.1	22.8	6.6	14.0	27.7	10.5
log K (MAC)	4.0	5.5	4.0	5.2	0.8	6.8

(log K$_1$ values from reference 27. Ionic radii (Å) for octahedral coordination from Ref. 89)

The low value for the macrocyclic effect for Hg(II) probably reflects the reluctance of Hg(II) to assume the higher coordination numbers required for low strain complexation with a ligand such as 15-aneN$_5$.

4.7. Macrocycles with Pendent Donor Groups

The chemistry of macrocycles with pendent donor groups has become of great interest because of their potential use in biomedical applications.[26,118-121] Some recent reviews[26,111,112,122-124] have covered aspects of macrocyclic ligands with pendent donor groups. Pendent donor groups have been added to triaza macrocycles, principally 9-aneN$_3$, which with full substitution on the nitrogen donors gives hexadentate ligands best suited for coordinating to smaller octahedral metal ions. Addition of pendent acetate groups to the nitrogens of tetraazamacrocycles, principally 12-aneN$_4$, gives octadentate ligands, best suited for coordinating larger metal ions. In addition, ligands with pendent donors have been generated by substitution on the nitrogens of mixed donor macrocycles to give differing requirements of geometry and coordination number. A further class of macrocycles with pendent donors is that where the pendent donor is attached to the ethylene bridges of the macrocycle.

4.7.1 Triaza macrocycles with pendent donor groups

Studies on ligands based on 9-aneN$_3$ (see Figure 4.22) have been carried out in aqueous solution with, for example, acetate (NOTA),[125-128] 3,5-dimethyl-2-hydroxybenzyl[129]

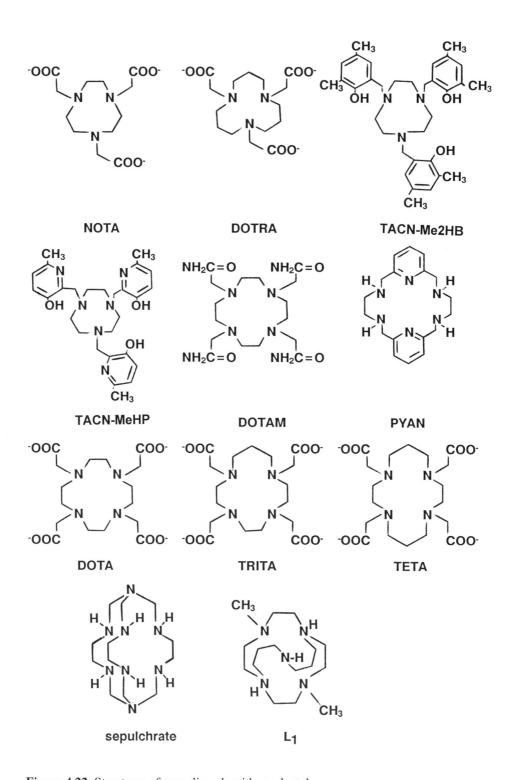

Figure 4.22. Structures of some ligands with pendent donor groups.

(TACN-Me2HB), methylphosphonate,[130] 2-hydroxyethyl,[131] 2-mercaptoethyl,[132] 3-hydroxy-6-methyl-2-pyridylmethyl,[133] and N,N-dimethylacetamide[134] pendent donors, and based on 12-aneN$_3$ with acetate pendent donors,[135] have been reported[15] in aqueous solution. In general these pendent donor groups provide large increases in stability constant, particularly with octahedral metal ions of about the right size. Thus, there is a large increase in log K$_1$ for NOTA complexes compared to those of the parent 9-aneN$_3$ macrocycle:

Metal ion	Cu(II)	Zn(II)	Cd(II)	Pb(II)
log K$_1$ (9-aneN$_3$)	15.5	16.2	9.5	11.1
log K$_1$ (NOTA)	21.6	21.1	16.0	16.6
increase in log K	6.1	4.9	6.5	5.5

(log K values from Ref. 15).

The effects of the various pendent groups is much as would be expected from their binding strength as simple unidentate ligands with metal ions, as exemplified by the log K values with Fe(III). Thus, the 3,5-dimethyl-2-hydroxybenzyl group has a more basic oxygen than does acetate, and so log K$_1$ for TACN-Me2HB is much higher at 51.3 than the NOTA complex[125] at 28.3.

An unusual feature of the complexes of NOTA with trivalent metal ions is that the Ga(III) complex is more stable[125] than that of Fe(III), whereas for virtually all other ligands the opposite is true. Crystallography[126,136] has shown (Figure 4.23) that the Fe(III) complex of NOTA has a structure distorted toward trigonal prismatic, whereas other metal ions such as Ga(III) or Ni(II) achieve more nearly regular octahedral geometry. The distorted geometry observed for the Fe(III) complex with NOTA suggests that the acetate arms are too short, and this factor would account for the lower stability of the Fe(III) complex as compared to the complex of the smaller Ga(III) ion. The six-membered chelate rings of TACN-MeHP the 9-aneN$_3$ derivative with 3-hydroxy-6-methyl-2-pyridylmethyl pendent groups, allow for adoption of regular octahedral coordination geometry[137] (Figure 4.23) in the Fe(III) complex. In complexes of ligands with 2-hydroxybenzyl groups that form six membered chelate rings, such as TACN-Me2HB, the stability of the Fe(III) complex (log K$_1$ = 51.3) is much higher than that of the Ga(III) complex (log K$_1$ = 44.2), suggesting[129] the importance for Fe(III) of being able to achieve more regular octahedral coordination geometry than allowed by NOTA.

If one is using a ligand to complex metal ions at a specific pH, for example the biological pH of 7.4, then the ideal ligand would not have donor groups numerically more basic than the required operating pH, since complex formation must otherwise occur in competition with the proton. Thus, a ligand such as TACN-MeHP has less basic 3-hydroxy-6-methyl-2-pyridylmethyl donor groups, which resemble TACN-Me2HB in having phenolate donor groups which form six membered rings, but of much lower basicity. Thus, although the formation constant of TACN-MeHP at[133] log K$_1$ = 49.98 is considerably lower than that of the naturally occurring siderophore enterobactin, at pH values below 9.0 TACN-MeHP complexes Fe(III) much more effectively than does enterobactin because the donor groups of enterobactin are extremely basic catecholate oxygens. The ligand TACN-MeHP represents from a thermodynamic standpoint possibly the ideal ligand for complexing Fe(III) at pH 7.4.

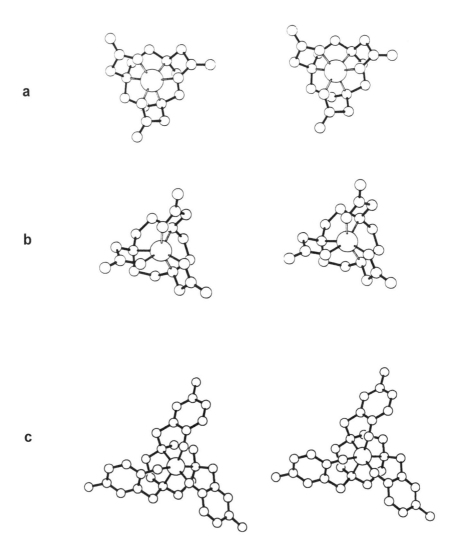

Figure 4.23. Stereoviews down the three-fold axes of (a) the Ni(II) complex of NOTA, (b) Fe(III) complex of NOTA, and (c) complex of Fe(III) with TACN-MeHP. The extent to which the projections of the N-M-O angles onto the plane at right angles to the three-fold axis of the complex is rotated away from the value of 60° for a regular octahedron is the twist angle. The twist angle is 0° for a regular octahedron, and 60° for trigonal prismatic coordination. The Ni(II) ion has short enough Ni-N and Ni-O bonds to allow fairly regular octahedral coordination, with[126] a twist angle of 6.9°. The acetate donor groups of NOTA are too short for Fe(III), thereby causing the trigonal prismatic distortion[136] of the complex (twist angle = 35°), whereas the six-membered rings of TACN-MeHP allow for more regular coordination geometry[137] (twist angle = 6°). Structures drawn from coordinates available in Refs. 126, 136, and 137.

A ligand based on 12-aneN$_3$ with acetate arms (DOTRA, Figure 4.22) has been studied,[135] and found to complex metal ions such as Ca(II), Mg(II), and Cd(II) only a little more weakly than does NOTA. DOTRA complexes the Zn(II) ion slightly better than does NOTA, which is surprising, because the parent 12-aneN$_3$ is a much weaker ligand than is 9-aneN$_3$. The reasonably strong complexation of metal ions by DOTRA may indicate that the added arms act to prevent the free ligand from achieving a low energy state attained by the free 12-aneN$_3$ ligand itself. One would surmise that part of why 12-aneN$_3$ is a weak ligand may relate to stabilization of the free ligand by internal hydrogen bonding of the N-H hydrogens to the lone pairs on other nitrogens within the macrocyclic cavity, which is denied to DOTRA because it has no N-H hydrogens.

4.7.2 Ligands based on cyclen, that have pendent donors

Solution chemistry of ligands based on the macrocycle cyclen has been reported with several donor groups, including acetate,[138-140] 2-hydroxypropyl,[141] methylphosphon-ate,[142,143] N,N-dimethylacetamide,[134] and acetamide.[144] Ligands based on 13-aneN$_4$ and 14-aneN$_4$ with a variety of pendent donors, such as acetate (TRITA and TETA, Figure 4.22) have also been studied.[15] Here the effect of increase of chelate ring size is much as would be expected, with larger metal ions showing larger decreases in log K$_1$ on increase of ring size in passing from DOTA to TETA, for example,[105] as seen in Figure 4.24.

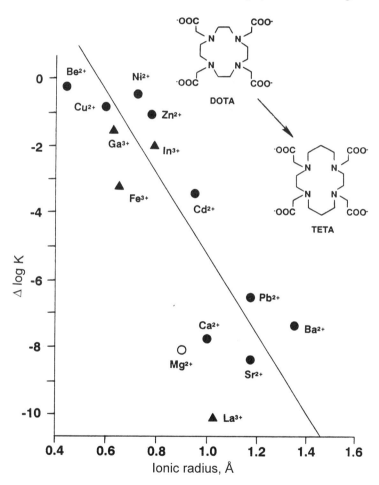

Figure 4.24. Effect of increase of chelate ring size on complex stability for tetraaza macrocycles with N-acetato pendent donor groups. The change in complex stability, Δ log K, in passing from DOTA to TETA (See Figure 4.22 for key to ligand abbreviations) is plotted as a function of metal ion radius for divalent (●) and trivalent (△) metal ions.

The type of structure afforded by cyclen with four pendent donor groups is particularly well suited to large metal ions, which MM calculations[145] suggest have optimally an ionic radius of 0.95 Å in the DOTA complexes. If one examines the increases in log K found in passing from complexes of cyclen to those of DOTA, it is evident that the greatest increase in log K occurs close to this metal ion radius:

M^{2+} ion	Cu(II)	Ni(II)	Zn(II)	Cd(II)	Ca(II)	Pb(II)
ionic radius	0.57	0.69	0.74	0.95	1.00	1.18
log K_1 cyclen	23.3	16.4	16.2	14.3	3.1	15.9
log K_1 DOTA	22.7	20.5	18.7	21.3	16.4	22.7
Δ log K	-0.6	+4.1	+2.5	+7.0	+11.3	+6.8

(log K_1 values from Refs. 15, 27, and 138-140, ionic radii (Å) from Ref. 89).

The greatest increase in complex stability occurs at Ca^{2+} with its ionic radius of 1.0 Å, fairly close to the radius which the MM calculations indicate would be optimal. The complex with DOTA is the most stable formed by Ca^{2+} with any ligand, showing the excellent geometry possessed by DOTA for complexing medium-large metal ions, which includes the Gd(III) ion.

The addition of pendent donor groups containing neutral oxygen donors to cyclen has also produced ligands of considerable complexing power.[141] Possibly even more remarkable has been the complexing ability of the ligand[144] DOTAM (Figure 4.22) which has amide donor groups. The structure of the Ca(II) complex is shown in Figure 4.25, along with the Zn(II) complex. The Ca(II) complex has close to regular square antiprismatic coordination geometry, typical of complexes with pendent donors based on cyclen with mean Ca-O lengths of 2.41 Å and Ca-N lengths of 2.56 Å. The Zn(II) is only six coordinate, with two Zn-O bonds at 2.10 Å, and the other two at beyond van der Waals contact distances at 3.14 Å. What is of interest here is the penalties that are associated with non-coordination of donor groups, as far as complex stability goes. The non-coordinated donor group does not only not contribute to the stability of the complex formed, but it also acts as a sterically hindering substituent. Thus, the Zn-N bonds to which the non coordinated amide groups are attached are fully 0.07 Å longer than those to the nitrogens which bear coordinated amide groups. This is typical of N-alkyl substitution, that steric crowding causes M-N bond lengthening. Accompanying these steric interactions is a drop in complex stability for the Zn(II) complex to a value of log K_1 nearly six log units lower than that for cyclen itself. In contrast, for the Ca(II) complex, which has all four amide donor oxygens coordinated, log K_1 for the DOTAM complex is over three log units higher than log K_1 for the parent cyclen complex.

4.7.3 Ligands based on hexadentate macrocycles, that have pendent donor groups

A variety of ligands based on 18-aneN_2O_4 with pendent N-donors have been studied.[27] These include hydroxyethyl, 2-hydroxypropyl, methoxyethyl, tetrahydrofuranyl, 2-pyridylmethyl, amide, and acetate donor groups.[145-151] The effect of these groups on complex stability is much as would be expected from their effects with triaza or tetraazamacrocycles. The presence of four neutral oxygen donors in the macrocyclic rings of these complexes leads to high selectivity for large metal ions such as Pb(II) or

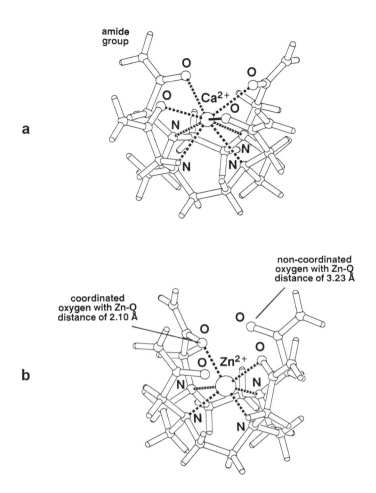

a

amide
group

b

coordinated
oxygen with Zn-O
distance of 2.10 Å

non-coordinated
oxygen with Zn-O
distance of 3.23 Å

Figure 4.25. The structure[144] of (a) the Ca(II) complex of DOTAM and (b) the Zn(II) complex of DOTAM, showing how the structure of the Zn(II) is affected by the Zn(II) being too small to coordinate in an octacoordinate fashion with the octadentate ligand. The coordination geometry around the Ca^{2+} is approximately square antiprismatic.

Ba(II) over smaller metal ions. Overall, however, the stability of the complexes of these ligands is not high, because of the preponderance of relatively weak neutral oxygen donors. More powerfully complexing ligands based on macrocyclic donors with six donor atoms have been developed for the ligand PYAN[153] (Figure 4.22). This study, and studies with pyridylmethyl pendent donor groups,[152] highlight the advantages of the pyridyl donor group as compared to the neutral oxygen donor, when the metal ion is large, and has a reasonably high affinity for nitrogen donor atoms. This is highlighted by log K values for the following N-substituted 18-aneN_2O_4 derivatives, where log K_1 with NH_3 is used as an indicator of the affinity of the metal ion for nitrogen donors:

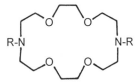

	log K_1 NH_3	log K_1 for: R = H	CH_2CH_2OH	CH_2CONH_2	$CH_2C_5H_4N$	CH_2COO^-
Ca(II)	-0.2	1.6	4.1	5.7	3.6	7.7
La(III)	(0.3)	3.0	3.2	-	3.5	11.7
Pb(II)	1.6	6.8	9.2	10.70	11.7	13.5
Cd(II)	2.7	5.3	7.6	8.6	11.0	11.9

log K_1 values from Refs. 27 and 152.

4.8   The Cryptands

The cryptands,[2] particularly the smaller cryptands, are generally more preorganized than simple monocyclic macrocycles. Smaller cryptands such as cryptand-111 (below) are fairly rigid,[25] and have a cavity very selective for the proton. This is illustrated by the protonation constants for cryptands of increasing cavity size[154] below. The protonation constant for triethanolamine is included as an example of a totally non-preorganized ligand with three groups consisting of an ethylene bridge to an oxygen, attached to the nitrogen donor. In Figure 4.26 is shown the variation[155] in log K_1 for alkali and alkali-earth metal ions as the number of ethereal oxygen donors, and therefore the size of the cavity in the cryptand, is increased. Here the size-match selectivity hypothesis does appear to give a reasonable account of the selectivity patterns observed, unlike the crown ethers and tetraaza macrocycles, with peaks in complex stability occurring for smaller metal ions with cryptands with smaller cavities. Both Na^+ and Ca^{2+} have ionic radii[89] of 1.00 Å, while the larger Ba^{2+} and K^+ ions are close to each other in size with ionic radii of about 1.35 Å. Thus, both Na^+ and Ca^{2+} in Figure 4.26 show a peak in complex stability with the same cryptand, cryptand-221, while both K^+ and Ba^{2+} peak at the larger cryptand-222.

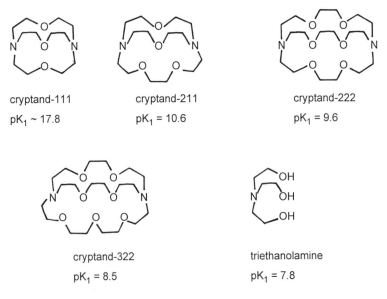

cryptand-111
$pK_1 \sim 17.8$

cryptand-211
$pK_1 = 10.6$

cryptand-222
$pK_1 = 9.6$

cryptand-322
$pK_1 = 8.5$

triethanolamine
$pK_1 = 7.8$

The benefit of high levels of preorganization is the sharp selectivity seen for the proton in the cavity of cryptand-111 below, and for the alkali and alkaline earth metal ions in Figure 4.26. The drawback to high levels of preorganization is[25] slow kinetics of complex formation. Thus, for cryptand-111, the proton initially attaches itself to the nitrogen outside of the cage of the cryptand, where the pK_a is only 7.1, and only slowly converts to a structure with the proton inside the cavity, where the pK_a is estimated to be 17.8.

MM calculations[156] have supported the idea of the rigid cavity in cryptand-111. Thus, the only low energy conformer of cryptand-111 is the endo-endo below, with lone pairs on the nitrogens directed into the cavity. In contrast, for cryptand-222 several conformers are of fairly low energy, and so cryptand-222 is not highly preorganized like cryptand-111 (see Figure 4.5).

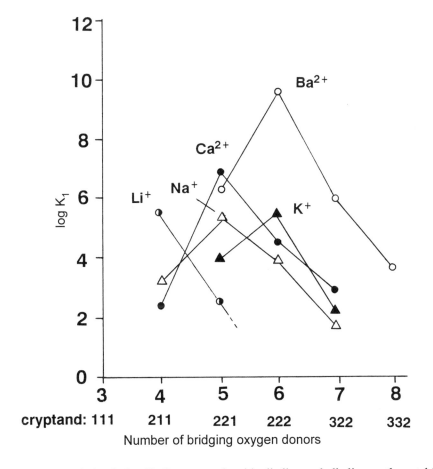

Figure 4.26. Variation in log K_1 for cryptands with alkaline and alkaline earth metal ions with number of oxygen donors in the groups bridging the two nitrogens of the cryptand. Formation constant data from references 15, 27, and 155. log K values are for Li+ (O), Na+ (Δ), K+ (Δ), Ca²+ (●), and Ba²+ (O).

| | endo-endo | endo-exo | exo-exo |

Cryptands appear to be highly preorganized, but what the primary effect is of the cage structure of cryptands on complex stability and metal ion selectivity can be seen by comparing two types of ligands that have exactly the same donor set, four nitrogens and four neutral oxygens, all sp³ hybridized, and see how the stability differs between the cryptand and macrocycle with pendent donors below:

cryptand-22-N2   THP-cyclen

Metal ion	Ionic radius	log K_1	log K_1	Δ log K_1
Cu(II)	0.57	12.7	19.5	6.8
Zn(II)	0.74	6.0	13.5	7.5
Cd(II)	0.95	12.0	17.5	5.5
Ca(II)	1.00	4.3	5.7	1.4
Pb(II)	1.18	15.3	15.1	-0.2
Ba(II)	1.35	6.7	3.7	-3.0

Ionic radii (Å) from Ref. 89, formation constants from Refs. 27 and 141.

The above results suggest that a primary effect of the larger cryptand type of structure compared with a tetraaza macrocycle with pendent donors is to depress the complex stability of smaller metal ions. The three dimensional cavity of the cryptand makes complexation of metal ions that are too small more difficult. The more highly preorganized cavity of the cryptand also produces a significant increase in log K_1 only for the very large Ba(II) ion relative to log K_1 for the pendent-donor macrocycle.

The sepulchrates (Figure 4.22) are cryptand-like ligands which generally have all nitrogen donor groups.[157] Ligands with a cryptand structure, containing all nitrogens donors have been prepared,[158] such as the cryptand-111 analogue with the oxygens replaced with N-methyl groups (L_1 in Figure 4.22). This ligand has a small cavity, associated with which is a high affinity for the Li⁺ ion, with log K_1 = 5.5. When some of the ethylene bridges in ligands of this type are replaced with trimethylene bridges, the affinity for Li⁺ drops dramatically, but the affinity for the proton rises rapidly.

4.9 Macrocycles with Pendent Donor Groups Attached *via* the Carbon Atoms of the Bridging Groups

Gokel and coworkers[124,159] have prepared a large number of crown ethers with pendent donor groups attached via a carbon of an ethylene bridge rather than the nitrogen donor atom, the "lariat ethers" (Figure 4.27).

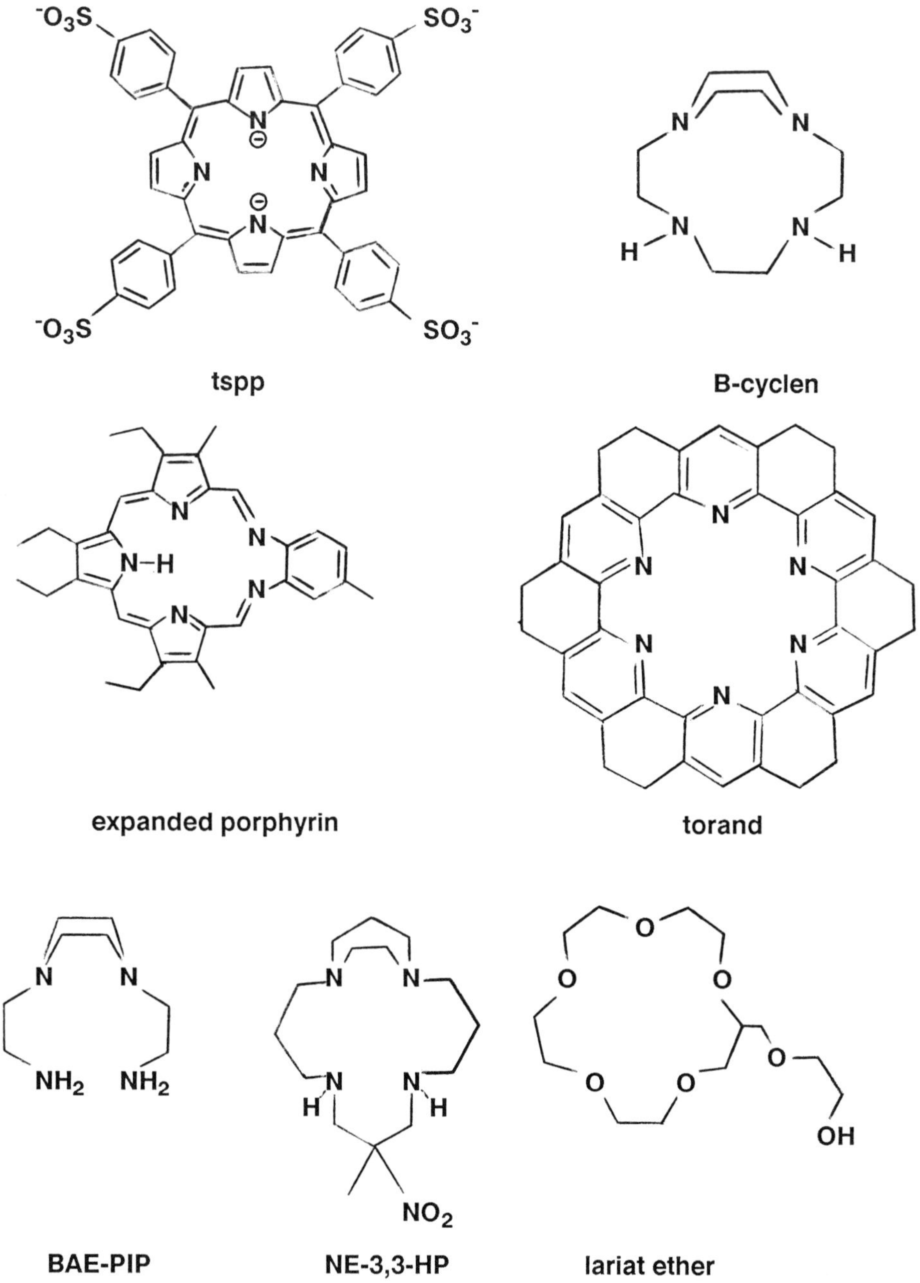

tspp

B-cyclen

expanded porphyrin

torand

BAE-PIP

NE-3,3-HP

lariat ether

4.27. Highly preorganized ligands discussed here.

The complexing strength of these C-pivot lariat ethers is somewhat disappointing, which may be traced to the fact that for the C-pivot donor group to coordinate to a metal ion, the pendent donor group has to move into what is effectively an "axial" position on the ethylene bridge to which it is attached. Ordinarily, the pendent donor group would occupy the energetically more favorable equatorial position, so that movement to the higher energy axial position so as to coordinate the metal ion carries a penalty in energy terms. Bartsch and coworkers[160] have studied crown ethers of the type I and II below:

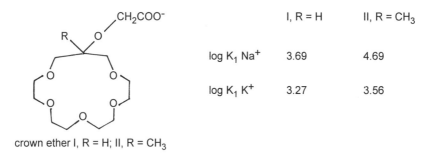

	I, R = H	II, R = CH$_3$
log K$_1$ Na$^+$	3.69	4.69
log K$_1$ K$^+$	3.27	3.56

crown ether I, R = H; II, R = CH$_3$

What one sees with the formation constant of Na$^+$ and K$^+$ with crown ethers I and II above is that the *gem* methyl group on II acts to preorganize the ligand for complexation as discussed in section 3.10. The C-methyl group increases complex stability of the Na$^+$ complex, and because Na$^+$ is smaller, promotes its selectivity over the larger K$^+$ ion.

4.10. More Highly Preorganized Macrocycles

In section 4.6 it was shown that crown ethers and tetraaza macrocycles are too flexible to exert the kind of size match selectivity that is apparent for the cryptands. Interesting ligands might be those consisting of a single ring that is more rigid. The prototypes of such ligands are the porphyrins.[161] Recently,[162] the stability constants of complexes of tetrakis(4-sulfophenyl)porphyrin (tspp^{6-}, Figure 4.27) were reported in 80:20 DMSO/water for Cu(II), Zn(II), and Mg(II). The use of DMSO/water as a solvent should not greatly affect the formation constants. The results demonstrate the enormous stabilization produced compared to the formation constants of tetraaza macrocycles such as 14-aneN$_4$:

Metal Ion	log K$_1$		
	tspp^{6-}	14-aneN$_4$	THEC
Cu(II)	38.1	27.2	15.7
Zn(II)	34.6	15.5	6.4
Mg(II)	28.8	(2)	1.9
protonation constants			
L + 2H = LH$_2$	32.8	22.0	17.0
LH$_2$ + 2H = LH$_4$	3.6	3.9	3.9

Abbreviations: tspp^{6-} = tetrakis(4-sulfophenyl)porphyrin, THEC = N,N',N'',N'''tetrakis(2-hydroxyethyl)cyclam. Formation constants for tspp^6 in 80:20 DMSO/water, from Ref. 162, other constants in water from Ref. 27. For the protonation constants L = ligand indicated. Charges are omitted from the protonation equilibria for simplicity, so that "H" is the proton.

The fact that the formation constants for tspp^{6-} with the above three metal ions are all closely bunched together suggests that the main source of the high complex stability is not the donor strength of the nitrogens of tspp^{6-} but rather its great degree of preorganization. In spite of the latter observation, MM calculations have shown[163] that porphyrins are surprisingly flexible. These MM calculations also support the idea that a fairly small metal ion with an M-N bond length of 2.04 Å would best fit into the cavity in the porphyrin ring.

In view of the high levels of preorganization apparent for porphyrins, it seems desirable to make ligands with similar structures, but cavities better suited to metal ions of other sizes. Thus, Bell *et al.*[164] and Sessler *et al.*[165] have synthesized the ligands shown in Figure 4.27. The torands of Bell *et al.*[164] have a high affinity for many metal ions, but show surprisingly little selectivity between alkali metal ions of different sizes. Thus, although the cavity of the torand in Figure 4.27 should fit K$^+$ best, the ligand complexes Li$^+$ almost as strongly. A possible explanation for this behavior is that the cavity suggests a stability order K$^+$ > Na$^+$ > Li$^+$, but the affinity for nitrogen donors is strongly the reverse of this order (see section 2.5.3). The cavities in the expanded porphyrins studied by Sessler *et al.*[165] are larger than those of regular porphyrins, and are well adapted to complexing metal ions of the size of the lanthanides.

The flexibility of tetraazamacrocycles has been reduced[166-168] by addition of extra bridges between adjacent nitrogen donors to give ligands of the type B-cyclen and NE-3,3-HP shown in Figure 4.27 to give *reinforced* macrocycles. The extra bridge greatly increases the level of preorganization of the ligand. Thus, the macrocyclic effect for the Ni(II) complex of B-cyclen compared to that of its open-chain analogue BAE-PIP is, at 9.6 log units, the largest known.[167] The rigid cavity in the macrocycle is also able to compress (Figure 4.28) the square planar low spin Ni(II) ion to the shortest Ni-N bonds

a

bond compressed to 1.84 Å from normal Ni-N length of 1.91 for S = O Ni(II)

b

bond stretched to 2.09 Å from normal length of 2.00 Å for square planar Cu(II)

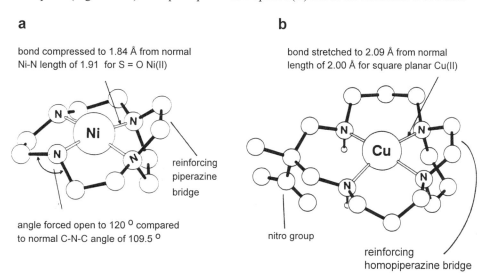

reinforcing piperazine bridge

angle forced open to 120 $^{\circ}$ compared to normal C-N-C angle of 109.5 $^{\circ}$

nitro group

reinforcing homopiperazine bridge

Figure 4.28. Structures of complexes of reinforced ligands. (a) shows the complex [Ni(B-12-aneN$_4$)]$^{2+}$ and structural features associated[167] with compression of the metal ion by the small cavity of the ligand, and (b) shows how the large cavity of the reinforced macrocycle NE-3,3-HP is able to stretch the Cu-N bonds in its complex with Cu(II).[168]

known, due to the fact that the macrocycle is unable to adopt other conformations that allow the metal ion to escape compression, as is the case for non-reinforced tetraaza macrocycles. At the other end of the cavity size range, the large cavity reinforced macrocycle NE-3,3-HP is able to stretch[168] the Cu-N bonds out to the very long value of 2.09 Å, as seen in Figure 4.28. The reinforcement of the macrocycle makes it extremely resistant to demetallation, and many of the complexes are able to survive 98% H_2SO_4 indefinitely, or survive refluxing in 0.1 M cyanide.

4.11 Binucleating Macrocycles

The interest in binucleating macrocycles has largely been the fact that they may serve as models for metalloproteins where the active site contains more than one metal ion.[169] Examples of such metalloproteins are hemocyanin,[170] tryrosinase,[171] and nitrogenase.[172] Macrocyclic complexes as models for metalloproteins have been covered in a recent review.[173] Research efforts have involved, for example, the synthesis of very large polydentate azamacrocycles[174] such as 36-aneN$_{12}$, which complexes three copper(II) ions, cyclam with four N-(2-aminoethyl) pendent donor groups which complexes two Cu(II) ions,[175] or large Schiff base macrocycles[176] capable of complexing two metal ions. The loading of dioxygen by BISTREN complexes containing two Co(II) ions coordinated within it (Figure 4.29) has been studied extensively.[177]

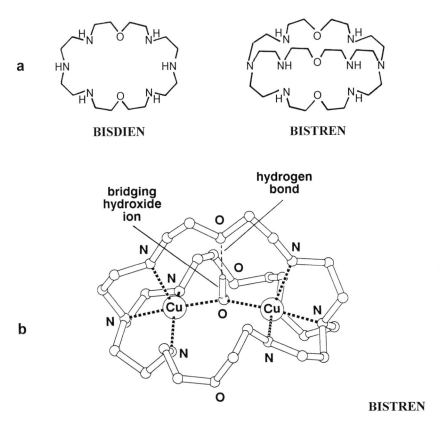

Figure 4.29. (a) The binucleating ligands BISTREN and BISDIEN, and (b) crystal structure of the hydroxide bridged dicopper(II) complex of BISTREN (Redrawn with coordinates available in Ref. 181.

The interest here is the effect that the presence of two metal ions held close together in a macrocyclic complex has on the ability of these two metal ions to bind small ligands between them. Is there, in fact, an enhancement of complex stability that might be referred to[177] as a "reverse chelate effect". A further point of interest is the size-based ligand selectivity that will be produced by the distance between the two metal ions in the minimum energy conformation for the binuclear complex.

For the di-copper(II) BISTREN complexes, there is a strong increase in log K_1 for the binding of simple unidentate ligands as compared to the binding of the same ligands to the simple Cu^{2+} aquo ion:[178]

log K_1 with:	Cu^{2+a}	$Cu(II)_2(BISTREN)^b$	$(Cu^{2+})_2^{c,d}$
OH^-	6.3	9.89	8.2
F^-	0.7	4.7	-
Cl^-	0.1	2.1	-

[a] Formation constants from Ref. 27 at 25 °C and ionic strength 0.1. [b] Formation constants from Ref. 180. [c] Ref. 27, temperature = 25 °C, and ionic strength 3.0. [d] The units are not strictly comparable.

There is a rise in log K_1 of from 2 to 4 log units for the binding of these simple unidentate ligands in Cu_2-BISTREN as compared to the simple Cu^{2+} aquo ion, showing that there is a reverse chelate effect operating. A single hydroxide ion is also bound more strongly by two Cu^{2+} aquo ions than by a single Cu^{2+}, in the formation of $[Cu_2OH]^+$, but this is less than the stability of the hydroxide complex with Cu_2-BISTREN. The Cu_2-BISTREN complex may provide preorganization of the complex for binding a bridging hydroxide ion, and also hydrogen bonding between the hydrogen of the bridging hydroxide group, and an oxygen of the ethereal bridge between the two halves of the Cu_2-BISTREN complex (Figure 4.29).

The bis-copper(II) complex of BISDIEN shows an ability to bind a wide variety of bridging ligands, ranging from hydroxide to pyrophosphate. How does the Cu_2-BISDIEN complex manage to complex well with ligands of such differing sizes? MM suggests here[182] that the Cu_2-BISDIEN complex is able to adopt several conformers of similar energies with differing Cu to Cu separations (Figure 4.30). Thus, the bowl-shaped conformer in Figure 4.30 has a Cu to Cu separation of 4.27 Å, and accommodates small bridging ligands such as OH^- and F^-. The intermediate conformer accommodates medium length ligands such as azide, while the extended conformer with its Cu to Cu separation of 7.31 Å accommodates longer bridging ligands such as pyrophosphate. In contrast, the three dimensional structure of the BISTREN ligand reduces the flexibility of its bis-copper(II) complex very considerably, and it accommodates short ligands such as OH^- (Figure 4.28) or F^- very well. Oxygen can be coordinated in the bis-Co(II) complex in the space below the hydroxide ion, but this produces very considerable distortion of the coordination sphere around the cobalt, and it is thus found that the oxygen is loaded weakly, and reversibly. Variation in the structure of the bridging groups in BISDIEN analogues, with, for example, replacement of the ethereal oxygens with furan groups has led to interesting alterations in the ability of dinuclear complexes to bind simple bridging ligands.[183]

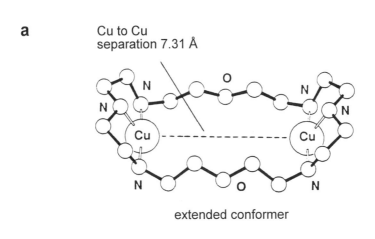

a   Cu to Cu
separation 7.31 Å

extended conformer

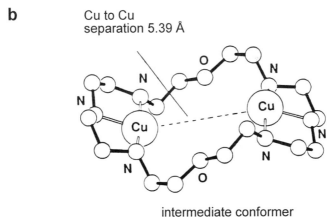

b   Cu to Cu
separation 5.39 Å

intermediate conformer

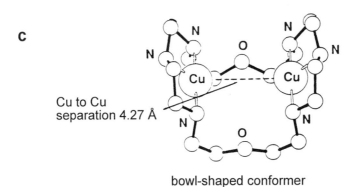

c

Cu to Cu
separation 4.27 Å

bowl-shaped conformer

Figure 4.30. Molecular mechanics generated structures of Cu_2-OBISDIEN complex showing how the binucleating complex can adjust the Cu to Cu separation to accommodate ligands of different sizes. Redrawn after Ref. 182

4.12   Conclusions

In this chapter a consistent theme has been that the selectivities displayed by macrocyclic ligands are more complex than might be supposed from the simple size match selectivity idea. The complexes of macrocyclic ligands are generally flexible enough to change conformation to accommodate metal ions of different sizes, or in the case of binucleating macrocycles, ligands of different size. In many cases the idea of a cavity in the macrocycle is inappropriate, since the metal ion lies out of the plane of the donor atoms. In this situation, the steric factors that govern selectivity are the same as for open-chain ligands, principally the size of the chelate ring. Much of the apparent size match selectivity of crown ethers reflects the fact that K^+ fits the five membered rings present in these ligands best. The cryptands as three dimensional ligands are more highly preorganized than monocyclic macrocyclic ligands, and their selectivities for metal ions approach more nearly the idea of size match selectivity, although the size of the chelate ring still plays an important part. Macrocyclic ligands may be made more preorganized by making their bridging structures more rigid, when something approaching size match selectivity might be achieved. The principal advantages of the macrocyclic structure are the increase in thermodynamic complex stability, and greater inertness to metallation and demetallation reactions, which might be advantageous in some applications, and detrimental in others. The best available information suggests that inductive effects caused by the alkyl bridges added to the open chain ligand to complete the macrocyclic ring cause greater M-L bond strengths, and account for much of the macrocyclic effect. Other contributions which may be important are related to the high energy state of the free macrocyclic ligand, which may be due to enforced electrostatic repulsion between lone pairs on the donor atoms in the cavity of the macrocycle, steric hindrance to solvation of the donor atoms in the cavity of the macrocycle, and greater preorganization of the macrocyclic ligand as compared to its open-chain analogue.

References.

1. C. J. Pedersen, *J. Am. Chem. Soc.*, **1967**, *89*, 2459.
2. J.-M. Lehn, *Acc. Chem. Res.*, **1978**, *11*, 49.
3. N. F. Curtis, *J. Chem. Soc.*, **1960**, 4409.
4. D. A. Busch, *Acc. Chem. Res.*, **1978**, *11*, 392.
5. D. B. Rorabacher, N. E. Meagher, K. L. Juntunen, P. V. Robandt, G. H. Leggett, C. A. Salhi, B. C. Dunn, R. R. Schroeder, L. A. Ochrymowycz, *Pure & Appl. Chem.*, **1993**, *65*, 573.
6. a) A. J. Blake and M. Schroder, *Adv. Inorg. Chem.*, **1990**, *35*, 1. b) G. Reid and M. Schroder, *Chem. Soc. Rev.*, **1990**, *19*, 239.
7. a) S. R. Cooper, *Acc. Chem. Res.*, **1988**, *21*, 141. b) S. R. Cooper and S. C. Rawle, *Struct. Bond. (Berlin)*, **1990**, *72*, 1.
8. G. A. Melson, Ed. *Coordination Chemistry of Macrocyclic Compounds*, Plenum, New York, 1980.
9. L. F. Lindoy, *The Chemistry of Macrocyclic Ligand Complexes*, Cambridge University Press, Cambridge, 1989.
10. R. M. Izatt and J. J. Christensen Eds., *Progress in Macrocyclic Chemistry*, Wiley, New York, 1987.
11. R. M. Izatt and J. J. Christensen Eds., *Synthetic Multidentate Macrocyclic Ligands*, Academic Press, New York, 1978.
12. a) G. W. Gokel and S. H. Korzeniowski, *Macrocyclic Polyether Synthesis*, Springer, Heidelberg and New York, 1982. b) G. W. Gokel and Y. Inoue, *Cation Binding by Macrocycles*, Marcel Dekker, New York and Basel, 1990. c) G. W. Gokel, *Crown Ethers and Cryptands*, The Royal Society of Chemistry, Cambridge, 1991.
13. a) A. Bianchi, M. Micheloni, P. Paoletti, *Coord. Chem. Rev.*, **1991**, *110*, 17. b) M. Ciampolini, N. Nardi, B. Vantancoli, M. Micheloni, *Coord. Chem. Rev.*, **1992**, *120*, 223.
14. S. R. Cooper Ed., *Crown Compounds, Towards Future Applications*, VCH Publishers, Deerfield Beach, Florida, 1992.

15. a) R. M. Izatt, J. S. Bradshaw, S. A. Nielsen, J. D. Lamb, J. J. Christensen, and D. Sen, *Chem. Rev.*, **1985**, *85*, 271. b) R. M. Izatt, K. Pawlak, J. S. Bradshaw, and R. L. Bruening, *Chem. Rev.*, **1991**, 91, 1721.
16. F. Vogtle, *Angew. Chem. Int. Ed. Engl.*, **1991**, *30*, 442.
17. D. J. Cram, *Science*, **1988**, *240*, 76-767.
18. D. K. Cabbiness and D. W. Margerum, *J. Am. Chem. Soc.*, **1969**, *91*, 6540.
19. D. A. Busch, K. Farmery, V. Goedken, V. Katovic, A. C. Melnyk, C. R. Sperati, and N. Tokel, *Adv. Chem. Ser.*, **1971**, *100*, 44.
20. L. Fabbrizzi, *Comments Inorg. Chem.*, **1985**, 4, 33.
21. A. J. Blake, R. O. Gould, A. J. Holder, T. J. Hyde, A. J. Lavery, M. O. Odulate, and M. Schroder, *J. Chem. Soc., Chem. Commun.*, **1987**, 119.
22. F. P. Hinz and D. W. Margerum, *Inorg. Chem.*, **1974**, *13*, 2941.
23. R. D. Hancock, *J. Chem. Ed.*, **1992**, *69*, 615.
24. A. Anichini, L. Fabbrizzi, P. Paoletti, R. M. Clay, *J. Chem. Soc., Dalton Trans.*, **1978**, 577.
25. R. D. Hancock, *Prog. Inorg. Chem.*, **1989**, *37*, 187.
26. R. D. Hancock and A. E. Martell, *Chem. Rev.*, **1989**, *89*, 1875.
27. A. E. Martell and R. M. Smith, *Critical Stability Constants*, Volumes 1-6, Plenum Press, New York, 1974-1989.
28. R. D. Hancock and A. Evers, *Inorg. Chim. Acta*, **1989**, *160*, 245.
29. M. Kodama and E. Kimura, *J. Chem. Soc., Dalton Trans.*, **1976**, 116 ; **1977**, 2269; **1977**, 1473.
30. L. Fabbrizzi, P. Paoletti, and R. M. Clay, *Inorg. Chem.*, **1978**, *17*, 1042.
31. D. C. Weatherburn, E. J. Billo, J. P. Jones, and D. W. Margerum, *Inorg. Chem.*, **1970**, *9*, 1557.
32. L. Fabbrizzi, R. Barbucci, and P. Paoletti, *J. Chem. Soc.*, **1972**, 1529.
33. B. L. Haymore, J. D. Lamb, R. M Izatt, and J. J. Christensen, *Inorg. Chem.*, **1982**, *21*, 1598.
34. F. Marsicano, R. D. Hancock, and A. McGowan, *J. Coord. Chem.*, **1992**, *25*, 85.
35. G. Anderegg, *Helv. Chim. Acta*, **1975**, *58*, 1218.
36. J. M. Lehn and J. P. Sauvage, *J. Am. Chem. Soc.*, **1975**, *97*, 6700.
37. F. Arnaud-Neu, B. Spiess, and M. J. Schwing-Weill, *Helv. Chim. Acta*, **1977**, *60*, 2633.
38. E. Kauffman, J. M. Lehn, and J. P. Sauvage, *Helv. Chim. Acta*, **1976**, 59, 1099.
39. V. M. Royal, R. Pizer, and R. G. Wilkins, *J. Am. Chem. Soc.*, **1977**, *99*, 7185.
40. R. J. Gillespie and R. J. Nyholm, *Q. Rev. Chem. Soc.*, **1957** *11*, 339 (,
41. B. P. Hay, J. R. Rustad, C. Hostetler, *J. Am. Chem. Soc.*, **1993**, 115, 11158.
42. C. Eaborn, P. B. Hitchcock, J. D. Smith, A. C. Sullivan, *J. Organomet. Chem.*, **1984**, *C23*, 263.
43. a) P. Groth, *Acta Chem. Scand., Ser. A.*, **1981**, *A35*, 460. b) J. Dale, J. Eggestad, S. B. Fredriksen, and P. Groth, *J. Chem. Soc., Chem. Commun.*, **1987**, 1391.
44. P. Luger, C. Andre, R. Rudert, D. Zobel, A. Knochel, and A. Krause, *Acta Crystallogr., Sect. B.*, **1992**, *B48*, 33.
45. a) G. Shoham, W. N. Lipscomb, U. Olsher, *J. Chem. Soc., Chem. Commun.*, **1983**, 208. b) J. D. Dunitz, M. Dobler, P. Seiler, and R. P. Phizackerley, *Acta Crystallogr., Sect B*, **1974**, *B30*, 2733.
46. A. Knochel, A. Haarich, P. Luger, T. Koritansky, and J. Buschmann, *Pure Appl. Chem.*, **1993**, *65*, 503.
47. R. D. Hancock and A. E. Martell, *Comments Inorg. Chem.*, **1988**, *6*, 237.
48. D. H. Busch, K. Farmery, V. Goedken, V. Katovic, A. C. Melnyk, C. R. Sperati, and N. Tokel, *Adv. Chem. Ser.*, **1971**, *100*, 52.
49. V. J. Thom, J. C. A. Boeyens, G. J. McDougall, and R. D. Hancock, *J. Am. Chem. Soc.*, **1984**, *106*, 3198.
50. R. D. Hancock and G. J. McDougall, *J. Am. Chem. Soc.*, **1980**, *102*, 6551.
51. V. J. Thom, C. C. Fox, J. C. A. Boeyens, and R. D. Hancock, *J. Am. Chem. Soc.*, **1984**, 106, 5947(
52. A. E. Martell, in *Development of Iron Chelators for Clinical Use*, A. E. Martell, W. F. Andersen, and D. G. Badman, Eds., Elsevier, North Holland, New York, 1981, pp 67-79.
53. B. Metz, D. Moran, and R. Weiss, *Acta Crystallogr., Sect. B*, **1973**, *B29*, 1377.
54. M. J. Bovill, D. J. Chadwick, I. O. Sutherland, and D. Watkins, *J. Chem. Soc., Perkin II*, **1980**, 1529.
55. G. Wipff, P. Weiner, and P. Kollman, *J. Am. Chem. Soc.*, **1982**, *104*, 3249.
56. G. Ranghino, S. Romano, J. M. Lehn, and G. Wipff, *J. Am. Chem. Soc.*, **1985**, *107*, 7873.

57. R. M. Clay, S. Corr., G. Keenan, and W. V. Steele, *J. Am. Chem. Soc.*, **1983**, *105*, 2070.
58. B. L. Haymore, J. D. Lamb, R. M. Izatt, and J. J. Christensen, *Inorg. Chem.*, **1982**, *21*, 1598.
59. V. F. Man., J. D. Lin., K. D. Cock, *J. Am. Chem. Soc.*, **1985**, 107, 4635.
60. I.-H. Chu, H. Zhang, and D. V. Dearden, *J. Am. Chem. Soc.*, **1993**, 115, 5736.
61. D. V. Dearden, H. Zhang, and I.-H. Chu, *Pure Appl. Chem.*, 65, 423 (**1993**,
62. J. S. Brodbelt and C.-C. Liou, *Pure Appl. Chem.*, **1993**, *65*, 409.
63. J. S. Brodbelt, S. Maleknia, C.-C. Liou, and R. Lagow, *J. Am. Chem. Soc.*, **1991**, *113*, 5913.
64. G. B. Reibnegger and B, M, Rode, *Inorg. Chim. Acta*, **1983**, *72*, 47.
65. T. Yamabe, K. Hori, K. Azaki, and K. Fukui, *Tetrahedron*, **1979**, *35*, 1065.
66. A. B. P. Lever, P. Paoletti, and L. Fabbrizzi, *Inorg. Chem.*, **1979**, *18*, 1324.
67. K. Wieghardt, W. Schmidt, W. Herrman, and H. J. Kuppers, *Inorg. Chem.*, **1983**, *22*, 2953.
68. L. Fabbrizzi, *Comments Inorg. Chem.*, **1985**, *4*, 33.
69. N. F. Curtis and D. F. Cook, *J. Chem. Soc., Chem. Commun.*, **1967**, 962.
70. W. Levason and M. D. Spicer, *Coord. Chem. Rev.*, **1987**, *76*, 45.
71. A. G. Lappin and A. McAuley, *Adv. Inorg. Chem.*, **1988**, *32*, 241.
72. D. C. Ohlson and J. Vasilevskis, *Inorg. Chem.*, **1969**, *8*, 1611.
73. D. P. Rillema, J. F. Endicott, and E. Papacostantinou, *Inorg. Chem.*, **1971**, *10*, 1973.
74. E. K. Barefield, F. V. Lovecchio, N. E. Yokel, E. Ochiai, and D. H. Busch, *Inorg. Chem.*, **1972**, *11*, 283.
75. D. G. Pilsbury and D. H. Busch, *J. Am. Chem. Soc.*, **1976**, *98*, 7836.
76. N. Jubran, G. Ginzberg, H. Cohen, and D. Meyerstein, *J. Chem. Soc., Chem. Commun.*, **1982**, 517.
77. N. Jubran, G. Ginzberg, H. Cohen, and D. Meyerstein, *Inorg. Chem.*, **1985**, *24*, 251.
78. M. Paik Suh, H. K. Kim., M. J. Kim, and K. Y. Oh., *Inorg. Chem.*, **1992**, *31*, 3620.
79. D. C. Olsen and J. Vasilevskis, Inorg. Chem., **1973**, 12, 463.
80. L. D. Diaddavio, W. R. Robinson and D. W. Margerum, *Inorg. Chem.*, **1983**, *22*, 1021.
81. J. Vasilevskis and D. C. Olsen, *Inorg. Chem.*, **1971**, *10*, 1228.
82. Y. Hung, L. Y. Martin, S. S. Jackels, A. M. Tait, and D. H. Busch, *J. Am. Chem. Soc.*, **1977**, *99*, 402.
83. M. O. Kostner and A. L. Allred, *J. Am. Chem. Soc.*, **1972**, *94*, 7198.
84. E. K. Barefield and M. T. Mocella, *Inorg. Chem.*, **1973**, *12*, 2289.
85. R. L. Deming, A. L. Allred, A. E. Dahl, A. W. Merlinger, and M. O. Kostner, *J. Am. Chem. Soc.*, **1976**, *98*, 4132.
86. T. W. Hambley, *J. Chem. Soc., Dalton Trans.*, **1986**, 565.
87. R. D. Hancock and B. S. Nakani, *S. Afr. J. Chem.*, **1983**, *36*, 117.
88. N. Herron and P. Moore, *Inorg. Chim. Acta*, **1979**, *36*, 89.
89. R. D. Shannon, *Acta Crystallogr., Sect. A*, **1976**, *A32*, 751.
90. H. L. Friedman and C. V. Krishnan, in *Water, a Comprehensive Treatise*, Vol. 3, F. Franks Ed., Plenum, New York, 1973, p 55.
91. L. Pauling, *The Nature of the Chemical Bond*, 3rd Edition, Cornell University Press, Ithaca, New York, 1960, p. 93.
92. V. J. Thom, M. S. Shaikjee, and R. D. Hancock, *Inorg. Chem.*, **1986**, 25, 2992.
93. T. Chantson and R. D. Hancock, *Inorg. Chim. Acta*, in the press.
94. V. J. Thom, G. D. Hosken, and R. D. Hancock, *Inorg. Chem.*, **1985**, 24, 3378.
95. P. A. Tasker and L. Sklar, *Cryst. Mol. Struct.*, **1975**, 5, 329.
96. B. Bosnich, C. K. Poon, and M. L. Tobe, *Inorg. Chem.*, 4, 1102 (1965)
97. N. W. Alcock, E. H. Curzon, N. Herron, and P. Moore, *J. Chem. Soc., Dalton Trans.*, **1989**, 1987.
98. R. G. Swisher, G. A. Brown, R. C. Smiercak, and E. L. Blinn, *Inorg. Chem.*, **1981**, 20, 3947.
99. N. F. Curtis in Coordination Chemistry of Macrocyclic Compounds, G. A. Melson, Ed., Plenum Press, New York, 1979.
100. R. Luckay, J. Reibenspies, and R. D. Hancock, *J. Chem. Soc., Dalton Trans*, in the press.
101. T. Ito, M. Kato, and H. Ito, *Bull. Chem. Soc. Japan*, **1984**, 57, 2641.
102. N. W. Alcock, E. H. Curzon, and P. Moore, *J. Chem. Soc., Dalton Trans.*, **1984**, 2813.
103. P. Seiler, M. Dobler, and J. D. Dunitz, *Acta Crystallogr., Sect. B*, **1974**, B30, 2744.
104. R. D. Hancock, in *Perspectives in Coordination Chemistry*, A. P. Williams, C. Floriani, and A. E. Merbach, Eds., Verlag Helv. Chim. Acta, Basel, and VCH Publishers, Weinheim, 1992, p 129.
105. R. D. Hancock, *Pure Appl. Chem.*, **1986**, *58*, 1445.
106. R. D. Hancock, *J. Incl. Phen., Mol. Recogn. Chem.*, **1994**, *17*, 63.

107. M. Dobler, and R. P. Phizackerley, *Acta Crystallogr., Sect. B*, **1974**, *B30*, 2748.

108. M. Dobler, J. D. Dunitz, and P. Seiler, *Acta Crystallogr., Sect. B*, **1974**, *B30*, 2744.

109. N. W. Alcock, M. Ravindran, and G. R. Willey, *J. Chem. Soc., Chem. Commun.*, **1989**, 1063.

110. M. A. Busch and M. R. Truter, *J. Chem. Soc., Perkin Trans*, **1972**, 2, 345.

111. P. Chaudhuri and K. Wieghardt, *Prog. Inorg. Chem.*, **1986**, *35*, 329.

112. R. Bhula, P. Osvath, and D. C. Weatherburn, *Coord. Chem. Rev.*, **1988**, *91*, 89.

113. R. D. Hancock, S. M. Dobson, and J. C. A. Boeyens, *Inorg. Chim. Acta*, **1987**, *133*, 221.

114. K. Wieghardt, M. Kleine-Boymann, B. Nuber, J. Weiss, L. Zsolnai, and G. Huttner, *Inorg. Chem.*, **1985**, *25*, 1647.

115. P. M. Schaber, J. C. Fettinger, M. R. Churchill, D. Nalewajek, and K. Fries, *Inorg. Chem.*, **1988**, *27*, 1641.

116. E. Kimura, T. Shiota, T. Koike, M. Shiro, and M. Kodama, *J. Am. Chem. Soc.*, **1990**, *112*, 5805.

117. G. Bombieri, E. Forsellini, A. del Pra, C. J. Cookesy, M. Humanes, and M. L. Tobe, *Inorg. Chim. Acta*, **1982**, *61*, 43.

118. P. J. Sadler, *Adv. Inorg. Chem.*, **1991**, *36*, 1.

119. S. Jurisson, D. Berning, W. Jia, and D. Ma, *Chem. Rev.*, **1988**, *6*, 237.

120. O. A. Gansow, *Nucl. Med. Biol.*, **1991**, *18*, 369.

121. R. B. Lauffer, *Chem. Rev.*, **1987**, *87*, 901.

122. R. A. Bulman, *Struct. Bonding*, **1987**, *67, 91.*

123. T. Kaden, *Top. Curr. Chem.*, **1984**, *121*, 157.

124. G. W. Gokel and J. E. Trafton, in *Cation Binding by Lariat Ethers*, Y. Inoue and G. Gokel Eds., Marcel Dekker, New York, 1990, p 253.

125. E. T. Clarke and A. E. Martell, *Inorg. Chim. Acta.*, **1991**, 181, 273.

126. M. J. van der Merwe, J. C. A. Boeyens, and R. D. Hancock, *Inorg. Chem.*, **1985**, *24*, 1208.

127. A. Bevilaqua, R. I. Gelb, W. B. Hebard, and L. Zompa, *Inorg. Chem.*, **1987**, *26*, 2699.

128. H. Hama, and S. Takamoto, *Nippon Kagaku Kaishi*, **1975**, 7, 1182.

129. E. T. Clarke and A. E. Martell, *Inorg. Chim. Acta*, **1991**, 186, 103.

130. T. Ya Medved, M. I. Kabachnik, F. I. Belskii, S. A. Pisareva, *Izv. Akad. Nauk. SSSR, Ser. Khim.*, **1988**, 2103.

131. B. A. Sayer, J. P. Michael, and R. D. Hancock, *Inorg. Chim. Acta*, **1983**, *77*, L63.

132. R. Ma, M. J. Welch, J. Reibenspies, and A. E. Martell, *Inorg. Chim. Acta*, **1995**, *236*, 75.

133. A. E. Martell, R. J. Motekaitis, and M. J. Welch, *J. Chem. Soc., Chem. Commun.*, **1990**, 1748.

134. R. Kataky, K. E. Matthes, P. E. Nicholson, D. Parker, and H. J. Buschmann, *J. Chem. Soc., Perkin Trans.* **1990**, 2, 1425.

135. S. Cortes, E. Brucher, C. F. G. C. Geraldes, A. D. Sherry, *Inorg. Chem.*, **1990**, *29*, 5.

136. K. Wieghardt, U. Bosseck, P. Chaudhuri, W. Herrman, B. C. Menke, and J. Weiss, *Inorg. Chem.*, **1982**, *21*, 4308.

137. A. E. Martell, *Mater. Chem. Phys.*, **1993**, *35*, 273.

138. a) H. Stetter and W. Frank, *Angew. Chem. Int. Ed. Engl.*, **1976**, *15*, 686. b) H. Stetter, W. Frank, and R. Mertens, *Tetrahedron*, **1981**, *37*, 767.

139. R. Delgado and J. J. R. Frausto da Silva, *Talanta*, **1982**, *29*, 815.

140. E. T. Clarke and A. E. Martell, *Inorg. Chim. Acta*, **1991**, *190*, 37.

141. R. D. Hancock, M. S. Shaikjee, S. M. Dobson, and J. C. A. Boeyens, *Inorg. Chim. Acta*, **1988**, *154*, 229.

142. C. F. G. Geraldes, A. D. Sherry, and W. P. Cacheris, *Inorg. Chem.*, **1989**, *28*, 3336.

143. R. Delgado, L. C. Siegfreid, and T. A. Kaden, *Helv. Chim. Acta*, **1990**, 73, 140.

144. L. Carlton, R. D. Hancock, H. Maumela, and K. P. Wainwright, *J. Chem. Soc., Chem. Commun.*, **1994**,

145. T. C. Chantson and R. D. Hancock, to be published.

146. S. Kulstad and S. Marmsten, *J. Inorg. Nucl. Chem.*, **1981**, *43*, 1299.

147. V. J. Gatto and G. W. Gokel, *J. Am. Chem. Soc.*, **1984**, *106*, 8240.

148. P. Corbaux, B. Spiess, F. Arnaud, and M. J. Schwing, *Polyhedron*, **1985**, *4*, 1471.

149. R. D. Hancock, R. Bhavan, P. W. Wade, J. C. A. Boeyens and S. M. Dobson, *Inorg. Chem.*, **1989**, *28*, 187.

150. M. Tazaki, K. Nita, M. Takagi, and K. Ueno, *Chem. Lett. (Japan)*, **1982**, *571.*

151. K. V. Damu, R. D. Hancock, P. W. Wade, J. C. A. Boeyens, D. G. Billing and S. M. Dobson, *J. Chem. Soc., Dalton Trans.*, **1991**, 293.

152. K. V. Damu, M. S. Shaikjee, J. P. Michael, A. S. Howard, and R. D. Hancock, *Inorg. Chem.*, **1986**, *25*, 3879.

153. G. L. Rothermel, L. Miao, A. L. Hill, and S. C. Jackels, *Inorg. Chem.*, **1992**, *31*, 4854.

154. P. B. Smith, J. L. Dye, J. Cheney and J. M. Lehn, *J. Am. Chem. Soc.*, **1981**, *103*, 6044.

155. J. M. Lehn and J. P. Sauvage, *J. Amer. Chem. Soc.*, **1975**, *97*, 6700.

156. R. Geue, S. H. Jacobsen, and R. Pizer, *J. Am. Chem. Soc.*, **1986**, *108*, 1150.

157. A. M. Sargeson, *Pure Appl. Chem.*, **1986**, *58*, 1511.

158. A. Bencini, A. Bianchi, S. Chimichi, M. Ciampolini, P. Dapporto, M. Micheloni, N. Nardi, P. Paoli, and B. Vantancoli, *Inorg. Chem.*, **1991**, *30*, 3687.

159. G. W. Gokel, D. M. Dishong, and C. J. Diamond, *J. Chem. Soc., Chem. Commun.*, **1980**, 1053.

160. R. A. Bartsch, *Pure Appl. Chem.*, **1993**, *65*, 399.

161. W. R. Scheidt and Y. J. Lee, *Struct. Bonding*, **1987**, *64*, 1.

162. H. R. Jiminez, M. Julve, and J. Faus, *J. Chem. Soc., Dalton Trans.*, **1991**, 1945.

163. O. Q. Munro, J. C. Bradley, R. D. Hancock, H. M. Marques, F. Marsicano, and P. W. Wade, *J. Am. Chem. Soc.*, **1992**, *114*, 7230.

164. T. W. Bell, P. J. Cragg, M. G. B. Drew, A. Firestone, A. D.-I. Kwok, R. T. Ludwig, A. T. Papoulis, *Pure Appl. Chem.*, **1993**, *65*, 361.

165. J. L. Sessler and A. K. Burrell, *Top. Curr. Chem.*, **1991**, *161*, 177.

166. K. P. Wainwright and A. Ramasubbu, *J. Chem. Soc., Chem. Commun.*, **1982**, 277.

167. R. D. Hancock, S. M. Dobson, A. Evers, P. W. Wade, M. P. Ngwenya, J. C. A. Boeyens, and K. P. Wainwright, *J. Am. Chem. Soc.*, **1988**, *110*, 2788.

168. R. D. Hancock, G. Pattrick, and G. D. Hosken, *Pure Appl. Chem.*, **1993**, *65*, 941.

169. a) D. E. Fenton, *Pure & Appl. Chem.*, **1989**, *61*, 903. b) D. E. Fenton, *Adv. Inorg. Bioinorg. Mech.*, **1984**, *2*, 187.

170. N. Kitajima, K. Fujisawa, C. Fujimoto, Y. Moro-oka, S. Hashimoto, T. Kitagawa, K. Toriumi, K. Tatsumi, and A. Nakamara, *J. Am. Chem. Soc.*, **1992**, *114*, 9232.

171. K. D. Karlin, and Y. Gultneh, *J. Chem. Ed.*, **1985**, *62*, 983.

172. J. Kim and D. C. Rees, *Science*, **1992**, *257*, 1677.

173. V. McKee, *Adv. Inorg. Chem.*, **1993**, *40*, 323.

174. A. Bencini, A. Bianchi, P. Paoletti, P. Paoli, *Pure & Appl. Chem.*, **1993**, *65*, 381.

175. A. Evers, R. D. Hancock, and I. Murase, *Inorg. Chem.*, **1986**, *25*, 2160.

176. M. G. B. Drew, P. C. Yates, J. Trocha-Grimshaw, A. Lavery, K. D. McKillop, S. M. Nelson, and J. Nelson, *J. Chem. Soc., Dalton Trans.*, **1988**, 347.

177. R. J. Motekaitis and A. E. Martell, *J. Am. Chem. Soc.*, **1988**, *110*, 7715.

178. R. D. Hancock and A. E. Martell, *Adv. Inorg. Chem.*, in the press.

179. a) A. E. Martell, *Adv. Supramolec. Chem.*, **1990**, *1*, 145. b) A. E. Martell, R. J. Motekaitis, D. Chen, and I. Murase, *Pure & Appl. Chem.*, **1993**, *65*, 959. c) A. E. Martell, in *Crown Compounds, Towards Future Applications*, S. R. Cooper Ed., VCH Publishers, Deerfield Beach, Florida, 1992, pp 99-134.

180. R. J. Motekaitis, A. E. Martell, I. Murase, J. M. Lehn, and M. W. Hosseini, *Inorg. Chem.*, **1988**, *27*, 3630.

181. R. J. Motekaitis, A. E. Martell, P. Rudolf, and A. Clearfield, *Inorg. Chem.*, **1988**, *28*, 3630.

182. P. E. Jurek, A. E. Martell, R. J. Motekaitis, and R. D. Hancock, *Inorg. Chem.*, **1995**, *34*, 1823.

183. Q. Lu, R. J. Motekaitis, J. J. Reibenspies and A. E. Martell, *Inorg. Chem.*, **1995**, *34*, 4958.

CHAPTER 5

MEDICAL APPLICATIONS OF METAL COMPLEXES

It is the purpose of this chapter to use the principles that have been described in previous chapters on metal complexes, chelates, and macrocyclic compounds to develop an understanding of the use of metal complexes or ligands for medical purposes. Previous chapters have emphasized the factors involved in designing metal complexes with maximum stability, and with a large degree of selectivity. Both of these factors are important in biological systems because of the fact that all metal ions are subject to interaction with the natural ligands that are present in body fluids for the purpose of storage, transport and the regulation of the activity of natural metal ions that are needed for various metabolic purposes. Therefore, in order for a ligand to be effective in biological systems, or a metal complex to retain its integrity in competition with natural carriers, the thermodynamic stability of the complex in question must be maximized. When a ligand is designed to remove a certain metal ion because it has reached toxic levels, it must be as selective as possible for that metal ion so as not to disturb the metals that are naturally present. However, high thermodynamic stability and selectivity are not the sole requirements of the ligand or of the metal complex being considered, because there are many other factors in biological systems which must be taken into consideration. Such factors include the method of administration (oral, sub-cutaneous, intravenous injection, etc.), bioavailability, membrane permeability, toxicity, and rapid elimination of the ligand and its metal chelate without spreading the undesired metal to other organs throughout the body. It will be the philosophy of this chapter that the thermodynamic stability of a metal chelate should be maximized, and the principles described in this book will be used for this purpose. While ligand design to achieve these purposes seems fairly straightforward, the other factors of importance in biological systems such as toxicity, bioavailability and membrane permeability are not as easily predicted. Therefore, many complexes that meet the stability requirements cannot be used in biological systems because they fail to meet the other biological requirements described above.

A few examples of metal complexes will serve to illustrate this point. Macrocycles such as cyclam which form very stable copper complexes (with stability constants of the order of ~10^{28}) were found to be considerably less effective in enhancing urinary excretion of copper than were the polyamines (with copper stability constants of ~10^{20}-10^{24}). This is possibly due to differences in bioavailability of the two types of complexing ligands, as well as the fact that the *rate* at which Cu(II) cyclam complexes form is very slow. Another example is the catechol group containing ligands such as the synthetic siderophores synthesized by Raymond and coworkers,[1] which cannot be used for the removal of iron in cases of iron overload because of toxicity, while the microbial iron binding agent Desferal with a somewhat lower affinity for Fe(III) is more effective and is currently employed for that purpose.[2,3] Another example is the ligand N,N',N''-tris-(3-hydroxy-6-methyl-2-pyridylmethyl)-1,4,7-triazacyclononane, which has a higher binding constant for Fe(III) than any other known ligand,[4,5] but which preliminary tests indicate may have toxicity problems in removal of iron from the body. The applications of metal chelates and chelating agents in biological systems for medical purposes have been covered in a number of reviews.[6-11] The material covered in this chapter will deal with the application of the principles of ligand design and selectivity described previously, to the development of new metal complexes and complexing agents for biological systems. In addition to the principles of ligand design the requirements of biological systems must also be met, and the more successful of these metal complexes and complexing agents will be described.

One should point out that when designing ligands, formation of the complex virtually always is accompanied by displacement of protons. Thus, the fact that a ligand may have a very high formation constant can be misleading if the pK_a values of the ligand are very high:

$$ML^{m+} + nH^+ \rightleftharpoons M^{m+} + LH_n^{n+}$$

Much more indicative of how a ligand will perform[12,13] at biological pH is the pM value, which takes into account the protonation constants of the ligand, as well as the dilution effects due to the fact that ligands and metals in biomedical applications are present at low concentrations. The pM value is defined as the negative log of the concentration of the free metal ion, -log [M]. It is seen from the above equilibrium that as the hydrogen ion concentration increases, competition with the metal ion for the ligand also increases, the metal chelate becomes more dissociated and the metal ion concentration also increases (the pM decreases). Therefore the more basic is the ligand (the higher are its pK's) the more the hydrogen ion competes with the metal ion for the ligand. The quantitative treatment of the equilibria involving ligand, hydrogen ion, metal ion, and other solution species, are described in detail in the next chapter.

When the ligand is in excess of the metal complex formed, the system may be considered a metal ion buffer, consisting of metal complex, complexing agent and a low concentration of free metal ion.

$$K = [ML]/[M][L]$$

$$1/[M] = K([L]/[ML])$$

In the presence of equal concentrations of metal complex and excess ligand,

$$[ML] = [L]$$

$$\text{Thus pM} = -\log [M] = \log K$$

and the metal ion concentration is maintained relatively constant by the buffer system consisting of the metal ion chelate and excess chelating agent.

In physiological systems the pH is frequently taken as pH 7.4 and pM is calculated at that pH. If one looks ahead to the data in Table 5.2, some pM values for Fe(III) together with the appropriate log K_1 values are listed. It is seen that although DFO has a much lower log K_1 than many of the other ligands in Table 5.2, the pM value of DFO is very adequate because of the lower total of the protonation constants for DFO.

5.1. Applications of Ligand Design Principles

In Figure 5.1 are shown some metal ions of biomedical interest,[8,14-16] and their applications, indicating the metal ions that should primarily be considered in discussing ligand design. The properties of metal ions that are useful in ligand design, as outlined in Chapters 1 to 4, are:
1) HSAB character of the metal ion[17-19]
2) size of the metal ion[20]
3) coordination number and geometry of the metal ion
4) formation constant of metal ion for particular types of ligand which are good indicators[19] of the general affinity of the metal ion for ligands with that donor atom type, e.g. OH-, NH_3, $HOCH_2CH_2S^-$,

In addition, the architecture of the ligand must be considered as it relates to such factors as chelate ring size, denticity, and the presence of sterically bulky groups on the ligand.[19]

5.1.1. Hard metal ions of biomedical interest (Li+, Rb+, Sr²⁺, Y³⁺, Sm³⁺, Gd³⁺, Dy³⁺, Ho³⁺, Yb³⁺, Fe³⁺, Pu⁴⁺, Al³⁺, Ga³⁺)

In relation to point number 1, if the metal ion is hard, then hard oxygen donor ligands are to be preferred. If the metal ion is hard, and strongly acidic (e.g., Ga^{3+}, Fe^{3+}, Pu^{4+}, Al^{3+}), then very basic negative oxygen donors such as phenol, catechol, hydroxamate, hydroxypyridinone, or methylphosphonate are indicated. From section 2.5.2 it is clear that selectivity for acidic metal ions over less acidic metal ions is promoted by strongly basic negatively charged oxygen donors. It is hard to draw a line between acidic and less acidic, but very acidic metal ions are probably best regarded as those with log $K_1(OH^-)$ greater than 9.0. If the hard acidic metal ion is very small (Al^{3+}, Ga^{3+}, ionic radius[20] less than 0.6 Å) then one or two six membered chelate rings will promote selectivity over larger metal ions. In addition, for very small metal ions, C-methyl groups (Section 3.8) tend to promote selectivity over larger metal ions. Pu^{4+} is too large to benefit from six-membered chelate rings. Fe^{3+} is small enough when six coordinate to benefit from six membered chelate rings, but when seven coordinate (e.g. as in $[Fe(EDTA)H_2O]^{3+}$) is too large.

If the metal ion is hard and of very low acidity (log $K_1(OH^-) < 2.0$), and also large (Na+, K+, Rb+, Cs+, Ca²⁺, Sr²⁺, Ba²⁺), then neutral oxygen donors are beneficial. Again, one somewhat arbitrarily defines a large metal ion as one having an ionic radius[20] greater than 1.0 Å. The neutral oxygen donors should be part of five membered chelate rings to maximize stability, and selectivity against small metal ions (Section 4.2). For these low acidity metal ions strongly basic negative oxygen donors will greatly reduce selectivity against more acidic metal ions, so that where negative oxygens must be used, e.g. to produce overall neutrality of the complex, or increase complex stability, low basicity groups such as carboxylates should be used. For small hard metal ions of low acidity (e.g. Li⁺) the considerations are the same as for larger ions of low acidity, except that six membered chelate rings will promote selectivity.

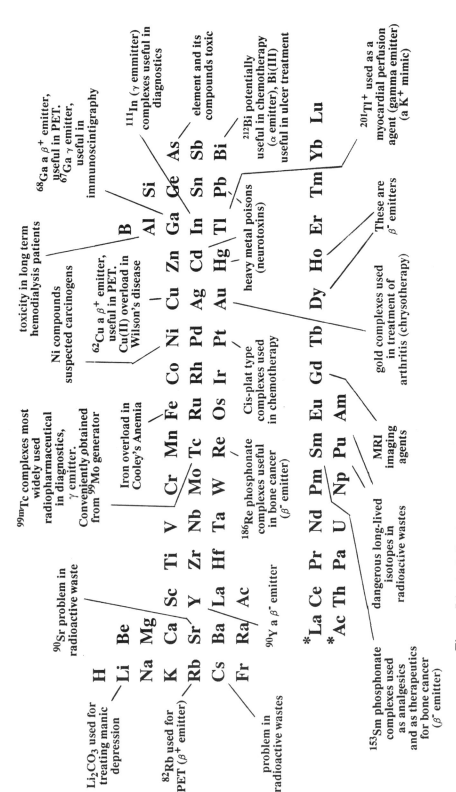

Figure 5.1. An indication of biomedical applications[8,14-16] of some metal ions in the Periodic Table.

For hard metal ions (Y^{3+}, La^{3+} - Lu^{3+}) of intermediate acidity (log $K_1(OH^-)$ between 2.0 and 9.0), an intermediate type of strategy which depends on the metal ions to be selected against is followed. It happens that all the metal ions that fall into this category here are large, or nearly so, and of high coordination number around eight or nine. One or two neutral oxygen donors can serve to produce selectivity against smaller metal ions, and in this situation ligands of high denticity are indicated. These ions are generally better Lewis acids than the hard large metal ions of low acidity (e.g. K^+, Sr^{2+}), so that virtually any ligand that does not consist mainly of neutral oxygen donors will produce adequate selectivity.

The next hardest donor atom is the nitrogen donor. The presence of the nitrogen donor will often decrease selectivity for hard metal ions relative to softer metal ions. Table 2.5 should be consulted for the affinities of metal ions for ammonia as an indicator of their affinity for nitrogen donors in general. The saturated nitrogen donor is important in ligand design as it can serve to connect up to three bridges to other donor groups together. The negative oxygen donor cannot serve to connect any bridges together. This is a weakness of ligands such as MECAM, **1**, where long chains are required to connect all of the negative oxygens of the catecholate groups to a central structural unit, and still allow for correct orientation so as to coordinate to Fe(III). The long chains of MECAM presumably lead to unfavorable entropic contributions to complex formation, and possibly also lead to steric problems. In contrast, the nitrogens of N,N'-bis(*o*-hydroxybenzyl)ethylenediamine-N,N'-diacetic acid, HBED, **2,** allow for a very compact ligand with the two nitrogens connecting together four negative oxygen donors. Metal ions such

1 MECAM   2 HBED

as Fe^{3+}, Ga^{3+}, and Pu^{4+} have (Table 2.5) high affinity for nitrogen, so that nitrogens are indicated. Metal ions such as the lanthanides have low affinity for nitrogens, so that nitrogens should be used sparingly if selectivity against softer metal ions is to be achieved, while the alkali and alkaline earth metal ions have very low affinity for nitrogens. Nitrogen donors can be used very profitably in conjunction with low basicity carboxylates to achieve compact ligands of high denticity such as diethylenetriaminepentaacetic acid, DTPA, **3**, and 1,4,7,10-tetraazacyclododecane-N,N',N'',N'''-tetraacetic acid, DOTA, **4**, which will complex these large weak Lewis acids very well.

The negative sulfur donor is not soft in HSAB as classified originally,[18] but is best regarded as intermediate in character, and forms stable complexes with In^{3+}, which is also intermediate in character. Negative sulfur donors also form stable complexes with Ga^{3+}, which is close to being intermediate in character (Table 2.3). For ease of reference, some log K_1 values for $HOCH_2CH_2S^-$ are shown in Table 5.1. It is seen here that the measured value of log K_1 for In(III) is 9.1 and the estimated value for Ga(III) is 8.7.

3 DTPA

4 DOTA

Table 5.1. Formation constants of mercaptoethanol with metal ions of biological and biomedical interest.

Metal Ion	log K_1 $HOCH_2CH_2S^-$	Metal Ion	log K_1 $HOCH_2CH_2S^-$
Ca(II)	-0.55[b]	Mg(II)	-1.42[b]
Zn(II)	5.7[b]	Cu(II)	8.1[b]
Cu(I)	16.7[b]	Ni(II)	3.9[b]
Fe(III)	8.6[b]	Gd(III)	0.1[b]
H^+	9.72[a]	Ga(III)	8.7[b]
In(III)	9.1[a]	Bi(III)	13.8[a]
Cd(II)	6.1[a]	Pb(II)	6.6[a]
Hg(II)	25.0[b]	CH_3Hg^+	15.9[a]

[a] From Martell and Smith, Ref. 21. [b] Estimated from equation 2.8 and parameters in Tables 2.3 and 2.4.

5.1.2. Metal ions of biomedical interest, that are intermediate in HSAB (Ni^{2+}, Cu^{2+}, In^{3+}, Pb^{2+}, Bi^{2+})

Here neutral nitrogen and negative sulfur donor atoms are indicated. Again size is an important factor. The Bi^{3+} ion, and to a lesser extent,$^{3+}$, have a high affinity for RS^- (Table 5.1) donor ligands. Smaller metal ions such as Ni^{2+} and Cu^{2+} will be favored by possibly one six membered chelate ring. The medium size In^{3+} and large Pb^{2+} and Bi^{3+} complexes are greatly destabilized by the presence of any six membered chelate rings. In^{3+} and, particularly, Bi^{3+} are of high acidity, so that very basic negative oxygen donors are suggested. Of these ions that are intermediate in HSAB (ionic radius[20] 1.19 Å) only Pb^{2+} is large enough to benefit from neutral oxygen donors, but for this ion this is a very viable ligand design strategy. Cu^{2+} is easily reduced to Cu^+ by RS^- groups, which are to be avoided, and nitrogen donor ligands are recommended.

5.1.3. Soft metal ions of biomedical interest (Cd(II), Hg(II), Tl(I), Pt(II), Au(I), Au(III), and possibly Tc in its +1 to +7 oxidation states)

For all these ions N, S, and soft donors such as P, As or C are useful. Complex formation of Pt(II), Au(III), and Tc in its different complexes is kinetically controlled, so that considerations of thermodynamic selectivity are probably not important for these ions. Au(I) is the softest of metal ions (Table 2.3), and so only the softest donor atoms are suitable here - nitrogen is probably too hard. For Cd(II) and Hg(II) which are labile, design considerations are important. Hg(II) is very soft, but with a tendency to form linear complexes that confuse ligand design considerations. Thus, although it is a large metal ion, it is reluctant to form complexes of high coordination number, and many of the design strategies that might be expected to apply to large metal ions, such as addition of neutral oxygen donors, do not work well with Hg(II). Hg(II) has a very large affinity for RS⁻ ligands (Table 5.1), and these are best suited for complexation of Hg(II). Cd(II) is just below what might be considered large (ionic radius[20] 0.95 Å) but does respond well to neutral oxygen donors.

5.2. Iron Overload

Iron is an essential element which is used by the body for the production of hemoglobin, which absorbs oxygen and distributes it to the tissues. Non-heme iron is stored in the body by the protein ferritin and transported by the protein transferrin both of which are not saturated with Fe(III). In cases of iron overload the natural storage and transport proteins are overwhelmed and the iron spills over into other tissues and organs such as the muscle, spleen and the liver.[22] As it accumulates, difficulties arise because of the toxicity of the metal ion. These difficulties can be relieved by the administration of an appropriate chelating agent which can combine with the iron and increase its rate of excretion, generally by the formation of water soluble complexes which are excreted in the urine, or other complexes which are excreted in the feces. The most common form of iron overload disease is β-thalassemia, otherwise known as Cooley's anemia, and hundreds of millions of people are affected by this disease to a greater or lesser degree.[23] Cooley's anemia is a genetic disorder in which a defect in the hemoglobin β-chain prevents it from being effective in the binding and transport of oxygen. Its treatment has generally been blood transfusions but the lifetime of a red blood cell is ninety to one hundred days. The transfusions therefore must be given on a regular basis to replace red blood cells that are destroyed. The general effect is the release of the iron when the life of the red blood cell is over and its elimination becomes necessary to maintain the health of the individual.

Iron(III) is a hard very acidic metal ion, and ligand design for its selective complexation involves use of negatively charged oxygen donors of high basicity (catechol, hydroxamate, hydroxypyridinone, methylphosphonate, 2-hydroxybenzylate, etc.). Fe^{3+} has a reasonable affinity for saturated nitrogen donors (Table 2.5), so that saturated nitrogens can be used where necessary. As discussed above, saturated nitrogens allow three organic bridges to be attached to them, and so lead to much more compact, and therefore entropically more favorable, ligands. Thus, HBED, **2**, does not suffer from the same entropic and steric drawbacks as do the long organic chains necessary to hold the catechol groups of enterobactin, **5**, together. Fe^{3+} is too small to coordinate well with neutral oxygen donors, which should be avoided. The coordination number of Fe(III) is commonly six, or occasionally seven.

The main compound used clinically for the purpose of removing iron from the body has been desferriferrioxamine-B, (DFO), **6**,[2,3] a trishydroxamate synthesized by a microorganism. While its principal use is to remove iron in cases of Cooley's anemia, it can also be used for the effective control of other diseases resulting in iron overload. While the clinical use of desferriferrioxamine-B has made possible the control of iron

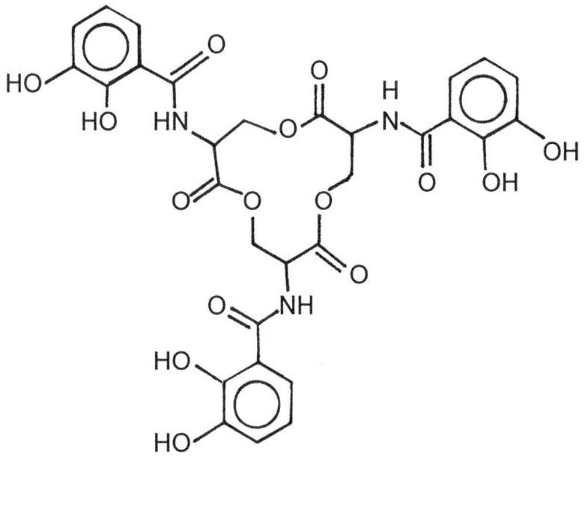

5 Enterobactin

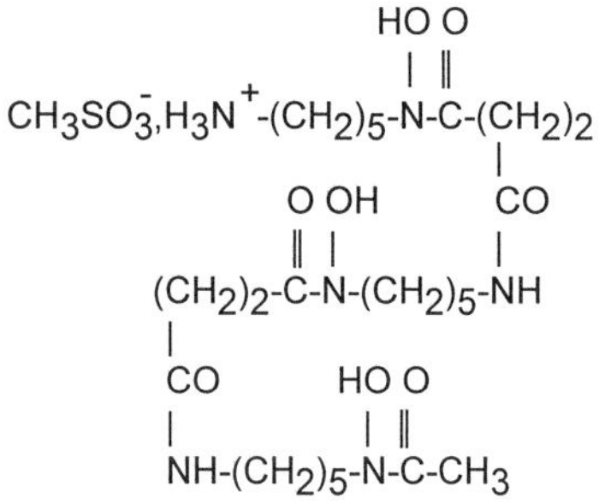

6 DFO

levels, it involves many disadvantages. First it must be administered parenterally since it is not absorbed through the gut. It is rapidly metabolized and only a small fraction of the dose is effective in the removal of iron. It is also very expensive since it is a microbial chelating agent and must be synthesized biologically. Its use can be accompanied by side-effects such as changes in renal function.[24] These problems have led to considerable research for the possible replacement of desferriferrioxamine-B for the treatment of iron overload. The types of compounds which have been investigated as potential replacements of DFO include natural siderophores and their synthetic models, hydroxamic acids, phenolic analogs of EDTA, and the 3-hydroxy-4-pyridinones. The main objective of such research has been the development of iron chelating drugs that are effective orally since the method of administration of desferriferrioxamine-B is an important factor. The compounds that were considered effective as substitutes for Desferal (DFO) were reviewed a few years ago.[25] Of these chelating agents the esters of 1,2-ethylenebis(*o*-hydroxyphenylglycine), EHPG, **5**, seemed like good substitutes but further research and testing of these compounds was halted because of their toxicity. Another compound, the methyl ester of HBED, **2**, now seems to offer considerable promise since it can be administered orally and is quite effective in the removal of iron from the body.[26,27] Another chelating agent which is claimed to be very effective on a number of human (volunteer) patients is 1,2-dimethyl-3-hydroxy-4-pyridinone, DMHP or L1, **8**. This has been reported to be effective even though it is a bidentate agent and its 3:1 iron chelates should dissociate considerably at low concentration. The following is a description of the various types of iron chelating agents that have been investigated as possible substitutes for desferriferrioxamine-B.

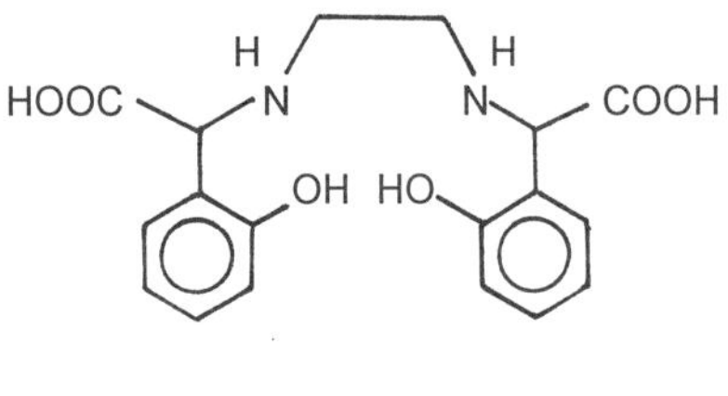

7 EHPG

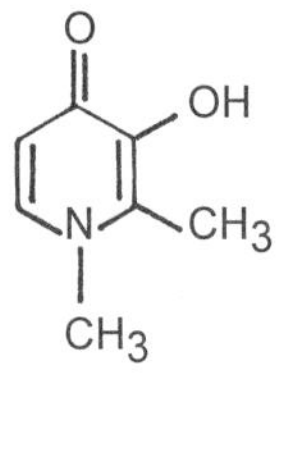

8 DMPH or L1

156

The amino carboxylic acids represented by the first three members of the series nitrilotriacetic acid, NTA, **9**, ethylenediaminetetraacetic acid, EDTA, **10**, and DTPA, **3**, are relatively poor chelating agents for iron(III). In spite of their seemingly high stability constants (with 10^{25} for Fe(III)-EDTA, and 10^{28} for Fe(III)-DTPA) they do not prevent Fe(III) from precipitating as ferric hydroxide from mildly alkaline solutions, indicating relatively weak affinity for the ferric ion. The introduction of the very basic phenolic donor into the EDTA framework, to produce ligands such as EHPG, **7**, and HBED, **2**, has been fairly successful in creating chelating agents which have high Fe(III) affinity and do not allow the precipitation of ferric hydroxide from mildly or even fairly strongly alkaline solutions. The relative advantages and disadvantages of EHPG and HBED are described above. These ligands are poorly absorbed through the gut so that their use as a possible substitute for DFO is based on the methyl, ethyl and other esters which are soluble in vegetable oil and can therefore be administered as solutions in such a medium. The esters are quickly converted to EHPG and HBED by enzymes (and hydrolysis is probably catalyzed by metal ions) in blood serum.

Variations of HBED are N,N'-bis(pyridoxyl)-N,N'-ethylenediaminediacetic acid, PLED, **11**, and N,N'-bis(3-hydroxy-6-methyl-2-pyridylmethyl)ethylenediamine-N,N'-diacetic acid, EDDA-MeHP, **12**. These ligands have somewhat lower Fe(III) ion affinity because of the electron-withdrawing effect of the pyridine nitrogen atoms. However, they have lower proton affinities (lower pK's) of the hydroxyl groups (with protonation occurring first at the pyridine nitrogens rather than at the hydroxyl groups) with the result that at the physiological pH of 7.4 the pM values are only a little less than that of HBED.

9 NTA

10 EDTA

11 PLED

12 EDDA-MeHP

The siderophore enterobactin, **5**, has a very high affinity for Fe(III), and other trivalent metal ions, by virtue of its three catechol donor groups. Many models of enterobactin prepared by Raymond and coworkers[28,29,30] have high affinities for Fe(III) but not as high as enterobactin, as indicated in Table 5.2. Examples are 3,4-LICAM-C, **13**, TRENCAM, **14**, MECAM, **1**, and the cryptand, bicapped TRENCAM, **15**. Most of the synthetic tris-catechols are too toxic to be considered as substitutes for DFO.

The ligands PLED, **11**, and 3,4-LICAM-C, **13**, which tested well in the mouse screen[25] and have suitably low toxicities, have not been studied in detail, but they offer good possibilities as effective chelators for Fe(III) and for the treatment of Cooley's anemia. The ligand EDDA-MeHP, **8**, should be very similar to PLED in its effectiveness is, in fact, a little better than PLED in complexing Fe(III). Further testing of this ligand is awaited with interest.

Table 5.2.   Formation Constants and pM values for Iron(III) chelating agents.

Ligand	log β	Σ pK's	pM[a]
L1[b]	35.88	13.45	19.3
DTPA	28.0	27.72	24.7
Transferrin[c]	20.2, 39.3[d]	-	21.2
Trishydroxamate cryptand	29.12	29.16	12.3
DFO	30.7	26.83	27.0
TRENCAM	43.6	66[e]	27.8
MECAM	46	58.5[e]	19.2
Bicapped TRENCAM	43.1	54.0[e]	30.7
HBED	39.7	36.85	31.0
Enterobactin[f]	~49[e]	~58.5[e]	33.5
TACN-Me2HB[g]	51.3	50.9	33.0
TACN-MeHP	49.98	51.3	40.3

[a] pM = -log [M] at pH 7.4, 10^{-6} M [ML] and 10^{-5} M [L]. [b] 3:1 constant. [c] Conditional constant at pH 7.4. [d] 2M:1L conditional constant. [e] Estimated. [f] Ref. 29. [g] 75% ethanol, 25% water.

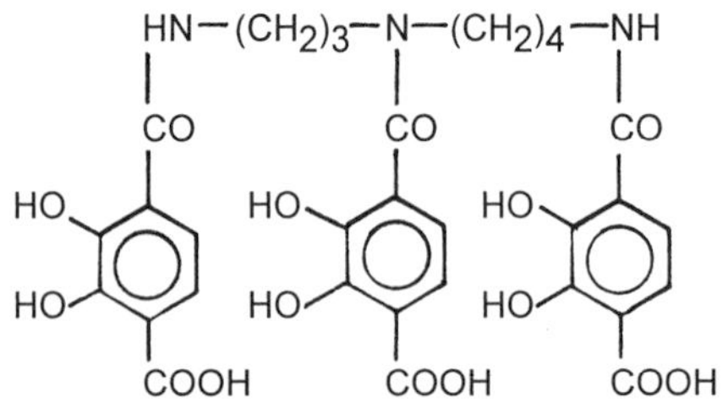

13 3,4-LICAM-C

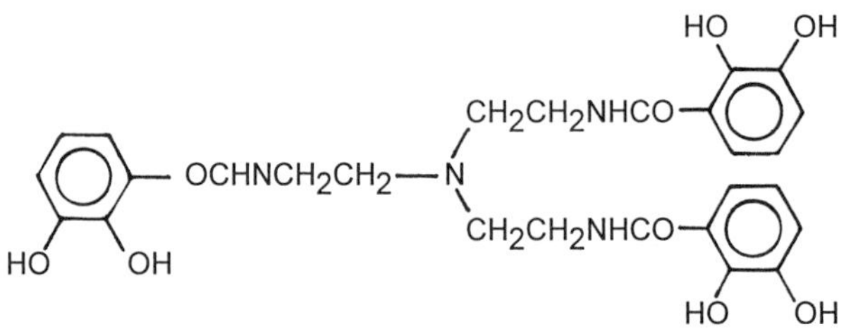

14 TRENCAM

15 Bicapped TRENCAM

5.2.1 The Hydroxypyridinones. The 3-hydroxy-4-pyridinones have recently been developed as oral chelating agents for the treatment of Cooley's anemia. These compounds were first described in a patent by Hider *et al.*[31] Kontoghiorghes and workers have developed 1,2-dimethyl-3-hydroxy-4-pyridinone (DMPA or L1, **8**) and related compounds.[32-36] Most of these studies were concerned with L1 which has been subjected to clinical trials. However, some of these analogs have been found to be somewhat less toxic.[37,38] Although all of the 3-hydroxy-4-pyridinones have high stability constants with Fe(III) (the 3:1 constants are about 10^{36}) their biological activity varies considerably with the substituent placed on the 1 and 2 positions of the ring.[39,40] A large number of 3-hydroxy-4-pyridinones have been studied by Hider and coworkers[40-44] who found that their ability to mobilize iron depends considerably on their membrane permeability, as indicated by their partition coefficients. In a review of drugs used for the treatment of iron overload, Hider and coworkers[39,42] showed that membrane permeability is important in providing access to intracellular iron deposits. The 3-hydroxy-4-pyridinones are suitable in this respect because they have zero charge at physiological pH and their 3:1 complexes involve displacement of three protons by ferric ion, so that the complexes themselves have zero charge.

It has been pointed out[45,46] that the hydroxypyridinones are bidentate and form stable 3:1 iron complexes, and as such they suffer from a strong dilution effect so that their complexes dissociate considerably at low concentration. For that reason it has been one of the objectives of ligand design to incorporate 3-hydroxypyridinones into a single covalent molecule, which will not have the dilution effect. Such ligands have been developed by Streater, *et al.*, N,N,N-tris[2-(3-hydroxy-2-oxo-1,2-dihydropyridin-1-yl)acetamido]ethylamine, TREN-2PO, **16**,[47] and by Sun, *et al.*, 1,1,1-tris[3-(3-hydroxy-2-oxo-1,2-dihydropyridin-1-yl)propoxymethyl]ethane, TRIS-2PO, **17**.[48] These ligands involve the 3-hydroxy-2-pyridinones which have been demonstrated in a qualitative way by Hider and coworkers to have lower stability than the 3-hydroxy-4-pyridinones.[48] The synthesis and testing of these ligands constitute an important development in the possible treatment of iron overload, if the hydroxypyridinone moieties behave in the manner indicated by the separate bidentate ligands.

16   TREN-2PO   **17**   TRIS-2PO

5.3   Aluminum

The Al^{3+} ion is a very small hard metal ion and has fairly high affinity for hard basic donors such as phenolate, hydroxamates and hydroxypyridinones. It is less acidic (log K_1 (OH^-) = 9.1) than Fe(III) (log K_1(OH^-) = 11.8), and also (Table 2.5) has a much lower affinity for nitrogen donors than does Fe(III), so that it will always be difficult to design ligands that will complex Al(III) selectively in the presence of Fe(III). Al(III) is, in general, a weaker Lewis acid than Fe(III). All that one has to work with is the smaller ionic radius[20] of Al(III) (0.54 Å) compared to Fe(III) (0.65 Å), so that six membered chelate rings might be of value.

The Al(III) ion is the third most abundant element in the earth's crust and a normal constituent of the human environment. It is rarely a cause for concern because it is absorbed relatively poorly from the gastrointestinal tract and is normally rapidly excreted in the urine.[49-51] Aluminum overload in the body can be the result of several special conditions including dialysis encephalopathy and other disorders involving the absorption and excretion of aluminum ion. It has been implicated in the senile and pre-senile dementia of Alzheimer's disease, but has not been proved to be the cause of this disease. Exposure to Al(III) may occur through intravenous introduction through medical treatment and dialysis. Also important in the introduction of Al(III) products is food processing, pharmaceuticals, and aluminum-containing antacids.[52,53] Exposure to low levels of aluminum in food and drinking water should also be mentioned. In cases of aluminum overload the use of desferriferrioxamine-B or other trivalent metal ion chelators for the treatment of iron overload have been recommended.[54-56] It may be that *in vivo*, the Fe(III) is so much more strongly bound in transferrin than Al(III) that such ligands can remove the Al(III) without too greatly depleting Fe(III). Various aspects of Al(III) exposure, toxicity, and elimination have been described.[57]

5.4   Nickel

Although nickel poisoning is not very common it is so toxic and carcinogenic[58,59] that its removal from the body should be considered. The Ni^{2+} ion is intermediate in hardness and has a high affinity for nitrogen bases although its affinity for carboxylate donor groups and mercaptide ligands has also been reported and is well known. A comparison of the relative effectiveness of a large number of amines, amino acids and other reagents showed that macrocycles such as cyclam and cyclam derivatives are superior to open chain compounds for the removal of nickel from the body. This type of ligand, as well as triethylenetetraamine, TREN, **18**, were found to be effective in reversing nickel induced alterations in body function.[60,61]

$$H_2NCH_2CH_2—N\overset{\displaystyle CH_2CH_2NH_2}{\underset{\displaystyle CH_2CH_2NH_2}{\big<}}$$

18 TREN

5.5   Copper

Although Wilson described the clinical aspects of Wilson's disease in 1911,[62] it was not until about forty years later that copper overload was implicated. The compound first used to treat the disease was BAL (British Anti–Lewisite), 2,3-dimercapto-1-propanol, **18**,[63,64] which was injected intramuscularly and was painful and disagreeable. This was soon replaced by D-penicillamine, **19**,[65] which could be administered orally. Although D-penicillamine is relatively non-toxic for most people, in some patients an allergy may develop after some time. Another chelating agent to which such individuals may be switched is triethylenetetramine, TRIEN, **20**, which is also effective in enhancing the urinary excretion of copper, but not as effective as D-penicillamine for this purpose. The toxic effects of both drugs have been described in detail.[66] For further information on Wilson's disease, the reader is referred to a general reference on this subject.[67]

$$\overset{\displaystyle SH \quad SH}{\underset{\displaystyle CH_2—CH—CH_2OH}{| \quad \ \ |}}$$

19 BAL

$$\overset{\displaystyle SH \ \ NH_2}{\underset{\displaystyle \underset{\displaystyle CH_3}{\overset{|}{CH_3—C—CH—COOH}}}{| \quad \ |}}$$

20 D-penicillamine

$$H_2N\text{-}(CH_2CH_2NH)_2\text{-}CH_2CH_2NH_2$$

21 TRIEN

5.6   Plutonium and the Actinides

The use of plutonium and other actinides in atomic weapons and in other nuclear applications has led to concern about the removal of plutonium and the actinide elements, when they are accidentally introduced into the body. Plutonium and other transuranium metals, particularly in the +4 state, have been shown to resemble iron in its body chemistry.[68,69] It is transported by transferrin and taken up by ferritin and very readily hydrolyzed to insoluble species which are deposited in various organs and in the skeleton. These metals are very dangerous and their deposits should be removed as soon as possible because they are generally alpha emitters and cause cancer in any location in which they are deposited. Because plutonium is a hard, very acidic, metal ion, in this respect resembling Fe(III), the search for ligands to form complexes generally follows the kind of donor group that has been found to be useful for complexing and removal of the ferric ion. Thus the donor groups considered to be effective are catecholate, hydroxypyridinonate, and hydroxamate. In addition, the aminopolycarboxylates, particularly DTPA, **3**, have been found to be surprisingly effective. The biochemistry of the actinides and the removal of the actinides, including plutonium from the body, have been reviewed by Duffield and Taylor[68] and by Taylor.[69] The use of zinc complexes of DTPA for the removal of actinides from the body has been suggested by Lushbaugh and Washburn.[70] The purpose of using the zinc chelate (also the calcium chelate has been employed) is to counteract the effects of zinc deficiency introduced by the ligand DTPA, which has a high affinity for transition metal ions and zinc ion in addition to the actinide elements.

Removal of plutonium and the actinides from the body has been accomplished by the use of catechol amides of the type indicated for **21** in which there are four catechol groups corresponding to a coordination number of eight, usually assumed for plutonium and other actinides in the +4 state. The various forms of **21** in which the benzene rings are substituted with sulfonate and carboxylate functions have been described by Durban, Raymond and coworkers.[71-73] The same authors have also used hydroxamate and 3-hydroxy-4-pyridinone ligands in addition to the polycatecholates.[72a]

22   The LICAM chelating agents

3,4,3-LICAM, $R_1 = R_2 = R_3 = H$; 3,4,3-LICAM(s), $R_1 = R_3 = H$, $R_2 = SO_3Na$;

3,4,3-LICAM(C) $R_2 = R_3 = H$, $R_1 = COONa$

Plutonium and the actinides are readily deposited in various parts of the body in forms which are very insoluble and very hard to remove. It is therefore necessary to remove these radioactive metals rapidly while they are still in labile form. Treatment with chelating agents is thus effective when administered very quickly but only small fractions of these radioactive metals can be mobilized when they have become firmly deposited in the body. Long treatment has been most effective with DTPA in the form of its calcium or its zinc salt, and DTPA in the form of its esters. Thus Markley[74] found that plutonium could be removed from mice by combined therapy with the calcium chelate and the pentaethyl ester of DTPA. The esters apparently have the advantage of penetrating cell membranes, and thus are more effective than ionic chelating agents in removing intercellular deposits of plutonium, actinides and other metal ions. Bulman and Griffin[75] used another approach to forming ligands with more lipophilic character, by attaching long alkyl chains to the chelating agent. Such approaches however, were associated with side effects which result in increased toxicity.

5.7   Uranium

The most stable form of uranium and the form that probably exists in most of the body fluids is dioxouranium(VI) commonly known as the uranyl ion. This is an ion of intermediate hardness with high affinity for carboxylate functional groups, although it has been reported that substituted catechols are also good ligands. The removal of uranium from the body has been studied by a number of investigators. The use of catechol derivatives as antagonists for uranyl acetate poisoning was reported by Basinger et al.,[76] who concluded that the best reagents under the conditions employed were Tiron (the sodium salt of catechol-3,5-disulfonate) and p-aminosalicylic acid. Examination of a large number of chelating agents for the treatment of uranyl ion intoxication show that the most effective compounds examined were Tiron, gallic acid, and DTPA.[77] A number of chelating agents were also examined by Xie and coworkers for uranium poisoning and removal of other metal ions from the body.[78-80] Examples of the compounds that they studied are catecholamic acid and quinamic acid. These ligands function as antagonists for a considerable number of metal ions. The derivatives of catechol seem to be less toxic than the parent compound.

5.8   Toxic Heavy Metal Ions

Metal ions which have a high degree of covalency in their coordinate bonds, such as Cd^{2+}, Pb^{2+} and Hg^{2+} have a relatively high affinity for soft donors, so that ligands with mercaptide donor groups form relatively stable complexes with these metal ions. Examples are the sulfur containing ligands **18,19,22-25**, which bind metal ions through S^- groups. One should state, here, however, that mercapto groups are not likely to produce selectivity for the heavy metal toxins Cd(II) and Pb(II) over Zn(II). Table 5.1 shows that the affinity of Cd(II), Pb(II), and Zn(II) for mercaptoethanol as a representative RS^- ligand, is rather similar. On the other hand, Table 5.1 shows a remarkably high affinity of Hg(II) for mercaptans, and for the removal of Hg(II) from the body, ligands with mercapto groups are likely to be the best.

SH   SH
| |
CH_2—CH—CH_2SO_3Na

23 DMPS

(sodium 2,3-dimercpatopropane-
1-sulfonate)

SH   SH
| |
HOOC—CH—CH—COOH

24 DMSA

(dimercaptosuccinic acid)

25 DDTC

(sodium diethyldithiocarbamate)

26

(sodium diethyldithiocarbamate derivative)

5.9   Cadmium

Cadmium accumulates in the body and is retained primarily in the liver and kidney in the form of a very stable complex of the metal binding protein metallothionein, which contains sulfhydryl groups. The stable complexes formed accumulate gradually reaching toxic levels,[81] and are not bound or removed by typical chelating agents such as EDTA or DTPA. The removal of cadmium from the body was first accomplished by **18**, BAL, and described in a report by Shaikh and Lucas.[82] Several sulfur containing ligands, in addition to BAL, were later reported to remove cadmium from the body and from metallothionein in vitro.[83-91] In addition to the use of BAL and other bidentate ligands containing two SH groups, such as the esters of DMSA, **24**, there should be mentioned the extensive studies of M. M. Jones and coworkers on diethyldithiocarbamate, **25**, and various derivatives, such as **26**, which proved to be very effective for the removal of cadmium from the body.[88-90,92]

5.10   Lead

It is well known that EDTA, DTPA and other aminopolycarboxylic acids have a high affinity for calcium ion and therefore cannot be used as free ligands in the body, because the depletion of calcium would be rapid and severe. The treatment of acute symptoms of lead poisoning in which part of the lead is in a fairly labile form in the serum has been accomplished by reaction with the calcium chelate of EDTA administered as the disodium calcium salt. The lead chelate of EDTA is much more stable than that of calcium so that the amount of lead complexed releases an equivalent amount of calcium ion but the ligand itself does not remove calcium from the serum. This technique for the treatment of lead poisoning was first described by Rubin et al.[93] Use of the calcium complex of EDTA for the treatment of lead poisoning is now the standard procedure. Both calcium zinc EDTA and zinc EDTA have been shown to be more effective than disodium calcium EDTA in enhancing pulmonary excretion of lead.[94] The replacement of EDTA by the use of ligands which have somewhat superior qualities with respect to the method of administration, high stability of its lead(II) chelates, and low toxicity have not been yet generally been adopted, partly because of the success of EDTA, its low price, and ready availability. In particular both DMSA and DMPS are more effective than EDTA in reducing lead levels in various organs in experimental animals.[95-97] Also clinical studies on orally administered DMPS proved to be effective[97a] but *meso*-2,3-dimercaptosuccinic acid, DMSA, is at least as effective when given orally and is definitely less expensive.[98-100] DMSA has recently been approved by the FDA for use as an orally administered drug for lead intoxication. D-penicillamine, **20**, has also been recommended for removal of lead from the body.

A recent development in the selective complexation of toxic metal ions is the report by Carlton et al.[101] and by Hancock et al.[102] that attaching four neutral acetamide donor groups to cyclen to give an octadentate ligand, 1,4,7,10-tetrakis(carbamoylmethyl)-1,4,7,10-tetraazacyclododecane, DOTAM, 23, greatly increases the affinity of the ligand for large metal ions relative to the parent macrocycle. Other weak neutral donors (such as hydroxyethyl groups) increase stability somewhat less when substituted in cyclen as additional pendent donors. The amide group, however, greatly increases the affinity of the ligand for Pb(II) and Cd(II), while apparently having little effect on small metal ions. Thus the log stability constant of Cu(II) complexes decreases from 23.3 to 16.4 on substitution of four acetamido groups, and the log stability constant of the Zn(II) complex decreases from 16.2 to 10.5. This decrease is probably due to steric effects of the substituents, and the fact that these metal ions have relatively low coordination numbers. The log stability constants of the Pb(II) and Cd(II) complexes, however, increase from 14.3 and 15.9, respectively, to very high values (>19) which are too high to be measured potentiometrically. The fact that the Pb(II) and Cd(II) stability constants are nearly nine orders of magnitude more stable than the Zn(II) complex indicates that pendent neutral amide donor groups may be effective in binding and removing toxic metal ions such as Cd(II) and Pb(II), while not affecting essential metal ions, such as Zn(II).

5.11 Arsenic

Arsenic(III) is a fairly soft metal ion in spite of its charge and combines most readily with soft ligands such as mercaptides. The use of 2,3-dimercaptosuccinic acid for treatment of arsenic poisoning was first investigated by Graziano and coworkers in 1978.[103] Subsequently Aposhian and coworkers investigated a group of vicinal dithiols and compared them with BAL for the treatment of arsenic poisoning.[104-106] These ligands, including *meso*-2,3-dimercaptosuccinic acid (DMSA), sodium-2,3-dimercaptopropane-1-sulfonate (DMPS) and N-2,3-dimercaptophthalamidic acid (DMPA) were shown to be more effective for the treatment of arsenic poisoning than BAL, and the order of effectiveness reported was DMSA > DMPS > DMPA > BAL.[107] The ligand DMSA will probably be shown to be a superior antidote for arsenic poisoning because of its greater availability, low toxicity, and oral administration.

5.12 Mercury

Both Hg(II) and Hg(I) are soft metal ions that are very strong Lewis acids, and are complexed very effectively by mercapto groups (Table 5.1), which are very strong bases. The toxicity of mercury is well known and its ingestion is much less widespread than it would otherwise be because of the reduction of the use of mercury in medicine. However, exposure due to industrial and commercial uses is quite common. The symptoms of mercury poisoning have been described in some detail in recent reviews.[108,109] The vicinal dimercapto ligands which have been found to be effective for the treatment of lead are even more effective for counteracting the effects of mercury. These include N-acetyl-*dl*-penicillamine, DMSA, and DMPS. These compounds have been introduced to replace the dimercapto compound first used for mercury poisoning, BAL.[110-113] Here again DMSA seems to one of the most effective antidotes for mercury poisoning because it can be administered orally, has few side effects, and is readily available.

5.13 General Summary

The above description of the removal of hard and soft metal ions from the body is based to a large extent on the concepts of hard and soft acids and bases, and the principles of coordination chemistry. When those concepts are applied to ligand design the result is the development of ligands that meet only part of the requirements for effective biological activity. The biological system itself, the method of administration and the essential clearance of the compound, imposes many more requirements for which there are only qualitative guidelines. In cases where unexpected toxicities have been encountered, the information available becomes quite empirical and the element of predictability becomes quite weak. Therefore the testing of the drug, first with animal models, and eventually with cautious clinical trials, has been and will continue to be, an important and essential part of the selection process. A few strategies for lowering toxicity have become apparent, such as the presence of several alcoholic oxygen groups on the ligand, which act to lower lipophilicity. Thus Jones has greatly reduced the toxicity of dithio-carbamates such as **25** by addition of sugar moieties to give ligands such as **26**.[88-90,92] Other strategies for getting the ligand to pass through the channels in the cell membrane, to get at metal ions within the cell, is the presence of only a single negative charge, which promotes passage through the porphyrin channels.[88-90,92]

The above discussion is not complete in that several metals were left out because exposure to them, and toxic effects, are less common. For example Be^{2+}, a very hard very small metal ion, is very toxic when it is absorbed (usually accidentally) because it interferes with other metal ions, such as Mg^{2+}, which have a regulatory function or activate certain enzymes.[114] Be^{2+} may also drastically affect membrane action and interfere with DNA transcription.[115] Chromium is also not mentioned since the compounds of Cr^{3+} are too insoluble or inert to generally cause trouble. However it has been considered to have serious toxic effects when it is introduced into cells, causing gene transformations and chromosomal damage. It may pass through cell membranes as chromate (Cr(VI)), using the anionic transport mechanism. Its subsequent reduction to Cr(III) results in Cr^{3+} being trapped in the cell where it reacts with components of the cell nucleus.

For further information on exposure and treatment of beryllium, chromium and other metal toxicities the reader is referred to a recent review paper by McKinney.[116]

5.14   Diagnostic Radiopharmaceuticals Based on Ga(III) and In(III) Complexes

The Ga(III) and In(III) ions are classified[18] as hard metal ions of high acidity, and their chemistry bears some resemblance to that of Fe(III). The Ga(III) ion is most similar to Fe(III), but is a slightly weaker Lewis acid than Fe(III), and is somewhat softer and a little smaller, as indicated by the data in Table 5.3

Table 5.3. A comparison of some properties of Ga(III), In(III), and Fe(III).

	Ga(III)	In(III)	Fe(III)
Ionic radius^a	0.62	0.80	0.65
$logK_1(OH^-)$	11.3	10.0	11.80
$logK_1(NH_3)$	4.1^b	4.0^b	3.8^b
$logK_1(RS^-)$	8.7^b	9.1	$(8.6)^{b,c}$

^{a} Ionic radii (Å) from Ref. 20, formation constants from Refs. 21, 21a. ^{b} Estimated as described in section 2.4 from equation 2.4.3. ^{c} Fe(III) is reduced to Fe(II) by mercapto groups.

In(III) is more different to Fe(III) in its chemistry than is Ga(III), being much larger, more weakly acidic, and significantly softer. The size differences mean that Ga(III) is found in crystal structures to adopt coordination numbers of four to six, but never seven. The larger In(III) frequently adopts coordination numbers of seven, and may be able to achieve coordination numbers of eight in complexes with DTPA, **3**, for example, although this is not known for certain. Some representative complexes of In(III) are seen in Figure 5.2.

Gallium-67 and indium-111 with half-lives of 77.9 hours and 67.9 hours, respectively, are currently utilized for such applications as tumor scanning,[117,118] abscess localization,[119,120] and thrombus detection.[121] Indium-111 is the radionuclide of choice for the labeling of monoclonal antibodies for diagnosis.[122] Positron emitting radionuclides are of particular importance because the biodistribution can be determined *in vivo* by positron emission tomography (PET).[123-125] The application of ^{68}gallium, a positron emitter, would be greatly increased if agents could be developed which would readily penetrate the blood brain barrier (BBB) and quantitate cerebral blood flow (CBF). There is need for development of agents labeled with metal radionuclides which cross the BBB and are trapped in the brain. Agents which have been developed include copper(II) pyruvaldehyde bis(N^4-methylthiosemicarbazone) [CuPTSM],[126,127] as well as a series of

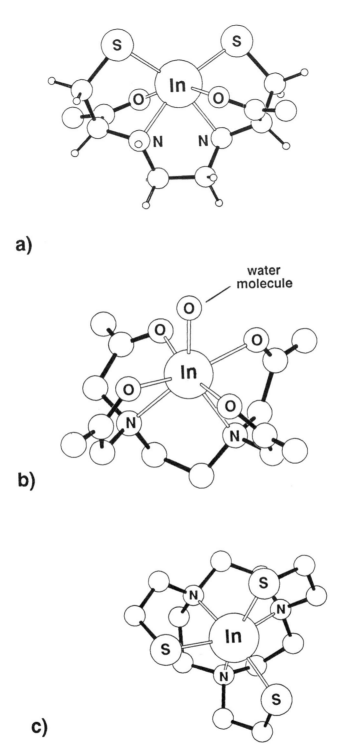

a)

b)

water
molecule

c)

Figure 5.2. Structure of some representative complexes of In(III). a) is the EC (N,N'-ethylenedi-l-cysteine) complex,[128] b) structure of the EDTA complex,[129] showing the seven coordinate structure, and c) structure of the N,N',N''-tris(2-mercaptoethyl)-1,4,7-triazacyclononane complex.[130] Redrawn with coordinates from the original papers.

technetium agents: ^{99m}Tc-hexamethylpropyleneamine oxime (HMPAO),[131] ^{99m}Tc-L,L-ethyl cysteinate dimer (ECD),[132] [99m]Tc-[bis(2,3-butanedionedioximato(1-)-0][2,3-butane-dionedioximato(2-)-0]2-methyl-1-propylborato-(2)-N,N',N'',N''',N'''',N''''']-chloro-technetium (SQ32,097).[133] The cerebral extraction and retention of all of these agents in the primate brain,[134,135] have been studied, and it was found that although they do cross the blood brain barrier they give low values of CBF at high flow rates.[136]

5.14.1 Synthesis and evaluation of new low molecular weight ligands

In this area, work has been carried out on the development of ligands with two types of structures. One type is based upon the HBED/PLED series and the second is based on substituted triazacyclononanes. Following the synthesis and evaluation of substituted PLED and HBED complexes[137] three derivatives were synthesized where the charge was anticipated to be significantly different.[138] The formal charges on these ligands after deprotonation vary from -4 to -2 (formulas 24-26). With the +3 charge of the metal ion the net charges would be -1, 0 and +1. It was anticipated that with the varying charges, different biodistributions would be observed. Although the molecules have different formal charges, in biological systems behavior of all three were found to be very similar.[138] This result may be due, at least in part, to the fact that various specific groups in the molecule are very polar and highly solvated. Also at physiological pH, species with different degrees of protonation exist in aqueous solution. Thus biological distribution appears to depend on several factors in addition to the overall charge. Another variation of the HBED structure involved the synthesis of Ga(III) and In(III) complexes of HBED in which the complexes were made more lipophilic by substituting alkyl groups on the benzene rings of the ligand, as in **27a** and **27b**.[5] The In(III) complex of HBED substituted with tertiarybutyl groups was found to be the most lipophilic of the complexes investigated. This was also reflected in the extraction coefficients of the complexes.

Alkyl derivatives of HBED
27a R_1 = R_2 = CH_3; **27b** R_1 = CH_3, R_2 = $C(CH_3)_3$

An early approach to reducing the charge of the complexes involving different coordinating groups was the preparation of a hexachelating ligand with an $N_3O_3^{3-}$ core which would produce neutral species with trivalent metals,[139] resulting in the synthesis of N,N',N''-tris(3,5-dimethyl-2-hydroxybenzyl)-1,4,7-triazacyclononane (TACN-Me2HB, **28b**). N,N',N''-tris(2-mercaptoethyl)-1,4,7-triazacyclononane (TACN-TM, **29**),[140] is an example of a ligand which should be less hydrophilic. However the data (Table 5.4), show that the complexes do not behave *in vivo* as lipophilic complexes. *In vivo* behavior of the indium(III) and gallium(III) complexes is probably due to the presence of the facial arrangement of thiolates about the metal center, which may create a substantial dipole moment for the complex and account for the relatively high solvation and observed hydrophilicity. The affinities of TACN-TM for indium(III) and gallium(III) were determined by potentiometric techniques.[141] At physiological pH = 7.4, 100% excess ligand, and at a total concentration of 1×10^{-6} M. The pM values of Ga^{3+} and In^{3+} are 26.2 and 27.9, respectively. These compare with Ga^{3+} values of 26.8 and 24.5 and In^{3+} values of 18.7 and 18.8 for the hexadentate ethylenediamine-based ligands, SHBED and PLED, respectively. The potentiometric data show that the complexes exist as neutral species, where the ligand is completely deprotonated and the metal is fully complexed.

29 TACN-TM

N,N',N''-tris(2-mercaptoethyl)-1,4,7-triazacyclononane

28a   R = R' = H   **28b** R = R' = CH$_3$
Alkyl derivatives of N,N',N''-tris(2-hydroxy-benzyl)-1,4,7-triazacylononane

Table 5.4.   Biodistribution of Complexes with TACN-TM

| | % ID/g (X s.d.) | | | | | |
| | [111]In | | | [68]Ga | | |
	1 min	5 min	1 hr	1 min	5 min	1 hr
Blood	1.76 ± 0.01	1.12 ± 0.07	0.19 ± 0.02	4.34 ± 1.23	1.98 ± 0.48	0.68 ± 0.28
Liver	2.08 ± 0.17	2.61 ± 0.22	0.77 ± 0.16	1.60 ± 0.21	2.44 ± 0.24	1.14 ± 0.15
Kidney	5.23 ± 1.16	6.86 ± 0.55	1.48 ± 0.18	5.95 ± 1.77	9.73 ± 2.18	3.19 ± 0.62
Muscle	0.25 ± 0.02	0.26 ± 0.03	0.24 ± 0.04	0.40 ± 0.07	0.38 ± 0.04	0.22 ± 0.03
Heart	0.77 ± 0.09	0.60 ± 0.09	0.24 ± 0.02	1.38 ± 0.28	0.79 ± 0.06	0.30 ± 0.07
Brain	0.07 ± 0.02	0.04 ± 0.01	0.01 ± 0.03	0.11 ± 0.04	0.07 ± 0.01	0.02 ± 0.01

The ligands tetrakis-(2-hydroxy-5-methylbenzyl)ethylenediamine, THMBED, **30a**, and tetrakis(2-hydroxy-3,5-dimethylbenzyl)ethylenediamine, THM$_2$BED, **30b**,[4] (Table 5.4) show significant differences between the *in vivo* behavior of Fe(III), In(III) and Ga(III) complexes. Also, with ^{67}Ga-THM$_2$BED (Table 5.5) a heart to blood ratio of ca. 2.8 is observed 30 minutes after administration and the myocardial activity remains high over long time frames. The myocardial uptake and retention of the compound indicate that this agent shows promise as a myocardial pervasion agent with properties superior to any other gallium-68 chelates described in the literature.[142,143] The differing properties of Fe(III), In(III), and Ga(III) complexes of THM$_2$BED warrant the investigation of similar effects for these metal complexes of related ligands.

30a THMBED R = H   **30b** THM$_2$BED R = CH$_3$

Table 5.5.   Biodistribution of THM$_2$BED Complexes at 1h.

	% ID.g (Xs.d.: n = 5)		
	Fe(III)	In(III)	Ga(III)
Blood	0.53 ± 0.08	3.39 ± 1.27	0.30 ± 0.02
Lung	1.91 ± 0.29	1.72 ± 0.40	1.74 ± 0.37
Liver	4.83 ± 0.52	0.94 ± 0.19	6.53 ± 0.66
Spleen	1.51 ± 0.18	0.68 ± 0.25	2.84 ± 1.02
Kidney	1.11 ± 0.13	3.99 ± 1.52	0.83 ± 0.10
Heart	0.72 ± 0.2	1.03 ± 0.03	0.71 ± 0.20
Brain	0.05 ± 0.005	0.10 ± 0.04	0.023 ± 0.003

An example of ligand design is a series of ligands containing a new type of donor group having high affinity for trivalent metal ions (Ga(III), In(III), Fe(III), Al(III), Gd(III), etc.), the 3-hydroxy-6-methyl-2-pyridylmethyl donor group, giving rise to the new ligands EDDA-MeHP, **31**, DTTA-MeHP, **32**, and TACN-MeHP, **33**. The development of these new ligands represents an important breakthrough in the achievement of both exceptionally high stability constants and very high metal binding efficiency. In fact TACN-MeHP, **32** binds the ferric ion more strongly than any other known ligand, including enterobactin, at physiological pH. This unprecedented coordinating effectiveness can be explained by several favorable factors: 1, the coordinating hydroxyl groups have low pK's because of the competing protonation of the pyridine nitrogens; 2, there is no coordination inhibition of the hydroxyl groups by the sterically-demanding methyl groups present in analogous pyridoxyl derivatives such as PLED; 3, the ring nitrogens (in TACN-MeHP) are part of the tightly-fitting coordinating framework giving rise to six chelate rings, and greatly assist metal coordination,[5,144] in contrast to the macrocyclic rings in enterobactin models, and enterobactin itself, which serve merely as frameworks on which to attach pendent coordinating groups. The fact that the coordinate bonds formed by the hydroxypyridine donor groups involve six-membered chelate rings also contributes to the high stability of the metal complexes with these small metal ions (see section 4.2, and Fig. 4.23). Whereas the short acetate pendent donor groups of NOTA lead to distortion to near trigonal prismatic coordination around the Fe(III), the six membered chelate rings of the TACN-MeHP complex allow the Fe(III) to adopt a coordination geometry much closer to octahedral, with presumably a decrease in steric strain.

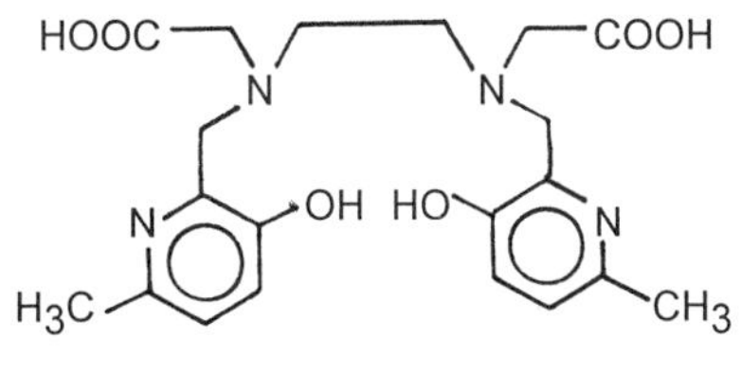

31 EDDA-MeHP

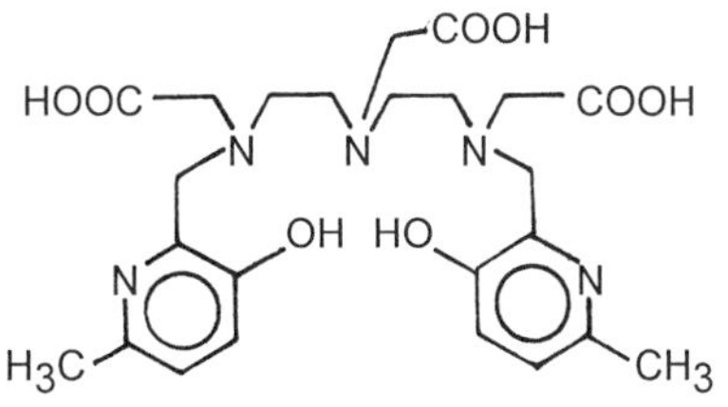

32 DTTA-MeHP

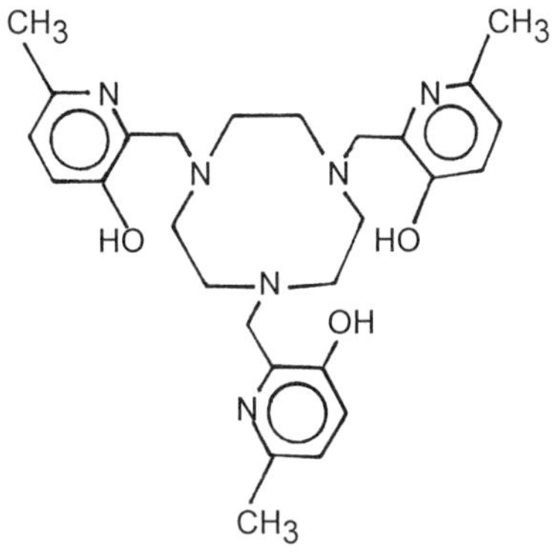

33 TACN-MeHP

The design of new ligands for improving the performance of metal chelate diagnostic agents involved the following guidelines: 1, the charge on the complex should be very small or zero; 2, thermodynamic stability should be maximized; and 3, hydrophilicity should be decreased. Also, with the working hypothesis that carboxyl groups are probably metabolized and destroyed *in vivo*, non-carboxylic donors, such as phenolate and mercapto groups will be preferred, as well as negative oxygen donors based on hydroxypyridines, hydroxamates, and hydroxypyridinones. The octahedral coordination model is selected for Ga(III) and In(III) complexes (with one exception). While the polyacetates **34a,b,c** have been known for some time, the hydroxypyridyl (A) and phenolic (B) analogs need considerable study of their metal binding affinities and physical properties. The hydroxypyridyl containing ligands have the advantage of providing phenolate-type oxygen donors with lower pK's, as noted above.

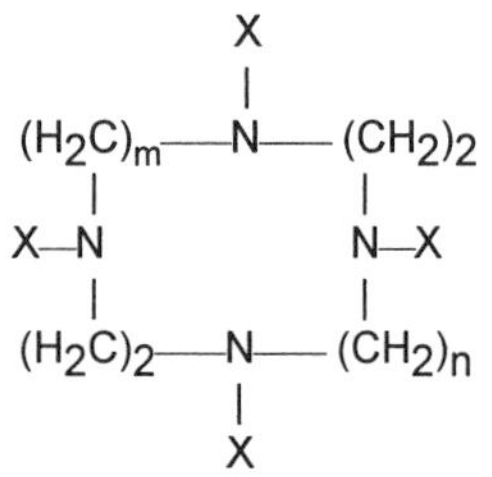

34 Tetrasubstituted tetraazacycloalkane

    a   DOTA m = n = 2; X = CH$_2$COOH

    b   TRITA m = 2, n = 3; X = CH$_2$COOH

    c   TETA m = 3, n = 3; X = CH$_2$COOH

    d   m = 2, n = 2; X = A

    e   m = 2, n = 3; X = A

    f   m = 3, n = 3; X = A

    g   m = 2, n = 2; X = A

    h   m = 2, n = 3; X = B

    i   m = 3, n = 3; X = B

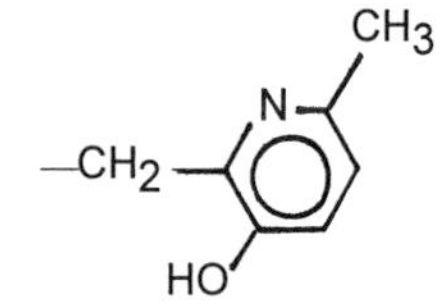

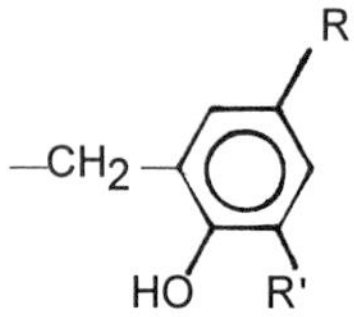

The effect of increasing the number of neutral oxygen donors in the macrocyclic ring was investigated recently by Delgado *et al.*,[145] by the study of the stabilities of 1-oxa-4,7,10-triazacyclododecane-N,N',N''-triacetic acid, (N3O-Ac) **35**, 1,7-dioxa-4,10,13-tri-azacylcopentadecane-N,N',N''-triacetic acid, N3O2-Ac, **36**, and 1,7,13-trioxo-4,10,16-triazacyclooctdecane-N,N',N''-triacetic acid, N3O3-Ac, **37**, with divalent and trivalent metal ions. Taken with 1,4,7-triazacyclononane-N,N',N''-triacetic acid, NOTA, **38**, a series of ligands containing three aminoacetate donor groups, and from 0 to 3 neutral oxygen donors is available for comparison. The results show that the hexadentate ligand of this series is by far the most effective for small trivalent metal ions such as Fe(III) and Ga(III), and expanding the numbers of donor groups does not have a significant effect on increasing the stabilities of complexes of larger metal ions such as In(III). It is noted, however, that the ratio of stabilities of In(III)/Ga(III) complexes is somewhat higher for **35**. For larger metal ions, such as Gd^{3+} (ionic radius[20] 0.94 Å), nonadentate **37** is superior since it matches the high coordination number of the metal ion. The available formation constants for **35** and **38** show very clearly the effect of metal ion size on the ability of the metal ion to adapt to the presence of the neutral oxygen donor of N3O-Ac (see section 2.5.1):

Metal ion	Ga(III)	Fe(III)	Mg(II)	In(III)	Ca(II)	Sr(II)	Pb(II)	Ba(II)
Ionic Radius (Å)	0.55	0.65	0.72	0.80	1.00	1.18	1.19	1.35
log K$_1$ NOTA , **38**	30.98	28.3	9.69	26.2	8.92	6.83	16.6	5.14
log K$_1$ N3O-Ac, **35**	21.3	26.8	10.25	25.48	12.98	11.37	19.27	9.91
Δ log K	-8.7	-1.5	+0.6	-0.6	+4.1	+4.6	+2.7	+4.8

(Formation constants from Refs 21 and 145)

35   1-oxa-4,7,10-triazacyclododecane-N,N',N''-triacetic acid (N3O-Ac)

36   1,7-dioxa-4,10,13-triazacyclopentadecane-N,N',N''-triacetic acid (N3O2-Ac)

37   1,7,13-trioxa-4,10,16-triazacyclooctadecane-N,N',N''-triacetic acid (N3O3-Ac)

38   1,4,7-triazacyclononane-N,N',N''-triacetic acid (NOTA)

A new phenolic ligand, HPED, has a structure very similar to that of HBED, except that the phenolic ring is bound directly to the aliphatic nitrogen, as indicated by **39**.[146] This ligand forms very stable complexes with Fe(III) and Ga(III) ions, as one would anticipate because it contains phenolate donor atoms. Because of the fact that it forms five five-membered chelate rings when it utilizes six donor atoms to form an approximately octahedral metal complex, it should be especially effective for binding somewhat larger metal ions, such as In^{3+}. This is borne out by the high stability constant, log K, and the high pM value at physiological pH and 100% excess ligand for the In(III) complex.

A search of the literature reveals that the mercaptoethyl donor group may be effective for In(III) (see Table 5.1). Thus mercaptoethylamine seems to bind strongly to In^{3+}, partly because of the great base strength of the sulfur donor atom, which forms a somewhat larger five membered ring (compared to, for example, the glycinate anion). Preliminary work on the In(III) complex of, EDDA-SS, **40** indicates that it has a stability constant higher than 10^{33}, and a pM value higher than those of EDTA and DTPA complexes.[147,148] It seems, therefore that mercaptoethyl groups should be introduced into the structures of multidentate ligands in order to increase the affinity (and perhaps selectivity) for the In(III) ion. This conclusion is further supported by the high affinity of **29**, TACN-TM for indium(III), and other chelating agents analogous to **40**.[149,150]

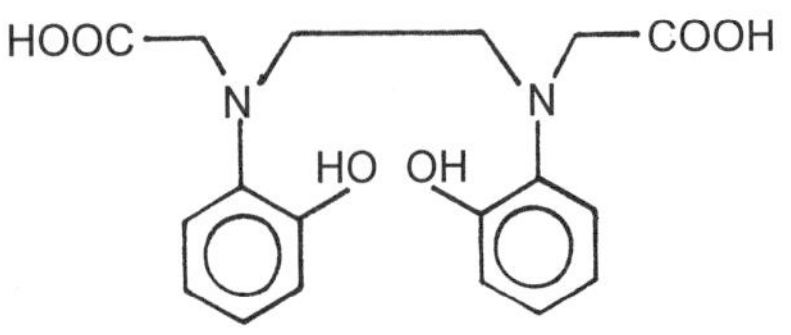

39   HPED, H$_4$L

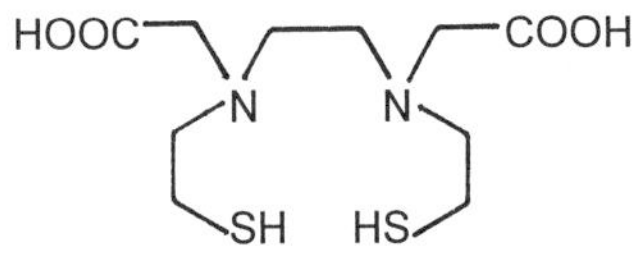

40   EDDA-SS, H$_4$L

5.14.2 Hydroxypyridinones

The seemingly successful application of 1,2-dimethyl-3-hydroxy-4-pyridinone, DMHP or L1, **8**, **41b**) for the treatment of iron(III) overload[32,151] has resulted in the development of considerable interest in the study of 3-hydroxy-2- and 4-pyridinones, **41** and **42**, as well as the cyclic hydroxamic acids, the 1-hydroxy-2-pyridinones, **43**. The ability of these bidentate ligands to form stable tris complexes of Fe(III)[152] Ga(III),[153] and In(III)[153] suggests that they should be seriously considered for the formation of complexes of Ga(III) and In(III) as radiopharmaceuticals. It has been shown[154] that bidentate ligands with very high intrinsic metal ion affinities form 3:1 (completely coordinated) metal complexes that tend to dissociate in dilute solution because their complex formation equilibria are highly concentration dependent. This problem was first pointed out by Martell[45,155] for the complexation of Fe(III) by **8**, and the incorporation of three molecules of the bidentate ligand into a single hexadentate 6-coordinate ligand was suggested. A hexadentate ligand containing three 3-hydroxy-2-pyridinone groups, **17**, was later synthesized.[48] A similar observation was made by Streater and coworkers,[47] who recently reported the synthesis of **16**. The bidentate units in **41-43** have very effective oxygen donors with metal ion affinities between those of hydroxamates and catecholates, with the compensating factor of binding only one hydrogen ion per bidentate unit, with moderate pK's. A hexadentate ligand incorporating these functional groups, such as **16** and **17** would be expected to have these advantages, without the adverse dilution effect.

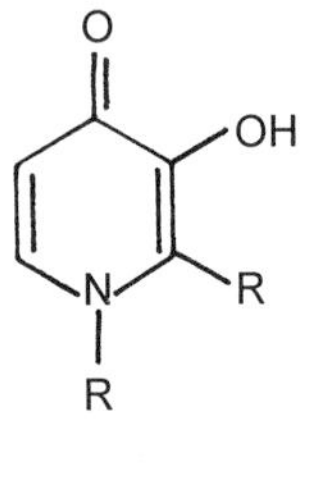

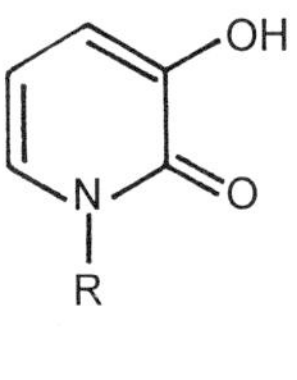

42   R = H; 3-hydroxy-2-pyridinone

41a   R = H; 3-hydroxy-4-pyridinone

41b   R = CH$_3$; 1,2-dimethyl-3-hydroxy-4-
        pyridinone

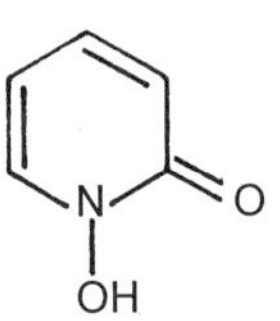

43   1-hydroxy-2-pyridinone

5.14.3 Bifunctional chelates of Ga(III) and In(III).

Because of the high stability constant of the transferrin complexes[156-158] and the large amount of transferrin in human plasma (0.2 mg/100mL), indium and gallium complexes must have very high stability constants in order to survive in the plasma. Sundberg *et al.*[159] prepared the initial bifunctional chelate 1-(*p*-benzenediazonium)ethylenediamine-N,N,N',N'-tetraacetic acid, which bonds the metal-ligand complex to a protein via the diazo group. The approaches to labeling proteins and antibodies with metal complexes generally required the attachment of either DTPA, **3**, or EDTA, **10**.[160-164] One of the major problems is that a significant fraction of the radioactivity is retained in the liver.[165,166] This is in contrast to[164] I-labeled antibodies where the radioactivity is rapidly excreted. Several mechanisms have been postulated including: 1, the complex undergoes exchange with an intracellular binding site which retains the metal; 2, the bond with which the chelate is attached to the antibody is resistant to enzymatic degradation and so the chelate remains attached to a relatively large peptide fragment which cannot escape from the liver cell; 3, portions of the chelate (possibly the acetate groups) are themselves enzymatically degraded in the liver and the strongly bonding chelate ceases to exist; 4, low molecular weight metal-labeled chelates are not retained in the liver, which suggests that the "trapping" phenomenon is a function of the antibody.

Meares and coworkers, using EDTA bound to antibodies suggested that 2 is the correct explanation.[167] When double labeled In-DTPA (^{111}In and ^{14}C-DTPA) was employed, ^{14}C from the DTPA was found in the urine in forms other than DTPA and the ^{111}In and ^{14}C have different biodistributions.[168] Thus, in view of these results, either mechanism 2 or 3 has been suggested. Other workers have carried out *in vivo* studies where In(III) and antibodies have very similar biodistributions and therefore favor mechanism 4.[169,170] Enhanced non-target organ clearance resulting in overall reduction of radiation exposure is needed. It is suggested that high thermodynamic stability may be essential for small metal chelate molecules, used directly, or formed by the complex which breaks away from the antibody, possibly by metabolism (degradation) of the covalent linkage.

Examples of metabolizable linkages for bifunctional chelating agents that may be used to radiolabel an antibody have been described by Deshpande *et al.*[167] In the examples given, the metabolizable groups used to covalently link the [111]In(III) chelate of 1-*p*-aminobenzyl EDTA and the mouse monoclonal antibody Lym-1 were thiourea, thioether, peptide and disulfide, as indicated in Chart I. All four linkages were cleaved and the radioactive indium was eliminated as the *p*-aminobenzyl EDTA chelate. The disulfide was cleaved very rapidly in the blood (presumably by glutathione), while the others were cleaved much more slowly, probably by enzymes in the liver. Although most of the reactivity was cleared, the liver was found to contain 10-20 percent of the ^{111}In after three days, the amount varying somewhat with the bifunctional group that was metabolized.

The antibodies labeled with the bifunctional chelates BrMe$_2$HBED, **44**, and BrϕHBED, **45**, appear to have significant advantages over bifunctional chelates currently in use.[171-173] Unfortunately these chelates still do not enable the In(III) to be cleared from the liver. A plan for testing antibody 1A3 labeled with In-111 using BrϕHBED, and clinical studies are in progress. The long-range objective remains: the development of a site-directed bifunctional chelate of In(III) which will not be retained in the liver.

Chart 1

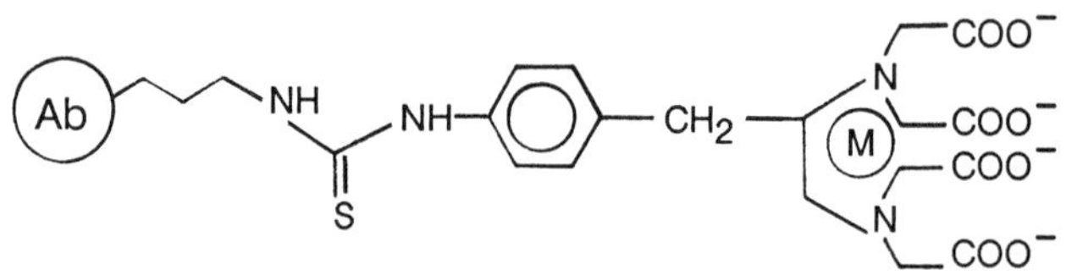

Lym-1-CITC (thiourea)

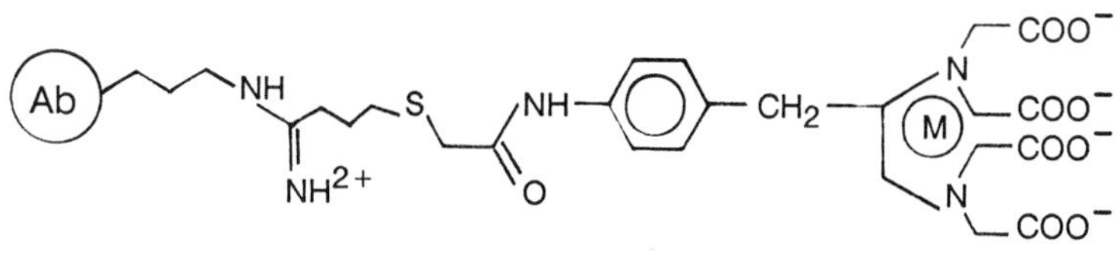

Lym-1-2IT-BABE (thioether)

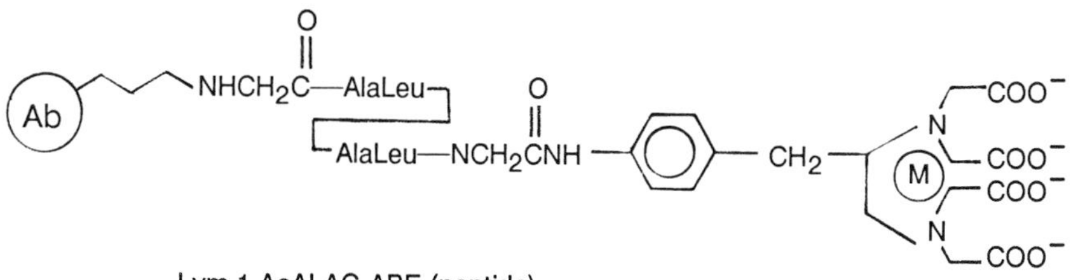

Lym-1-AcALAG-ABE (peptide)

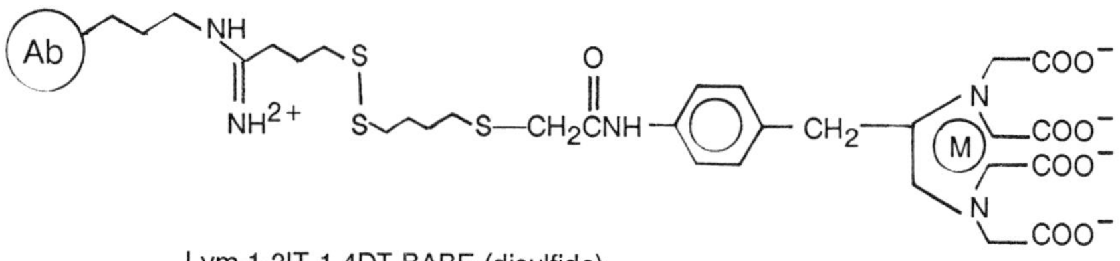

Lym-1-2IT-1,4DT-BABE (disulfide)

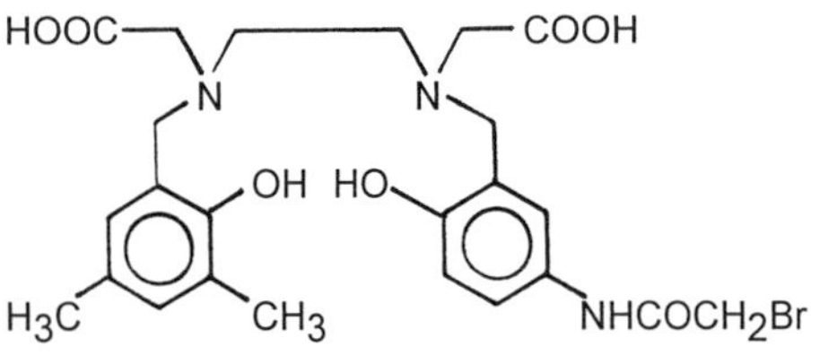

44 BrMe$_2$HBED

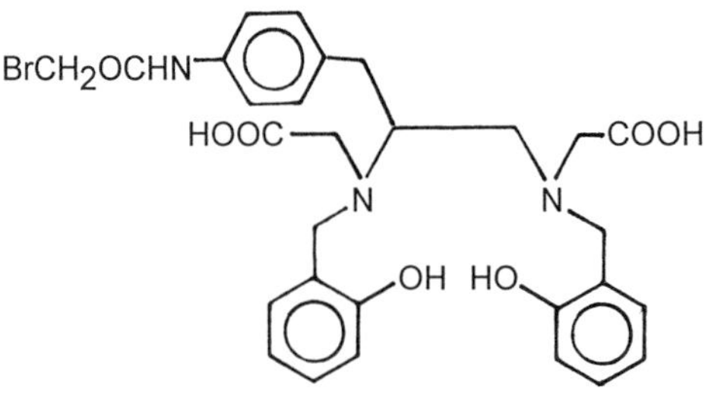

45 BrϕHBED

A recent development in the synthesis of bifunctional chelating agents for the radiolabeling of proteins is the report by Gansow and coworkers[174] of the use of *p*-benzylthiocyanate group to establish a covalent linkage with chelating agents such as

NOTA, DOTA, and TETA. The stability of the metal chelate was maximized by attaching the covalent linkages to a carbon atom of the macrocyclic ring, so as not to eliminate a coordinating group, as would occur if the linkage were to amino nitrogen or a carboxyl group.

A bifunctional reagent that forms a stable ^{111}In(III) complex has been made by covalently linking DTPA to octreotide,[175-177] a cyclic octapeptide, which is a selective and potent analog of somatostatin. The bifunctional ligand, **46**, was synthesized by reaction of the DTPA dianhydride directly with the octapeptide, giving the compound.

HOOCCH$_2$ CH$_2$CH$_2$ CH$_2$CH$_2$ CH$_2$COOH
N   N   N
HOOCCH$_2$ CH$_2$OOH   CH$_2$CO-D-Phe-Cys-Phe-D-Trp-Lys-Thr-Cys-Thr(d)
⌐————————— S – S —————————⌐

46 [DTPA-D-Phe]octreotide

The octapeptide is cyclic, the two cysteine residues being connected by a disulfide linkage. The ^{111}In complex of this bifunctional complexing agent was then made from ^{111}InCl$_3$ in acetic acid. The In(III) complex should be as stable, or nearly as stable as the In(III)-DTPA chelate itself, since it has all of the original coordinating groups of the DTPA ligand, except for one carboxyl group which has been converted to an amide group. The bifunctional complex is thought to retain most of the selectivity of the original octapeptide, octreotide, from which it is derived, and is now undergoing clinical trials.

5.15   Technetium

Technetium, which is element number 43 in the Periodic Table, does not occur naturally, since all of its isotopes are sufficiently radioactive to have decayed completely in the earth's crust. The longest-lived isotope, ^{99}Tc ($T_{1/2} = 2.1$ x 10^5 years) is used to study the chemical and physical properties of the element at conventional chemical concentrations. A short-lived nuclear isomer ^{99m}Tc ($T_{1/2} = 6.03$ hr) is used widely in nuclear medicine, because it has several favorable characteristics: a favorable half life, it emits gamma rays but no alpha particles and low energy Auger electrons, and easy analysis by a thallium-doped sodium iodide scintillation counter.

Little is known about the metabolism and ultimate fate of the various ^{99m}Tc-containing radiopharmaceuticals injected into the body. There is no unifying theme in the nature of the complexes formed, except for the successful use of technetium complexes for imaging purposes. At the present time more than three quarters of the routine medical diagnostic applications are based on technetium compounds. The highly radioactive ^{99m}Tc complexes decay to the ^{99}Tc nuclear isomer with low radioactivity, presumably giving rise to complexes with the same ligands. However, at the low (nanomolar) concentrations employed it is not possible to detect their fate or the metabolites formed. Of course it can be argued that at the low concentrations involved the degradation products can have little effect.

Technetium exists in eight oxidation states, from +7 to 0. Its chemical reactions parallel those of manganese, which lies above it in the Periodic Table. It resembles most closely rhenium which lies below it in the Periodic Table. The lanthanide contraction makes the radii of the two metals almost identical when they are in the same oxidation state. Therefore, because of this similarity the chemical reactions of rhenium are often taken as guidelines to the chemical reactions of analogous technetium compounds. Because of the complexity of Tc chemistry, and the lack of general guidelines, the following description is based on oxidation states, and only a few of the most important compounds for each oxidation state will be given as examples. One of the general characteristics of the metal is that Tc complexes are generally robust, although thermodynamically unstable. ^{99m}Tc complexes (with respect to dissociation or hydrolysis) may be used as imaging agents during the lifetime of the isotope before any appreciable change occurs. Some representative complexes of Tc are seen in Figure 5.3.

5.15.1 Technetium(VII)

Technetium 99m is usually made from the β-decay of molybdenum-99, along with a small amount (13%) of the more stable isotope, ^{99}Tc. Thus the molybdate anion is converted to the pertechnetate, with no breaking of metal-oxygen bonds.

$$\begin{bmatrix} & O & \\ & \| & \\ O-&Mo&-O \\ & | & \\ & O & \end{bmatrix}^{2-} \longrightarrow \begin{bmatrix} & O & \\ & \| & \\ O-&Tc&-O \\ & | & \\ & O & \end{bmatrix}^{-} + e$$

While pertechnetate is usually reduced to a lower valence state to form complexes for diagnostic purposes, it is sometimes intravenously injected directly into the blood stream in a number of diagnostic procedures. The anion is certainly rapidly bound to various serum proteins. Although the nature of the complexes formed is not known, it is almost certainly reduced to some lower valence state, with the reduction process being somewhat slower than the initial binding of Tc(VII).

5.15.2 Technetium(VI) compounds

These are subject to disproportionation reactions, which give mixtures of Tc(VII) and Tc(V) complexes, so that examples of stable Tc(VI) compounds are not as numerous as are complexes of other valence states. Tc(VI) may be formed by reduction of pertechnetate (or by oxidation of Tc(V) compounds) in the presence of a coordinating ligand. The tetrachloro-nitridotechnetate anion, **47**, was reported,[178] along with the corresponding bromo analog, whereby the high positive charge of the metal center is partly compensated for by the trinegative nitrido group. When the ligand has sufficient negative charge to balance the charge of the metal ion, octahedral complexes may form without oxo or nitrido groups associated with the metal center. Thus 3,6-ditertiarybutylcatechol forms a neutral 3:1 complex of Tc(VI), indicated by **48**.[179] Analogous compounds are the neutral complexes formed from the dithio analog of the 3,6-ditertiarybutylcatechol[180] and from the o-aminobenzenethio analog.[181] The formation of a neutral complex with the latter ligand requires the displacement of a proton from each of the amino groups.

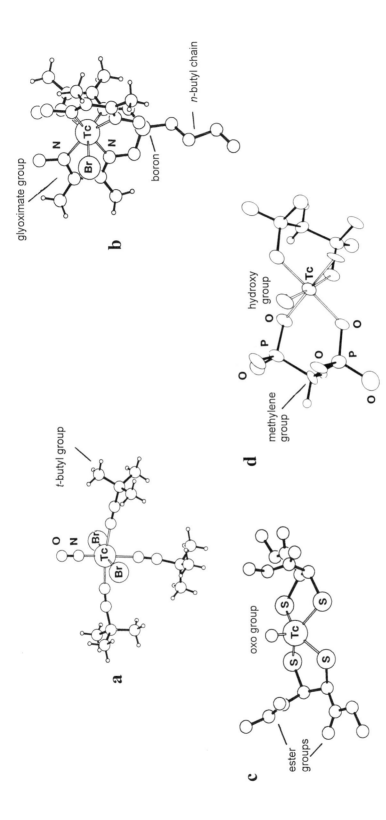

Figure 5.3. Structures of some representative complexes of technetium. a) An octahedral *t*-butylisonitrile complex of Tc(I) which has a nitroso group and two bromides coordinated to it.[182] b) A complex of a BATO ligand,[183] which consists of a triester of three glyoximates with borate, complexing with Tc(III), which has a coordinated bromide to give a coordination number of seven. c) a square-pyramidal complex of oxo-Tc(V) with two dimercaptodimethylsuccinate ligands coordinated to it.[184] d) Complex of Tc(IV) with two methylenediphosphonates and two hydroxy groups coordinated to it.[185] These complexes are bone seeking and are useful for imaging bone cancers. Redrawn with coordinates available in the original papers.

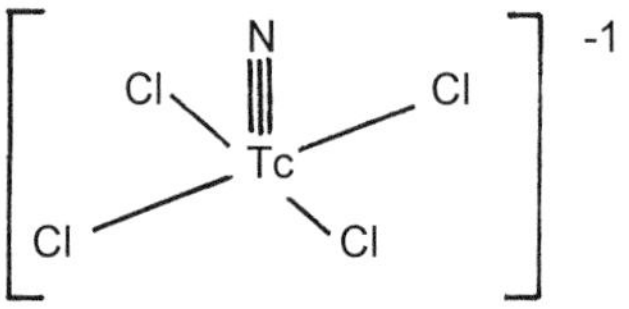

47

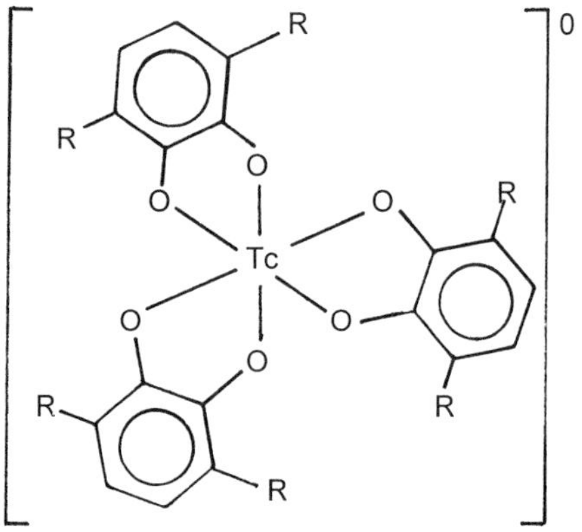

48   R = C(CH3)3

5.15.3 Technetium(V)

Typical Tc(V) complexes of tetradentate ligands are illustrated by formulas **49-53**.[186-193] They all contain an oxo- or nitridotechnetium core with the closely and strongly associated nitrido and oxo group providing a reduction of the effective charge of the technetium center. With neutral donors the most stable complexes seem to be those with the strongly negative nitrido or *trans* dioxo groups. When the donor groups of the ligands have negative charge the central metal may have a single oxo group, as in **52**[190-192] and **53**,[186,187] or be an oxohydroxo species, as in **50**.[189] The equatorial donor groups are more easily replaced than are the central (axial) nitrido and oxo groups. In fact many Tc(V) complexes have been synthesized by replacement of the halogens of the square pyramidal OTcCl$_4^-$ or OTcBr$_4^-$ complexes. Also a polyamine, such as cyclam, can replace a weaker nitrogen donor such as a monodentate amine. The amino complexes of Tc(V) can be made by reduction of pertechnetate with SnCl$_2$ in alkaline solution in the presence of the amine ligand. The complexes of cyclam and its macrocyclic analogs seem to be the most stable, with additional stability and inertness supplied by the macrocyclic ring structure.

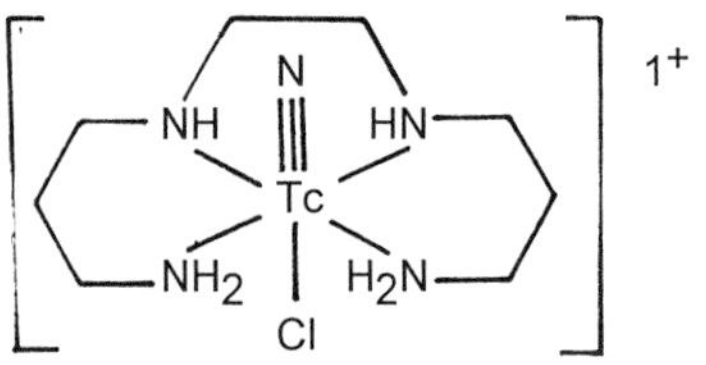

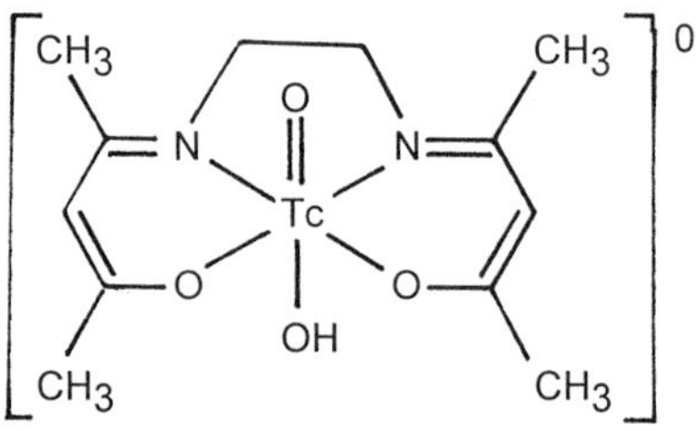

49

50

51 [+1]

52 [0]

53 [-1]

The highly charged metal center may also promote deprotonation of a coordinated amide or an amino group adjacent to other electron-withdrawing functional groups. Thus in the complex **52**, there is a hydrogen bond between the two oxime oxygens of the ligand but more importantly a proton is displaced from each of the amino donor groups, rendering them trigonal, negative, and very effective donors. The resulting complex has a hydrogen bond between the oxygens, which are partially dissociated. The amino groups, rendered somewhat more acidic because of the adjacent oxime groups, each lose a proton through the influence of the metal center, so that the complex formed is neutral, Many analogs of **52** have been made, since the complex and some of its analogs pass through the blood-brain barrier. This property is ascribed to its zero charge, low molecular weight, and lipophilicity. The methyl derivatives are also given credit for steric effects which slow the dissociation of the complex. It is noted that the negative charge on the ligand limits the metal center to one oxo group. Amides also undergo proton dissociation and become very strong donors under the influence of the Tc(V) center. Complex **53** is such an example. The ligand has two amide groups and two more acidic mercaptoethyl groups. The ligand therefore has a charge of 4-, which when combined with monooxo-technetium(V) forms a complex with a net -1 charge. Many Tc(V) complexes have been prepared by ligand exchange. Examples are complexes **50-53**, in which the ligand anion replaces the more weakly coordinating chloride ion. There are many other examples, with the ligands varying from polyamines, polyaminopolycarboxylic acids, mercaptoalkyl-amines, aminophenols, and many others. In Figure 5.3(c) is shown a typical complex of Tc(V) with an oxo group and two dimercapto ligands coordinated to it.

5.15.4 Technetium(IV)

Technetium(IV) complexes are formed by ligand exchange reactions of the chloride and bromide complexes of Tc(IV) such as K_2TcCl_6 and K_2TcBr_6. The complexes formed are generally octahedral, or distorted octahedral, as indicated by formulas **54-56**.[194-195] Complex **54**[196] is binuclear and can be described as trigonal prismatic with four mercapto bridges, but the coordination sphere around each metal center may be considered distorted octahedral. In this case eight mercaptophenyl donor groups provide six-coordination to two metal centers by having two of the 1,2-mercaptophenyl donors serve as bridging groups. The complexes of Tc(IV) formed with polyamino polycarboxylate ligands are not well characterized, frequently polynuclear, and partially hydrolyzed with oxotechnetium cores or hydroxo groups coordinated with the metal centers. An important characteristic of Tc(IV) complexes is a strong tendency to hydrolyze completely to form the hydrous oxide $TcO_2\text{-}XH_2O$. This oxide, in the freshly precipitated form, is a convenient starting material for preparing many technetium complexes. It is also conveniently prepared by the reduction of TcO_4^- in aqueous solution.

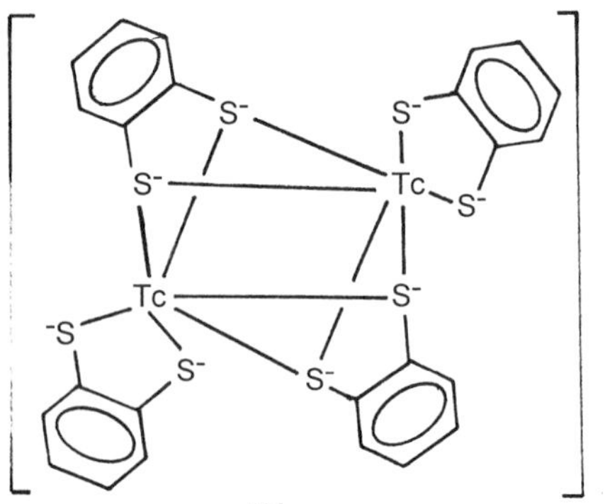

54 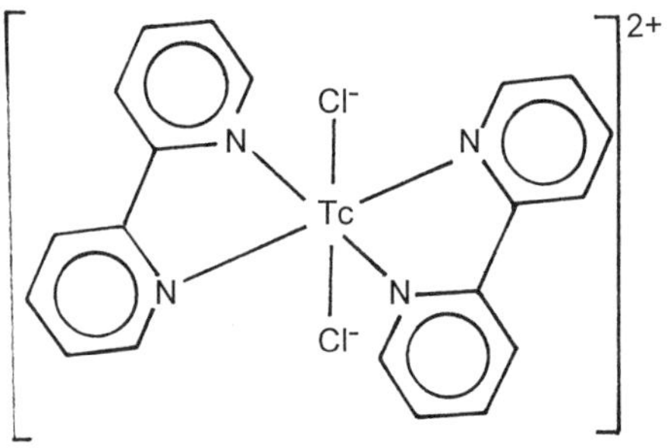   55

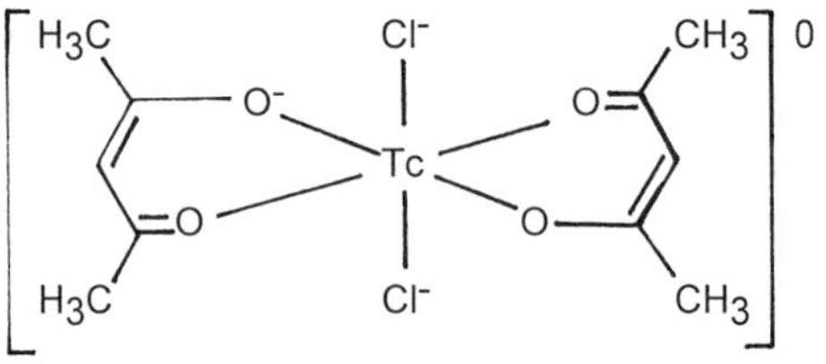

56

5.15.5 Technetium(III)

Many technetium(III) complexes have been prepared by the facile displacement of the monodentate thiourea donors from the hexakis(thiourea-S)technetium(III) chloride (or fluoborate). The latter complex is obtained by the reduction of pertechnetate with thiourea in ethanol. The complexes of Tc(III) can also be prepared from the tetravalent halide, such as $TcCl_4$ or K_2TcCl_6, where the ligand also acts as a reducing agent. A large number of phosphine, polyphosphine, and the corresponding arsine complexes are known, of which **57**,[197] **58**,[198] and **59**[199] are examples. Various mixed carbonyls are also known. They can be prepared by carrying out the synthesis under an atmosphere of carbon monoxide. An example is complex **57**.

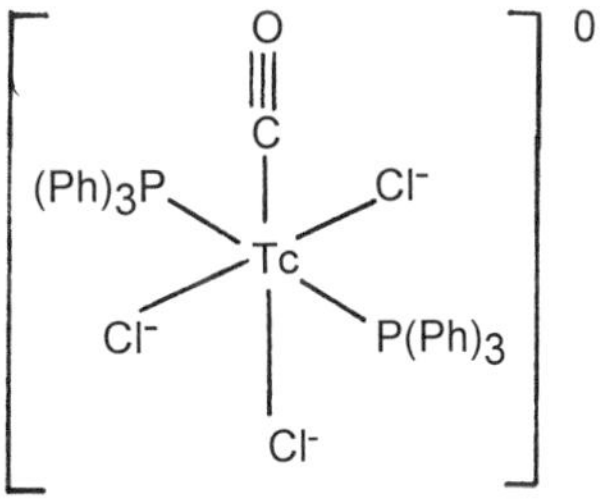

57

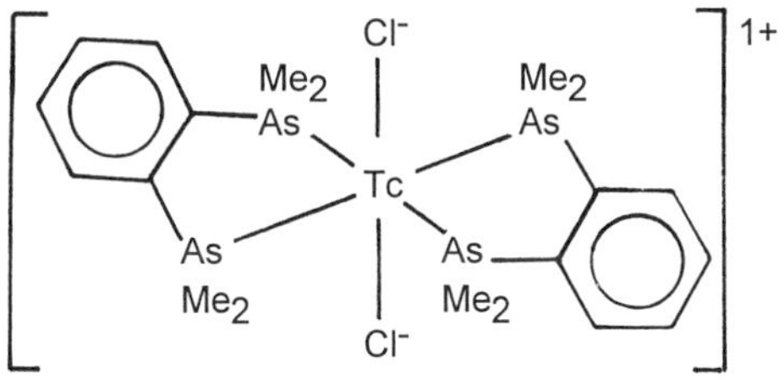

58

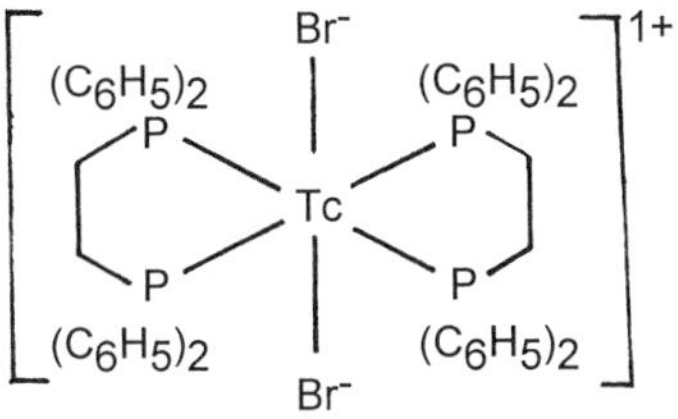

59

The trisacetylacetonate octahedral complex **60** was synthesized by heating TeCl₄ in acetylacetone. The reduction to a Tc(III) complex occurred in the solvent. It can be converted to the Tc(IV) trisacetylacetonate complex by gentle oxidation. Tris-oxime complexes of Tc(III) which show brain and heart uptake have been developed by the Squibb Institute for Medical Research.[183,200] They are called BATO's because they are boronic acid adducts. These compounds contain hydrogen bonds involving two hydrogen ions between three oxime oxygens on one side of the complex, and a boronic acid cap on the other. The oximes used are 1,2-dicyclohexanedioxime and dimethylglyoxime. An example of a seven coordinate BATO complex of Tc(III), which includes a coordinated bromide ion, is seen in Figure 5.3(b).

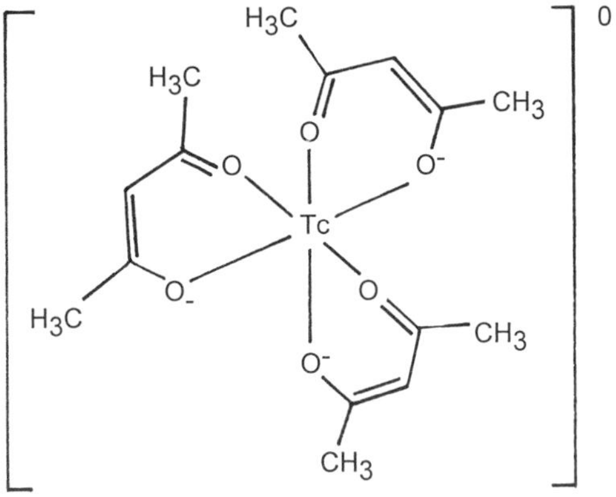

60

5.15.6 Technetium(II)

Technetium(II) complexes of π-acceptor ligands such as NO, CO, and phosphorus and arsenic compounds are stabilized by π backbonding from the metal center because of its low oxidation state. These complexes are generally six-coordinate and octahedral. An example of a Tc(II) complex of a ligand which generally favors low oxidation states is the trisbipyridyl Tc(II), illustrated by formula **61**.[201]

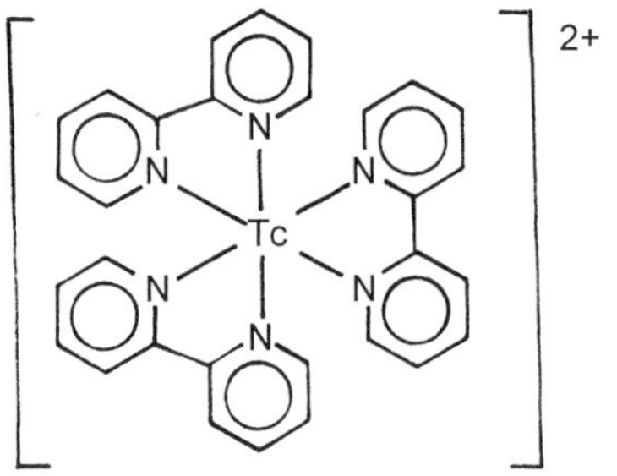

61

5.15.7 Technetium(I)

Technetium(I), even more than Tc(II), forms octahedral complexes with π-acceptor donor ligands such as carbon monoxide and phosphines, which can undergo backbonding from the metal center. Examples of such complexes contain as ligands phosphines, arsines, isonitriles, carbon monoxide, nitrous oxide, aromatic organic molecules and anions, and cyanide ion. Some of the complexes thus formed are quite stable and inert. An example of the complexes formed are the isonitriles represented by formula **62** (R = *tert*-butyl, methyl, cyclohexyl, or phenyl).[202] They may be synthesized directly from pertechnetate anion by reduction with dithionite in the presence of ligand in alcohol-water. An example of a sandwich compound, **63**,[203] was obtained by reduction of TeCl$_4$ with benzene, and precipitated as the hexafluorophosphate salt. The structure of a Tc(I) complex of *t*-butylisonitrile is shown in Figure 5.3(a).

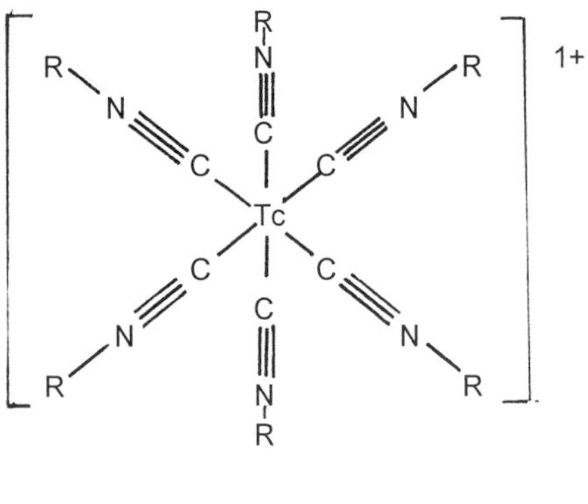

62

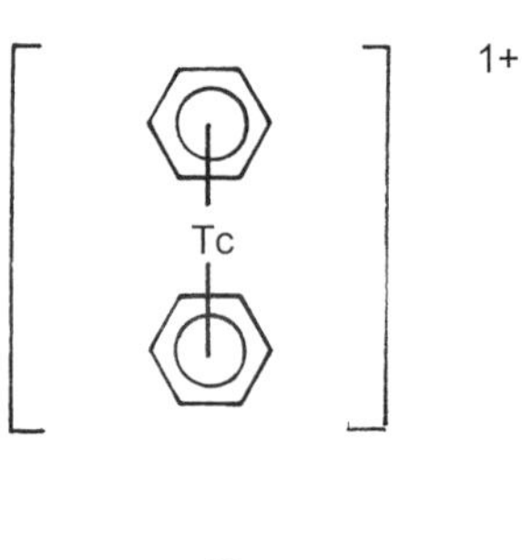

63

5.16 Rhenium

Similarities between rhenium and technetium chemistry have long been known. Structurally, the two metals in all oxidation sates form the same types of complexes. In the oxidation state (V), with analogous ligands, both rhenium and technetium exhibit the tendency to form the same oxo cores: tr-$M(V)O_2^+$, $M(V)O^{3+}$, or $M(V)O(OH)^{2+}$, depending on the ability of the ligand donor groups to neutralize the charge on the metal center. Also, because of lanthanide contraction the ionic radii of Re(V) and Tc(V) are almost the same, 0.72 Å and 0.74 Å, respectively.[20]

Although structural aspects of Re(V) and Tc(V) complexes with similar ligands are virtually the same, other properties differ significantly. Rhenium compounds are harder to reduce and concomitantly easier to oxidize.[204] This affects reductive synthetic methods (reduction of $Re(VII)O_4^-$ in the presence of a ligand L to form Re(V)-L). Therefore, syntheses of Re(V)-complexes with multidentate ligands rely on ligand exchange methods from starting complexes such as $Re(V)OCl_4^-$ and $ReOCl_3(PPh_3)_2$. Ligand exchange methods have also been used in Tc(V) chemistry[205] but the methods utilized for forming chelate complexes for the two metal ions differ in that harsher reaction conditions are necessary for forming the desired Re(V) complexes. The tendency of third row metals to form more robust complexes than their second row congeners is well-known.

5.16.1 Development of rhenium-radiopharmaceuticals of therapeutic value

Two different types of radiopharmaceuticals are being developed for ^{186}Re. These are: 1) small discrete metal-ligand complexes that, by virtue of their physical properties (charge and lipophilicity) localize in certain organs in the body, and 2) radiolabeled biologically active molecules such as monoclonal antibodies that are specific for certain tumors. While ^{99m}Tc radiopharmaceuticals are useful for diagnostic imaging of certain organs, ^{186}Re radiopharmaceuticals are useful for actually treating the tumors that are located in that organ. The physical properties of certain isotopes of rhenium and technetium that make them desirable for clinical use are listed in Table 5.6.

Table 5.6. Properties of Important Isotopes of Rhenium and Technetium.

Isotope	Half-life	Emissions	Comments
^{185}Re	stable	none	naturally occurring isotope, 37.07%
^{187}Re	stable	none	naturally occurring isotope, 62.93%
^{186}Re	90 hours	γ, 137 KeV β-, 1.07, 0.93, MeV	reactor produced irradiation of enriched ^{185}Re (85%) in 10^{14} neutron/cm^{2}s flux
^{188}Re	17 hours	γ, 155 KeV^a β-, 2.12 MeV	^{188}W/^{188}Re generator produced as ^{188}ReO$_4^-$ no carrier added
^{99m}Tc	6 hours	γ, 143 KeV	^{99}Mo/^{99m}Tc generator produced as ^{99m}TcO$_4^-$ no carrier added
^{99}Tc^b	2.1 x 10^5 yr	β-, 0.292 MeV	available in gram quantities

^{a} Many gamma rays with higher energies are also emitted. ^{b} This isotope is used only for synthetic studies.

^{131}I is currently the only clinically used radiolabeled antibody but *in vivo* dehalogenation of the antibody radiolabel and other undesirable biological properties have prompted a search for other more efficient radiolabels. ^{67}Cu, ^{77}Br, ^{82}Br, ^{90}Y, ^{111}In, ^{99m}Tc, ^{186}Re, and ^{211}At contain desirable properties, but ^{186}Re and ^{90}Y have been shown to possess the best properties for use as therapy radiolabels.[206] Desirable properties include sufficiently long half-lives (necessary for tumor localization), singular imaging gamma emission, an intermediate beta energy, stable daughter products, and the ability to form stable chelates with functional groups attached to an antibody molecule (see Table 5.6). The conditions necessary to radiolabel a biologically active molecule containing DTPA as the chelating ligand, from ^{186}ReO$_4^-$ at physiological conditions have been examined.[207] Problems with low levels of radiolabeling occurred because of the difficulty in reducing Re(VII) by the same methods as those used for Tc(VII), and by hydrolysis of reduced species to non-binding species such as ReO$_2$, before formation of stable chelate complexes could occur.

Re(V) forms stable chelate complexes under a variety of conditions by the ligand exchange method, especially when starting from an intermediate Re(V)O(citrato)$^{n+}$ complex. This method is being investigated for formation of ^{99m}Tc(V) imaging agents with gluconate or citrate to stabilize the Tc(V) from disproportionation in water before chelation with the desired ligand.[208,209]

Currently, a portable generator is not available for production of ^{186}Re, as is also true of ^{99m}Tc and ^{188}Re. The ^{186}Re must be specially ordered from a reactor facility and is only available in carrier-added form (carrier added = containing the daughter products ^{185}Re and ^{187}Re). This means that the total rhenium concentration in a clinical preparation would be about 10^{-3} M to achieve the desired amount of radioactivity as compared to 10^{-6} to 10^{-7} M for ^{99m}Tc and ^{188}Re.[210] This inconvenience has not, however, discouraged development of ^{186}Re radiopharmaceuticals. For example, clinical trials on humans of ^{186}Re-Sn-HEDP (hydroxyethylidenediphosphonate), a polymeric complex for treatment of skeletal metathesis, are being carried out with positive results.[211] Also, a recent study comparing the biodistribution of the well-known kidney imaging agent ^{99m}Tc-CO$_2$DADS with the analogous ^{186}Re-CO$_2$DADS (CO$_2$DADS = 2,3-dimercaptoacetamidopropanoate), showed them to behave identically *in vivo*.[212]

5.17 Magnetic Resonance Imaging (MRI)

It is well known that proton NMR provides a highly-effective non-invasive technique for diagnostic imaging and has sufficient sensitivity to detect a wide variety of lesions[213-215] including liver disease, cerebral and vascular abnormalities, kidney malfunctions, and tumors in several organs and tissues. Nearly all NMR clinical imaging has involved proton magnetic resonance, and the principal proton species that generates the NMR signal is the water molecule. Water is the predominant molecular species throughout the body, in all types of organs and tissues, with, in many cases, rather small variations in the percent water, thus usually producing only small variations in the intensity of the NMR water signal. MR image contrast arises from three magnetic resonance properties of the hydrogen nucleus of water: 1) its concentration or spin density (SD); 2) the spin-lattice or T$_1$ relaxation time; and 3) the spin-spin or T$_2$ relaxation time. (T$_1$ and T$_2$ are related to molecular mobility and are strongly influenced by the localized physiochemical environment.) With the appropriate MRI techniques, each of these parameters can be measured as a function of spatial location to create a weighted image, whose contrast results primarily from that single parameter. Differences in any or all of the parameters in different tissues or between normal and pathological states, provide anatomical and/or morphological differentiation.[213-215]

In an effort to increase the applicability and diagnostic potential of MRI, agents have been developed which can be used to selectively enhance image contrast. Several approaches to this problem are available, with the most promising being the development of paramagnetic metal ion chelates.[14] Highly paramagnetic ions alter the magnetic properties of the tissues by adding large increments to the normal relaxation rates ($1/T_1$ and $1/T_2$) of the water protons. This effect, which essentially involves relaxation of the water signal, is roughly proportional to the number of unpaired electrons responsible for the magnetic moments of the solvated metal ions in solution. The effect cannot take place, however, without direct exposure of the paramagnetic metal ion to the bulk solvent water, and this is generally achieved by exchange of solvated (coordinated) water around the metal ion with the water in the tissue or medium being considered. Exchange of inner sphere coordinated water is generally much more effective than loosely bound outer sphere molecules. The contrast in the NMR signal is achieved and amplified through the fact that the paramagnetic metal is carried (and protected) by an organic ligand which forms metal chelates which themselves vary greatly in tissue penetration and membrane permeability, thus achieving variations in the intensity of the NMR signal. Chelation of the paramagnetic metal ion to suitable ligands reduces *in vivo* toxicity while still retaining the relaxivity properties of the paramagnetic center. A list of paramagnetic metal ions that may be considered for NMR contrast agents is presented in Table 5.7.

Table 5.7.   Magnetic Moments of Some Paramagnetic Metal Ions.

Ion	d-Electron Configuration (where applicable)	Number of Unpaired Electrons	Magnetic Moment (Bohr magnetons)
Ti^{3+}, V^{4+}, Cu^{2+}		1	1.7 -2.2
V^{3+}		2	2.6 - 2.8
Ni^{2+}		2	2.8 - 4.0
Cr^{3+}, V^{2+}		3	3.7 -3.9
Co^{2+}	high spin	3	4.1 -5.2
Mn^{3+}, Cr^{2+}	high spin	4	4.8 - 5.0
Fe^{2+}	high spin	4	5.1 - 5.5
Fe^{3+}, Mn^{2+}	high spin	5	5.8 -6.0
Gd^{3+}		7	8.0

From Ref. 14.

5.17.1 Factors influencing the water-relaxation ability of paramagnetic complexes

The changes of water relaxation rate caused by paramagnetic metal ions in metal complexes are dependent on many factors, which are outlined below. If the paramagnetic metal ions listed in Table 5.7 are to be employed, they must be strongly complexed, because their ionic salts are insoluble at physiological pH as well as highly toxic in most cases. Strongly chelating ligands reduce toxicity by preventing exchange with metal ion receptors in the body and by facilitating clearance and elimination. Strong complexing, however, reduces exchange with the solvent so that a balance must be achieved. The following are the most important factors that must be considered:

5.17.2 Concentration of the paramagnetic complex

High concentration requires low toxicity, and considerable aqueous solubility, both of which are strongly affected by ligand design. The nature of the ligand may be modified to increase membrane permeability, but such modifications usually decrease aqueous solubility and strongly influence toxicity.

5.17.3 Magnetic moment of the paramagnetic complex

From Table 5.7 it is seen that metal ions with large numbers of unpaired d or f electrons have the highest magnetic moments. Because of the necessity of having very stable metal complexes, and the fact that it is much easier to form very stable complexes of Fe(III) and Gd(III) than of Mn(II), it is suggested that priority be given to the design, synthesis, and evaluation of new iron(III) and gadolinium(III) chelates which have the molecular structures considered favorable for contrast imaging purposes.

5.17.4 Modulation of magnetic interactions between unpaired electrons and nuclei

This complicated effect, which is not completely predictable, influences the relaxation efficiency of a paramagnetic species. Detailed discussions of this phenomenon have been published.[14,216,217] The effect is strongly influenced by molecular motion, and hence on the size of the complex molecule. It may be enhanced by covalently attaching the paramagnetic complex to a large molecule such as a protein. The development of structure-activity relationships for this effect requires *in vivo* testing of a considerable number of paramagnetic complexes.

5.17.5 Number of coordinated water molecules

The increase in the water relaxation rate achieved in the bulk solvent is roughly proportional to the number of water molecules in the coordination sphere of the metal ion. Thus it is seen that the more completely coordinated is the paramagnetic metal ion by one or more ligands, the smaller is the number of remaining coordinated water molecules available for exchange with the solvent. Therefore stability of the complex and resistance to metal ion exchange *in vivo*, and effectiveness in paramagnetic relaxation of the NMR water signal, vary in opposite directions. Because of this factor, a balance must be achieved. As a working hypothesis, molecular design may be directed toward retaining at least one water molecule in the coordination sphere of the metal ion, even for the most stable metal chelates.   The structures of the complexes of DTPA,[218] DOTA,[219] and EDTA[220] with Gd(III) are seen in Figure 5.4, showing the coordinated solvent molecules. Gd(III) appears to prefer a coordination number of nine, so octadentate ligands such as DTPA and DOTA leave one coordinated water molecule, whereas a hexadentate ligand such as EDTA leaves three coordination sites for water molecules.

5.17.6 Rate of water exchange

The rate of exchange of coordinated water molecules with the bulk solvent must be rapid in order to achieve effective paramagnetic relaxation. It is therefore necessary to avoid metal complexes having slow exchange rates, such as Cr(III). This is not a problem with metal ions such as Fe(III) and Gd(III).

5.17.7 Metal-to-nucleus distance

The paramagnetic relaxation effects drop off rapidly with increasing distance between the paramagnetic metal ion and the protons of the coordinated water molecules. This factor will not change much in most complexes for which the lengths of the coordinate bonds to water are not expected to change to any significant extent as the coordinated ligand is varied. Although the distance between coordinated water and Gd(III) is somewhat greater than that for Fe(III), the difference is somewhat compensated for by the greater lability of water in the coordination sphere of the more basic Gd(III) ion.

5.17.8 Metal chelates recently investigated for magnetic resonance imaging

A number of paramagnetic metal complexes of readily available chelating agents have been employed for the development of contrast media for MRI. Early results have been described in several reviews.[221-223] The Cr(III) complex of EDTA, **10**, and the Gd(III) complex of DTPA, **3**, have been used as contrast agents for the liver and kidneys.[223] Recently, the Gd(III) chelate of a more sophisticated and complex ligand, EHPG, **7**, and its bromo derivative was employed by Lauffer and coworkers[224,225] for radio-contrast imaging. The most thoroughly characterized paramagnetic relaxation agent is Gd(DTPA).[14,214,215,223] This is a nonspecific agent which distributes passively in vascular and other extracellular tissues. It is currently undergoing clinical trials for uses including evaluation of blood-brain barrier integrity, renal function, and as a myocardial imaging agent. In Figure 5.4 is shown the structures of some representative Gd(III) complexes.

Tweedle and coworkers[215,226-228] investigated Gd(III) and other lanthanide complexes of a seven coordinate ligand, 1,4,7,10-tetraazacyclododecane-1,4,7-triacetic acid, **64a**, and of an eight coordinate ligand (in which the eighth coordinating group is a relatively weak donor) 10-(2-hydroxypropyl)-1,4,7,10-tetraazacyclododecane-1,4,7-triacetic acid, **64b**. The purpose was to prepare neutral Gd(III) complexes with a vacant coordination site (probably occupied by a water molecule, or a weakly coordinated group, the 2-hydroxypropyl donor) to promote exchange with the solvent, thus extending the paramagnetism of the metal ion throughout the solvent system. The omission of a pendent acetate donor resulted in the formation of a Gd(III) complex with a surprisingly high stability constant ($10^{21.0}$) which is only three log units lower than that of the Gd(III)-DOTA complex, $10^{24.0}$. Replacement of an acetate donor of DOTA by a 2-hydroxypropyl group gave a stability constant of the Gd(III) complex that was found to be considerably higher, $10^{23.8}$, only slightly lower than that of the DOTA complex.

64   Analogs of DOTA

  a) R = H   DO3A

  b) R = $CH_2CHOHCH_3$ DO3A-HP

  c) R = CH_2COOH DOTA

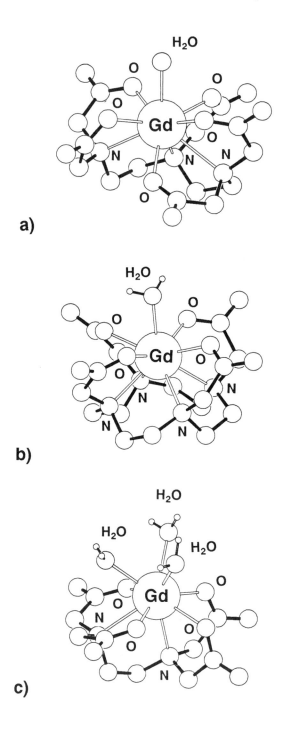

Figure 5.4. Structures of some representative complexes of Gadolinium(III), showing the preferred coordination number of nine, and the coordinated water molecules necessary for producing adequate relaxivity in MRI applications. a) shows the complex with DTPA[218] (diethylenetriaminepentaacetic acid), b) the complex with DOTA (1,4,7,10-tetraazacyclododecane-1,4,7,10-tetraacetate),[219] and c) the complex with EDTA.[220] Redrawn with coordinates available in the original papers.

5.17.9 Hepatobiliary MRI contrast agents

Use of nonspecific MR contrast agents to image the liver has generally proven disappointing in the diagnosis of liver diseases. This is likely due to the fact that these agents are cleared through the liver via the reticuloendothelial cells, which comprise only about two percent of the tissue volume. On a mole-to-mole basis these agents will only influence the relaxivity of a small fraction of the tissue's water, thereby providing poor or no resolution of pathological states. The Fe(III) chelate of EHPG, **7**, has recently been shown to be specific for liver hepatocytes, a cell type which comprises seventy eight percent of the tissue volume, and the associated biliary excretion pathway. By providing contact with a significantly greater fraction of tissue water, targeting of contrast agents through the hepatocytes is likely to significantly increase the diagnostic utility of MRI of the liver.[229-231] Fe(EHPG)⁻ has been shown to decrease liver T_1 and T_2 times by twenty-five and fifteen percent, respectively. This results in a three fold increase in MR image intensity in the normal liver. Using this agent, imaging of mice, intrahepatically inoculated with tumor cells, revealed significantly enhanced liver-to-tumor image contrast, thereby demonstrating the potential of this agent for early detection of liver metastases.

5.17.10 New MRI contrast agents

New relaxation agents currently being developed are designed to be more tissue specific. An approach to direct targeting of metal chelates is achieved by linking them to bioactive molecules (i.e., monoclonal antibodies) or through judicious modification of functional groups on the ligand in order to alter lipophilicity, pK_a, molecular weight, etc. This latter strategy has been effectively exploited to develop a class of relaxation agents specific for hepatocytes. An interesting use of paramagnetic complexes in MRI imaging is the labeling of the monoclonal antibody of myosin (which localizes in infarcted myocardium) with the Mn(II)-DTPA complex.

Recently Gansow and coworkers[232] described a new class of magnetic resonance imaging contrast agents: gadolinium complexes of DTPA covalently linked to Starburst™ dendrimers. The agents are built from the conjugation of the amino groups of the polyamidoamine form of the dendrimer to the chelator 2-(4-isothiocyanatobenzyl)-6-methyldiethylenetriaminepentaacetic acid. The formula of this chelator is

65 PANAM-TU-DTPA

The Gd(III) complexes of the new reagents which have as many as 170 Gd(III) ions bound to a single dendrimer molecule show large proton relaxation enhancements and high molecular relaxivities. This new class of contrast agents has the potential for diverse and extensive applications in MRI imaging.

References

1. K. N. Raymond, V. L. Pecoraro and F. L. Weitl, In: Development of Iron Chelators for Clinical Use, A. E. Martell, W. F. Anderson and D. G. Badman, Eds., Elsevier/North Holland, New York, 1981, pp. 165-187.
2. A. J. Jacobs, In: *Development of Iron Chelators for Clinical Use*, A. E. Martell, W. F. Anderson and D. G. Badman, Eds., Elsevier/North Holland, New York, **1981**, pp. 29-46.
3. W. R. Grady and A. J. Jacobs, In: *Development of Iron Chelators for Clinical Use*, A. E. Martell, W. F. Anderson and D. G. Badman, Eds., Elsevier/North Holland, New York, 1981, pp. 134-164.
4. S. L. Madsen, M. J. Welch, R. J. Motekaitis and A. E. Martell, *Nucl. Med. Biol.*, **1992**, *19*, 431.
5. R. J. Motekaitis, A. E. Martell, and M. J. Welch, *J. Inorg. Chem.*, **1990**, *29*, 1463
6. M. M. Jones, *Comments Inorg. Chem.*, **1992**, *13*, 91.
7. M. M. Jones, In: *Metal Ions in Biological Systems*, Vol.16, H. Sigel, Ed., Marcel Dekker, Inc., New York, 1983, p. 47.
8. R. A. Bulman, *Structure and Bonding*, **1987**, *67*, 91.
9. D. R. Williams and B. W. Halstead, *J. Toxicol.: Clin. Toxicol.*, **1982-83**, *19*, 1081.
10. S. W. Schwarz and M. J. Welch, In: *Applications of Enzyme Biotechnology*, . T. O. Baldwin and J. W. Kelly, Eds., Plenum Publishing Co., Inc., New York, 1992.
11. R. C. Hider and A. D. Hall, *Prog. Med. Chem.,* **1991**, *28*, 40.
12. A. E. Martell and M. Calvin, *Chemistry of Metal Chelate Compounds*, Prentice-Hall, Englewood Cliffs, N.J. 1952, p. 472.
13. A. E. Martell, *J. Chem. Ed.,* **1952**, *29*, 270-280.
14. R. B. Lauffer, *Chem. Rev.*, **1987**, *87*, 901.
15. P. J. Sadler, *Adv. Inorg. Chem.,* **1991**, *36*, 1.
16. S. Jurisson, D. Berning, W. Jia and D. Ma, *Chem. Rev.,* **1993**, *93*, 1137.
17. S. Ahrland, J. Chatt and N. R. Davies, *Q Rev. Chem. Soc.*, **1958**, 12, 321.
18. R. G. Pearson, *Chem. Ber.*, **1967**, *3*, 103.
19. R. D. Hancock and A. E. Martell, A. E. J. Chem. Ed., in press.
20. R. D. Shannon, *Acta Crystallogr. Sect. A*, **1976**, *A32*, 751.
21. A. E. Martell and R. M. Smith, *Critical Stability Constants*, Plenum Press, New York, Vols. 1-6, 1974-1989; a) R. M. Smith, A. E. Martell and R. J. Motekaitis, *Critical Stability Constants Database*, Beta Version, NIST, Gaithersburg, MD, USA, 1993.
22. D. J. Weatherall and J. B. Clegg, *The Thalassemia Syndromes*, 3rd Ed.; Blackwell Scientific Publications, Oxford, 1981.
23. D. J. Weatherall, In: Development of Iron Chelators for Clinical Use, A. E. Martell, W. F. Anderson and D. G. Badman, Eds., Elsevier/North Holland, New York, 1981, pp. 1-12.
24. G. Koren, Y. Bentur, D. Strong, E. Harvey, J. Klein, R. Baumal, S. P. Spielberg and M. H. Freedman, *Am. J. Dis. Child.*, **1989**, *143*, 1077.
25. A. E. Martell, R. J. Motekaitis; I. Murase, L. F. Sala, R. Stoldt, C. Y. Ng, H. Rosenkrantz and J. J. Metterville, *Inorg. Chim. Acta*, **1987**, *138*, 215.
26. R. Grady, In: *The Development of Iron Chelators for Clinical Use*, R. J. Bergeron and G. M. Brittenham, Eds. CRC Press, Boca Raton, FL, 1993, pp. 395-406.
27. R. Grady, 5th International Conference on Iron Chelation in the Treatment of Thalassemia and Other Diseases, No.12-13, 1993, University of Milan.
28. K. N. Raymond, S. S. Isied, L. D. Brown, F. R. Fronczek, J. H. Neibert, *J. Am. Chem. Soc.*, **1976**, *98*, 1767.
29. T. M. Garrett, T. J. McMurry, M. W. Hosseini, Z. E. Reyes, F. E. Hahn, and K. N. Raymond, *J. Am. Chem. Soc.*, **1991**, *113*, 2965.
30. T. M. Garrett, P. W. Miller and K. N. Raymond, *Inorg. Chem.*, **1989**, *28*, 128.
31. R. C. Hider, G. J. Kontoghiorghes and J. Silver, J. Pharmaceutical compositions, U. K. Patent GB 2 118 176, 1983; 2 136 807, 1984; 2 146 990, 1984.
32. G. J. Kontoghiorghes, *Lancet, I,* **1985**, 81.
33. G. J. Kontoghiorghes and A. V. Hoffbrand, *Br. J. Haematol.*, **1986**, *62*, 607.
34. G. J. Kontoghiorghes, *Mol. Pharmacol.*, **1986**, *30*, 670.
35. G. J. Kontoghiorghes, L. Sheppard and S. Chambers, *Arzneim Forsch./Drug Res.*, **1987**, *37*, 1099.
36. J. H. Brock, J. Liceaga, H. M. L. Arthur and G. J. Kontoghiorghes, *Am. J. Hematol.*, **1990**, *34*, 21.

37. R. C. Hider, S. Singh, J. B. Porter and E. R. Huehns, *N.Y. Acad. of Sciences*, **1992**.
38. R. Ma and A. E. Martell, *Inorg. Chim. Acta*, **1994**, *233*, 21.
39. J. B. Porter, E. R. Huehns, R. C. Hider, *Bailliere's Clin. Haematol.*, **1989**, *2*, 257.
40. E. R. Huehns, J. B. Porter and R. C. Hider, *Hemoglobin*, **1989**, *12*, 593.
41. M. Gyparaki, J. B. Porter, S. Hirani, M. Streater, R. C. Hider, and E. R. Huehns, *Acta. Haematol.*, **1987**, *78*, 217.
42. P. S. Dobins and R. C. Hider, *Chemistry in Britain*, **1990**, 565.
43. J. B. Porter, K. P. Hoyes, R. Abeysinghe, E. R. Huehns and R. D. Hider, *Lancet, ii*, **1989**, 156.
44. J. B. Porter, M. Gyparaki, L. C. Burke, E. C. Huehns, P. Sarpong, and R. C. Hider, *Blood*, **1988**. *72*, 1497.
45. R. J. Motekaitis and A. E. Martell, *Inorg. Chim. Acta*, **1991**, *183*, 71.
46. A. E. Martell, R. J. Motekaitis, E. T. Clarke and Y. Sun, *Drugs of Today*, **1992**, *28*, 11.
47. M. Streater, P. D. Taylor, R.C. Hider and J. Porter, *J. Med. Chem.*, **1990**, *33*, 1749.
48. J. B. Porter, L. C. Burke, R. C. Hider and E. R. Huehns, *Br. J. Haematol.*, **1988**, *69*, 88 (abstract). Y. Sun, R. J. Motekaitis and A. E. Martell, to be submitted.
49. M. Wilhelm, D. E. Jager and F. K. Ohnesorge, *Pharmacol. & Toxicol.*, **1990**, *66*, 4.
50. R. B. Martin, *Clin. Chem.*, **1986**, *32*, 1797.
51. R. B. Martin, *Metal Ions in Biol. Systems*, **1988**. *24*, 1.
52. J. L. Cannata and J. L. Domingo, *Veter. Hum. Toxicol.*, **1989**, *31*, 577.
53. B. Fulton and E. G. Jeffrey, *Fund. Appl. Toxicol.*, **1990**, *14*, 788.
54. D. J. Brown, K. N. Ham, J. K. Dawborn and J. M. Xipell, *Lancet, ii*, **1982**, *8294*, 343.
55. J. L. Domingo, M. Gomez, J. M. Llobet and J. Corbella, *J. Toxicol., Clin. Toxicol.*, **1988**, *26*, 67.
56. E. J. Shherrard, J. V. Walker and J. L. Boykin, *Am. J. Kidney Dis.*, **1988**, *12 (2)*, 126.
57. T. E. Lewis, *Evironmental Chemistry and Toxicology of Aluminum*, Lewis Publishers, Inc. Chelsea, MI, 1989.
58. T. P. Coogan, D. M. Latta, E. T. Snow and M. Costa, *CRC Critical Reviews in Toxicology*, **1989**, *19 (4)*, 341.
59. F. W. Sunderman, In: *Nickel in the Human Environment, IARC Monograph Series, #53*, IARC Scientific Publications, Lyon, France, 1988.
60. M. Athar, M. Misra and R. C. Srivastava, *Fund. Appl. Toxicol.*, **1987**, *9*, 26.
61. M. Misra, M. Athar, S. K. Hasan and R. C. Srinivastava, *Fund. Appl. Toxicol.*, **1988**, *11*, 285.
62. S. A. K. Wilson, *Brain*, **1912**, *34*, 295.
63. J. N. Cumings, *Brain*, **1951**, *74*, 10.
64. D. Denny-Brown and H. Porter, *New Eng. J. Med.*, **1951**, *245*, 917.
65. J. M. Walshe, *Lancet*, **1956**, 1, 25.
66. J. M. Walshe and M. Yealland, *Quartl. J. Med.*, **1993**, *86*, 197.
67. I. H. Scheinberg, *ACS Sym. Ser.*, **1980**, *140*, 373.
68. J. R. Duffield and D. M. Taylor, In: *Handbook on the Physics and Chemistry of the Actinides*, A. J. Freeman and C. Keller, Eds., Vol.4, North Holland Elsevier Science Publishers B. V., Amsterdam, 1986, 129.
69. D. M. Taylor, In: *Handbook on the Physics and Chemistry of the Actinides*, A. J. Freeman and C. Keller, Ed. Elsevier Science Publishers, 1990, Vol. 5.
70. C. C. Lushbaugh and L. C. Washburn, *Health Phys.*, **1979**, *36*, 472.
71. P. W. Durbin, N. Jueng, E. S. Jones, F. L. Weitl and K. N. Raymond, *Radiat. Res.*, **1984**, *99*, 85.
72. P. W. Durbin, E. S. Jones, K. N. Raymond and F. L. Weitle, *Radiation Res.*, **1980**, *81*, 170.
73. P. W. Durbin, D. L. White, N. Jeung, F. L. Weitle, L. C. Uhlir, E. C. Jones, F. W. Breunger and K. N. Raymond, *Health Physics*, **1989**, *56*, 839. a) P. W. Durbin, N. Jeung, S. J. Rodgers, P. N. Turowski, F. L. Weitle, D. L. White and K. N. Raymond, *Radiation Protection Dosimetry*, **1989**, *26*, 351.
74. J. F. Markley, *Int. J. Radiat. Biol.*, **1963**, 7, 405
75. R. A. Bulman and R. J. Griffin, *Naturwissenschaften*, **1981**, *67*, 483.
76. M. A. Basinger, R. L. Forti, L. T. Burka, M. M. Jones, W. M. Mitchell, J. E. Johnson and S. J. Gibbs, *J. Toxicol. Environ. Health*, **1983**, *11*, 237.
77. A. Ortega, J. L. Domingo, M. Gomez and J. Corbella, *Pharmacol. Toxicol.*, **1989**, *64*, 147.
78. W. -Z. Chen and Y. -Y. Xie, *Acta Pharmaceutica Sinica*, **1984**, *19*, 865.

79. Y. -Y. Hsieh, W. -Z. Chen, M. -Z. Xu, Z. -Q. Tao, Y. -Y. Liang, X- H. Xu, X. -M. Yan and J. -S. Zhang, *Plzen. Lek. Sbornik, Suppl.*, **1988**, *56*, 59.
80. Z. -Q. Tao, X. -H. Xu, X. -M. Yan, Z. -J. Chen, J. -S. Zhang, Y. -Y. Liang and Y. -Y. Xie, *Plzen Lek., Sborn. Suppl.*, **1988**, *56*, 95.
81. M. Piscator, In: Cadmium, E. C. Foulkes, Ed., Springer-Verlag, Berlin, 1986, p.179.
82. Z. A. Shaikh and L. J. Lucis, *Arch. Environ. Health*, **1972**, *24*, 410.
83. R. von Burg and J. C. Smith, *J. Toxicol. Environ. Health*, **1980**, *6*, 75.
84. M. G. Cherian, *Nature*, **1980**, *187*, 871.
85. M. G. Cherian, S. Onosaka, G. K. Carson and P. A. W. Dean, *J. Toxicol. Environ. Health*, **1982**, *9*, 389.
86. W. Rau and F. Planas-Bohne, *Biol. Trace Element Res.*, **1989**, *21*, 227.
87. G. R. Gale, A. B. Smith and E. M. Walker, Jr., *Ann. Clin. Lab. Sci.*, **1981**, *11*, 476.
88. S. G. Jones, M. M. Jones, M. A. Basinger, L. T. Burka and L. A. Shinobu, *Res. Commun. Chem. Pathol. Pharmacol.*, **1983**, *40*, 155.
89. G. R. Gale, L. M. Atkins, E. M. Walker, Jr., A. B. Smith and M. M. Jones, *Ann. Clin. Lab. Sci.*, **1983**, *13*, 474.
90. S. G. Jones, P. K. Singh and M. M. Jones, *Chem. Res. Toxicol.*, **1988**. *1*, 234
91. S. Kojima, K. Kaminaka, M. Kiyozumi and T. Honda, *Toxicol. Appl. Pharmacol.*, **1986**, *83*, 516.
92. P. K. Singh, S. G. Jones, G. R. Gale, M. M. Jones, A. B. Smith and L. M. Atkins, *Chem. Biol. Interactions*, **1990**, *74*, 79.
93. M. Rubin, S. Gignac, S. P. Bessman and E. L. Belknap, *Science*, **1953**, *117*, 659.
94. C. F. Brownie and A. L. Aronson, *Toxicol. Appl. Pharmacol.*, **1984**, *75*, 167.
95. U. Hofmann and G. Segewitz, *Arch. Toxicol.*, **1975**, *34*, 213.
96. E. Fiedheim, C. Corvi and C. H. Walker, *J. Pharm. Pharmac.*, **1976**, *28*, 711.
97. Z. -F. Xu and M. M. Jones, *Toxicology*, **1988**, *53*, 277. a) J. J. Chisolm, Jr. and D J. Thomas, *J. Pharmacol. Exp. Ther.*, **1985**, *235*, 665.
98. L. Fournier, G. Thomas, R. Garmier, A. Buisine, P. Houze, F. Pradier and S. Dally, *Med. Toxicol.*, **1988**, *3*, 499.
99. H. V. Aposhian, *Ann. Rev. Pharmacol. Toxicol.*, **1983**, *23*, 193.
100. H. V. Aposhian, *Ann. Rev. Pharmacol. Toxicol.*, **1990**, *30*, 279.
101. L. Carlton, R. D. Hancock, H. Maumela and K. P. Wainwright, *J. Chem. Soc. Chem. Commun.*, **1994**, 1007.
102. H. Maumela, R. D. Hancock, L. Carlton, J. Reibenspies and K. P. Wainwright, *J. Am. Chem. Soc.*, in press.
103. J. H. Graziano, D. Cuccia and E. Friedheim, *J. Pharmacol. Exp. Ther.*, **1978**, *207*, 1051.
104. C. H. Tadlock and H. V., *Biochem. Biophys. Res. Commun.*, **1980**, *94*, 501.
105. H. V. Aposhian, C. H. Tadlock and T. E. Moon, *Toxicol. Appl. Pharmacol.*, **1981**, *61*, 385.
106. E. R. Stine, C. -A. Hsu, T. D. Hoover, H. V. Aposhian and D. E. Carter, *Appl. Pharmacol.*, **1984**, *75*, 329.
107. H. V. Aposhian, D. E. Carter, T. D. Hoover, C. -H Hsu, R. M. Maiorino and E. Stein, *Fund. Appl. Toxicol.*, **1984**, *4*, S58.
108. L Magos, In: *Mercury, Handbook on the Toxicity of Inorganic Compounds*, H. G. Seiler, H. Sigel and A. Sigel, Eds., Marcel Dekker, Inc., New York, 1988, 419.
109. F. W. Sunderman, *Ann. Clin. Lab. Sci.*, **1988**, *18*, 89.
110. H. V. Aposhian and M. M. Aposhian, *J. Pharmacol. Exp. Ther.*, **1959**, *126*, 131.
111. L. Magos, *Br. J. Pharmacol.*, 1976, *56*, 478.
112. J. Aaseth, J. Alexander and N. Raknerud, *J. Toxicol. Clin. Toxicol.*, **1982**, *19*, 173.
113. M. E. Lund, T. W. Clarkson and M. Berlin, *J. Toxicol. Clin. Toxicol.*, **1984**, *22*, 31.
114. A. L. Reeves, In: *Handbook on the Toxicology of Metals*, Vol.II. L. Frigerg, G. F. Nordberg and V. B. Vouk, Eds., Elsevier, Amsterdam (1986).
115. A. L. Reeves, *J. Am. Coll. Toxicol.*, **1989**, *8 (7)*, 1307-1313.
116. M McKinney, *J. Toxicol. Environ. Chem.*, **1993**, *38*, 1.
117. C. L. Edward and R. L. Hayes, *J. Nucl. Med.*, **1969**, *10*, 103.
118. M. L. Thakur, M. W. Merrick and W. E. Ganasekera, In *Radiopharmaceuticals and Labelled Compounds*, Vol. 2; IAEA: Vienna, 1973; pp. 183-193.
119. J. P. Lavender, J. Lowe and J. R. Barker, *Br. J. Radiol.*, **1971**, *44*, 361.
120. M. L. Thakur, R. E. Coleman and M. J. Welch. *J. Radiology,* **1976**, *119*, 731.
121. M. L. Thakur, M. J. Welch and. J. H. Joist, *Thromb. Res.*, **1975**, *9*, 345.

122. See for example Zalutsky, M. R. Ed. *Antibodies in Radiodiagnosis and Therapy*, CRC Press; Boca Raton: Florida, 1989.

123. C. A. Burnham, S. Aronow and D. L. Brownell, *Phys. Med. Biol.,* **1970**, *15*, 517.

124. M. M. Ter-Pogossian, M. Phelps and E. Hoffman, *Radiology,* **1975**, *114*, 89.

125. P. V. Harper, *Int. J. Appl. Radiat. Isotop.* **1977**, *28*, 5.

126. M. A. Green, *Int. J. Radiat. Appl. Instrum, Part B,* **1987**, *14*, 59.

127. C. J. Mathias, M. J. Welch, M. R. Raichle, M. A. Mintun, L. L. Lick, A. H. McGuire, K. R. Zinn, E. K. John and M. A. Green, *J. Nucl. Med.*, **1990**, *31*, 351.

128. Y. J. Li, A. E. Martell, R. D. Hancock, J. H. Reibenspies, C. J. Anderson, and M. J. Welch, in press.

129. V. M. Agre, N. P. Kozlova, V. K. Trunov and S. D. Ershova, *Zh. Strukt. Khim.*, **1979**, *22*, 138.

130. U. Bosser, D. Hanke and K. Wieghardt, *Polyhedron*, **1993**, *1*, 12.

131. D. P. Nowotnik, L. R. Canning and S. A. Cumming, *Nucl. Med. Comm.*, **1985**, *6*, 499.

132. S. Vallabhajosula, R. E. Zimmerman and M. Picard, *J. Nucl. Med.*, **1989**, *30*, 599.]

133. K. Linder, T. Feld and P. N. Juri, *J. Nucl. Med.*, **1987**, *28*, 492.

134. M. R. Raichle, J. O. Eichling and M. G. Straatman, *Amer. J. Physiol.*, **1976**, *230*, 543.

135. D. D. Dischino, M. J. Welch and M. R. Kilbourn, *J. Nucl. Med.,* **1983**, *24*, 1030.

136. C. J. Mathias, M. J. Welch and L. Lich, *J. Nucl. Med.*, **1988,** *29*, 747.

137. Y. Sun, C.J. Mathias, M. J. Welch, W. L. Madsen and A. E. Martell, *Nucl. Med. Biol.*, **1991**, *18*, 289.

138. R. J. Motekaitis, Y. Sun, A. E. Martell and M. J. Welch, *Inorg. Chem.*, **1991**, *30*, 2737.

139. E. T. Clarke and A. E. Martell, *Inorg. Chim. Acta*, **1990**, *29*, 1463.

140. D. A. Moore, P. E. Fanwick and M. J. Welch, *Inorg. Chem.*, **1990**, *29*, 672.

141. R. Ma, J. Reibenspies, M. J. Welch and A. E. Martell, *Inorg. Chim. Acta*, **1995**, 236, 75.

142. M. A. Green, *J. Lab. Compd. Radiopharm.*, **1990**, *23*, 1227.

143. H. F. Kung, B. Liu, D. Mankoff, M. P. Kung and J. Billings, *J. Nucl. Med.*, **1990**, *31*, 727.

144. Y. Sun, A. E. Martell, J. H. Reibenspies and M. J. Welch, *Tetrahedron,* **1991**, *47*, 357.

145. R. Delgado, Y. Sun, R. J. Motekaitis and A. E. Martell, *Inorg. Chem.*, **1993**, *32*, 3320.

146. R. Ma and A. E. Martell, *Inorg. Chim. Acta*, **1993**, *209*, 71.

147. Y. Sun, R. J. Motekaitis, A. E. Martell and M. J. Welch, *Inorg. Chim. Acta*, 1995, *228*, 77.

148. Y. Sun and A.E. Martell, *J. Coord. Chem.*, in press.

149. C. J. Anderson, C. S. John, Y. J. Li, R. D. Hancock, T. J. McCarthy, A. E. Martell and M. J. Welch, *Nucl. Med. Biol.*, **1995**, *22*, 165.

150. Y. Sun, C. J. Anderson , T. S. Pajeau, D. E. Reichert, R. D. Hancock, R. J. Motekaitis, A. E. Martell, and M. J. Welch, *J. Med. Chem.*, in press.

151. D. M. Taylor and G. J. Kontoghiorghes, *Inorg. Chim. Acta*, **1986**, *125*, 35.

152. R. C. Scarrow and K. N. Raymond, *Inorg. Chem.*, **1988**, *27*, 4140, and references therein.

153. C. A. Matsuba, W. O. Nelson, S. J. Rettig and C. Orvig, *Inorg. Chem.*, **1988**, *27*, 3935.

154. A. E. Martell, In: *Development of Iron Chelators for Clinical Use*, A. E. Martell, W. F. Anderson and D. G. Badman, Eds., Elsevier/North Holland, New York, **1981**; p.67.

155. A. E. Martell and R. J. Motekaitis, *Proc. Int. Conf. on Oral Chelation in the Treatment of Thalassaemia and Other Diseases*, London, 1989.

156. W. R. Harris and V. L. Pecoraro, *Biochem.*, **1983**, *22*, 292.

157. M. J. Welch and T. J. Welch, In *Radiopharmaceuticals*; G. M. Subramanian, B. A. Rhodes, J. F. Cooper and V. J. Sodd, Eds.; New York Soc. of Nucl. Med., **1975**; pp. 73-79.

158. W. R. Harris, Y. Chen and K. Wein, *Inorg. Chem.*, **1994**, *33*, 4991.

159. M. J. Sundberg, C. F. Meares and D. A. Goodwin, *Nature,* **1974**, *250*, 587.

160. C. F. Meares, D. A. Goodwin and C. S. H. Leung, *Proc. Natl. Acad. Sci. USA,* **1976**, *73*, 3803.

161. G. E. Krejcarek and K. L. Tucker, *Biochem. Biophys. Res. Comm.*, **1977**, *77*, 581.

162. W. C. Eckelman, S. M. Karesh and R. C. Reba, *J. Pharm. Sci.*, **1975**, *64*, 704.

163. A. Najafi, R. L. Childs and D. Hnatowich, J. *Int. J. Appl. Radiat. Isotop.*, **1984**, *35*, 554.

164. C. F. Meares, M. J. McCall and D. R. Redan, *Anal Biochem.,* **1984**, *142*, 68.

165. G. L . DeNardo, S. J. DeNardo and J. S. Peng, *J. Nucl. Med.,* **1985**, *26*, 67.

166. F. L. Otsuka and M. J. Welch, *Int. J. Nucl. Med. Biol.*, **1985**, *12*, 331.

167. S. V. Deshpande, S. J. Denardo, C. F. Meares, M. J. McCall, G. P. Adams and G. L. Genardo, *Int. J. Radiat. Appl. Instrum.*, **1989**, *Part B 16(6)*, 587.

168. C. J. Mathias and M. J. Welch, *J. Nucl. Med.,* **1987**, *28*, 657.

169. M. W. Brechbiel, O. A. Gansow and R. W. Atcher, *Inorg. Chem.,* **1986**, *25*, 2772.

170. P. L. Carney, P. E.; Rogers and K. D. Johnson, *J. Nucl. Med.*, **1989**, *30*, 374.
171. C. J. Mathias, Y. Sun, J. M. Connet, G. W. Philpott, M. J. Welch and A. E. Martell, *Inorg. Chem.*, **1990**, *29*, 1475.
172. C.J. Mathias, Y. Sun, M. J. Welch, G. W. Philpott and A. E. Martell, *Bioconjugate Chem.*, **1990**, *2*, 204.
173. S. W. Schwarz, C. J. Mathias, Y. Sun, W. G. Dilley, S. A. Wells, A. E. Martell and M. J. Welch, *Nucl. Med. Biol.*, **1991**, *18*, 477.
174. M. W. Brechbiel, T. J. McMurry and O. A. Gansow, *Tetrahedron Lett.*, **1993**, *34*, 3691.
175. W. Bauer, U. Brimer, W. Doepfner, R. Haller, R. Huguenin, P. Marbach, T. J. Petcher and J. Pless, *Life Sciences*, **1982**, *31*, 1133.
176. W. H. Bakker, R. Albert, C. Bruns, W. A. P. Breeman, L. J. Hofland, P. Marbach, J. Pless, D. Pralet, B. Stolz, J. W. Roper, S. W. J. Lamberts, T. J. Biscer and E. P. Krenning, *Life Sciences*, **1991**, *49*, 1583.
177. S. R. Cooper, Paper No. 186, American Chemical Society, Division of Inorganic Chemistry, 208th ACS National Meeting, Washington, D.C., August 20-25, 1994.
178. J. Baldas, J. R. Boas, J. Bounyman and G. A. Williams, *J. Chem. Soc. Dalton Trans.*, **1984**, 2395.
179. L. A. Learie, R. C. Holtiwanger and C. G. Pierpont, *J. Am. Chem. Soc.*, **1989**, *111*, 4324.
180. M. Kawishima, M. Koyama and T. Fuginaja, *J. Inorg. Nucl. Chem.*, **1976**, *38*, 801.
181. B. Johannsen and H. Spies, *Akadimie der Wissenchaften der DDR, Central Institut fur Kernforschung DDR-5051, Dresden*, **1981**, 122.
182. K. E. Linder, A. Davison, J. C. Dewan, C. E, Costello and S. Maleknia, *Inorg. Chem.*, **1986**, *25*, 2085).
183. E. N. Treher, L. C. Francesconi, J. Z. Gougoutas, M. R. Malley and A. D. Nunn, *Inorg. Chem.*, **1989**, *28*, 3411.
184. G. Bandoli, M. Nicolini, U. Mazzi, H. Spies, and R. Munze, *Trans. Met. Chem.*, **1984**, *9*, 127.
185. K. Libson, E. Duetsch and B. L. Barnett, *J. Am. Chem. Soc.*, **1980**, *102*, 2476.
186. A. Davison, A. G. Jones, C. Orvig and M. Sohn, *Inorg. Chem.*, **1981**, *20*, 1629.
187. S. A. Zuckman, G. M. Freeman, D. E. Groutner, *et al.*, *Inorg. Chem.*, **1981**, *20*, 2386.
188. D. Brenner, A. Davison, J. Lister-Jones and A. G. Jones, *Inorg. Chem.*, **1984**, *23*, 3793.
189. S. Jurisson, L. F. Lindoy, K. P. Dancey, *et al.*, *Inorg. Chem.*, **1984**, *23*, 227.
190. S. Jurisson, E. O. Schlemper, D. E. Troutner, L. R. Canning, D. P. Nowotnik and R. D. Neirinckx, *Inorg. Chem.*, **1986**, *25*, 543.
191. S. Jurisson, E. A. Schlemper, D. E. Troutner, L. R. Canning, L. R.; D. P. Nowotnik and R. D. Neirinckx, In: *Technetium in Chemistry and Nuclear Medicine 2*; G. Bandoli and U. Mazzi, Eds., Cortina International, Verona, Italy, 1987, 37.
192. S. Jurisson, K. Aston, C. Fair, *et al.*, *Inorg. Chem.*, **1987**, *26*, 3576.
193. X. Marchi, P. Garuti, A. Duatti, *et al.*, *Inorg. Chem.*, **1990**, *29*, 2091.
194. S. F. Colmanet and M. F. Mackay, *Aust. J. Chem.*, **1988**, *41*, 269.
195. J. E. Ferguson and J. H. Hickford, *J. Inorg. Nucl. Chem.*, **1966**, *28*, 2293.
196. U. Mazzi, E. Roncair, G. Bandoli and L. Magon, *Transit. Met. Chem.* (Weinheim, Ger.), **1979**, *4*, 151.
197. R. M. Pearlstein, W. M. Davis, A. G. Jones and A. Davison, *Inorg. Chem.*, **1989**, *28*, 3332.
198. R. C. Elder, R. Whittle, K. A. Glavan, *et al.*, *Acta Crystallog.*, **1980**, *B36*, 1662.
199. K. Libson, B. L. Barnett and E. Deutsch, *Inorg. Chem.*, **1983**, *22*, 1695.
200. K. E. Linder, M. F. Malley, J. Z. Gougouts, S. E. Unger and A. D. Nunn, *Inorg. Chem.*, **1990**, *29*, 2428.
201. G. M. Archer and J. R. Dilworth, *J. Nucl. Med.*, **1989**, *30*, 939.
202. T. H. Tulip, J. Calabrese, J. F. Kronange, A. Davison and A. G. Jones, In: *Technetium in Chemistry and Nuclear Medicine 2*; G. Bandoli and U. Mazzi, Eds., Cortina International, Verona, Italy, 1987, 119.
203. E. O. Fischer and M. W. Schmidt, *Chem. Ber.*, **1969**, *102*, 1954.
204. K. Schwochau, In: *Technetium in Chemistry and Nuclear Medicine*, Vol. 2, p.13; M. Nicolini, L. G. Bandoli and U. Mazzi, Raven Press: New York, 1986.
205. J. Steigman and W. C. Eckelman, *The Chemistry of Technetium in Medicine*, National Academy Press, Washington, DC, 1992.
206. B. W. Wessel and R. D. Rogus, *Med. Phys.*, **1984**, *1*, 638.
207. S. M. Quadri and B. W. Wessel, *Nucl. Med. Biol.*, **1986**, *13*, 447.

208. W. A. Volkert, D. E. Troutner and R. A. Holmes, *Int. J. Appl. Radiat. Isotop.*, **1982**, *33*, 891.

209. M. Nicolini, G. Bandoli and U. Mazzi, *Technetium in Chemistry and Nuclear Medicine, Vol. 2*; Raven Press: New York, 1986; p.169.

210. E. Deutsch, K. Libson, J. -L. Vanderheyden, A. R. Ketring and H. R. Maxon, *Nucl. Med. Biol.*, **1986**, *13*, 465.

211. H. L. Maxon, S. R. Thomas, V. S. Hertzberg, L. E. Schroder, E. A. Deutsch, K. F. Libson, R. C. Samaratunga, C. C. Williams and J. S. Moulton, *J. Nucl. Med.*, **1989**, *30*, 837.

212. A. R. Fritzberg, J. -L. Vanderheyden, T. N. Rao, S. Kasina, D. Eshima and A. T. Taylor, *J. Nucl. Med.*, **1989**, *30*, 743.

213. J. A. Koutcher and C. T. Burt, *J. Nucl. Med.*, **1984**, *25*, 371.

214. J. A. Koutcher, C. T. Burt, R. B. Lauffer and T. J. Brady, *J. Nucl. Med.*, **1984**, *25*, 506, and references therein.

215. R. J. Alfidi, J. R. Haaga, Jr. and S. J. El Yousef, *Radiology*, **1982**, *142*, 175.

216. R. A. Dwek, A. *Nuclear Magnetic Resonance in Biochemistry. Applications to Enzyme Systems*; Oxford Clarendon Press: 1973; Chapters 9-11.

217. D. R. Burton, S. Forsen and G. Karlstrom, Prog. *NMR Spectroscopy*, **1979**, *13*, 1.

218. T.-Z. Jin, S.-F. Zhao, G.-X. Xu, Y.-Z. Han, N.-C. Shi, Z.-S. Ma, *Acta Chim. Sinica (Chin)*, **1991**, *49*, 569.

219. C. A. Chang, L. C. Francesconi, M. F. Malley, K. Kumar, J. Z. Gougoutas, M. F. Tweedle, D. W. Lee and L. J. Wilson, *Inorg. Chem.*, **1993**, *32*, 3501.

220. L. K. Templeton, D. H. Templeton, A. Zalkin, and H. W. Rubin, *Acta Crystallogr. Sect. B.*, **1982**, *B38*, 2155.

221. R. C. Barash, *Radiology*, **1983**, *147*, 781.

222. M. A. Mendonca-Dias, E. Gaggelli and P. Lauterbur, *Semin. Nucl. Med.*, **1983**, *12*, 364.

223. V. M. Runge, J. A. Clanton and D. M. Lukehart, *An. J. Roent.*, **1983**, *141*, 1209.

224. R. B. Lauffer, A. C. Vincent, S. Padmanabhan, A. Villringer, W. L. Greig, D. Elmaleh and T. Brady, *J. Inorg. Paper 0087*, 192nd ACS National Meeting, Anaheim, Sept. 7-12, 1986.

225. R. B. Lauffer, A. C. Vincent, S. I. Padmanabha and J. Meade, *J. Am. Chem. Soc.*, **1987**, *109*, 2216.

226. K. Kumar, C. A. Chang and M. F. Tweedle, *Inorg. Chem.*, **1993**, *32*, 587.

227. S. I. Kang, R. S. Ranganathan, J. E. Emswiler, K. Kumar, J. Z. Gougoutas, M. F. Malley and M. F. Tweedle, *Inorg. Chem.*, **1993**, *32*, 2912.

228. Z. Zhang, C. A. Chang, H. G. Brittain, C. M. Garrison, J. Tesler and M. F. Tweedle, *Inorg. Chem.*, **1992**, *31*, 5597.

229. R. B. Lauffer, W. L. Greif, D. D. Stark, A. C. Vincent, S. Saini, V. Wedeen and T. J. Brady, *J. Comput. Assist. Tomogr.*, **1985**, *9*, 431.

230. B. L. Engelstad, D. I. White, E. C. Ramos, T. R. Johnson, H. A. Macapinlac and M. E. Moseley, *Abstracts, Soc. Mag. Res. in Med.* 5th Annual Meeting, Aug. 19-22, 1986, SMRM Program, p.175.

231. F. Shtern, L. Garrido, C. Compton, J. K. Swiniarski, R. B. Buxton, R. B. Lauffer and T. Brady, *J. Soc. Magn. Res. in Med.*, 6th Annual Meeting, New York, 1987, SMRM Program, p.175.

232. E. C. Weiner, M. W. Brechbiel, H. Brothers, R. L. Magin, O. A. Gansow, D. A. Tomalia, P. C. Lauterbur, *Mag. Res. Med.*, **1994**, *31*, 1-8.

Chapter 6

The Selectivity of Ligands of Biological Interest for Metal Ions in
Aqueous Solution. Some Implications for Biology

In Chapters 1 to 4, the factors that control selective complexation of metal ions were discussed. Here factors that control strength of metal-ligand interactions as these might relate to metal ions in biology are examined. It is not possible here to give an exhaustive account of selective complexation of metal ions in biology. Rather, factors that control selectivity are examined as they relate to selected examples in biology, and it is hoped that readers will be able to identify these same principles acting in their own area of interest in metal ions in biology.

6.1   Significance of HSAB Ideas for Zinc-Containing Metalloenzymes

Biological processes use metals such as Mg, Ca, Zn, Fe, Cu, Co, Ni, and Mn, with Mo being the only metal used from the second row of transition elements.[1,2] How, then, is the chemistry of these metal ions altered to give required bonding properties? For example, Zn^{2+} as the aquo ion has a modest affinity for hydroxide ion, with log K_1 OH^- being[3] only 4.6. It occurs in many metalloenzymes such as carbonic anhydrase or alcohol dehydrogenase,[4] where Zn(II) has a log K_1 with hydroxide ion of over 6.0. This increase in acidity of Zn(II) is important for it to function at biological pH. Replacement of Zn(II) with less acidic Cd(II) (log K_1 OH^- = 3.8[3]) produces enzymes that function, but only at higher pH.[4] A further alteration of ligand binding properties of Zn(II) in metalloenzymes is that, whereas the Zn^{2+} aquo ion is classified as borderline[5] in the HSAB classification, and has log K_1 values as shown below, in carbonic anhydrase[6] the order is reversed to F^- < Cl^- < Br^- < I^-, so that Zn(II) in this metalloenzyme would be classified as a soft Lewis acid:

log K_1	F^-	Cl^-	Br^-	I^-	OH^-
$Zn^{2+}(aq)^a$	1.15	0.43	0.03	-0.9	4.6
Zn(II) in carbonic anhydrase^b	-0.07	0.72	1.19	2.07	6.2

^{a} Ref. 3. ^{b} For bovine carbonic anhydrase, from Ref. 6.

Binding in the metalloenzyme has altered the Lewis acid properties of Zn(II) so that it is a stronger acid, and has become soft instead of intermediate in HSAB. This seems to relate to the coordination geometry and coordination number forced on Zn(II) by the enzyme binding site, since these properties have been replicated in simple model Zn(II) complexes.[7,8] There may also be[1] a contribution from the hydrophobic cavity in which the Zn(II) occurs in these metalloenzymes to the "soft" behavior of the Zn(II), since this would make the Zn(II) more like a metal ion in the gas phase. However, duplication of the soft behavior in model compounds with no hydrophobic cavities indicates that it must be largely a function of the bonding properties of the Zn(II) itself. Also important is that Zn(II) in alcohol dehydrogenases appears to bind alcohols.[9] Ordinarily, metal ions bind alcohols only weakly in aqueous solution, so how does Zn(II) bind alcohols in these metalloenzymes?

An insight into metalloenzyme action is provided[10] by the work of Chung,[11-13] in studies of binding of ligands to Cu(II) complexes of the macrocycle tetb, discussed in

tetb

section 2.4. As with $Zn^{2+}(aq)$, $Cu^{2+}(aq)$ has low affinity for halide ions,[3] and is classified[5] as intermediate in HSAB. On incorporation into tetb, binding constants for simple unidentate ligands at the axial site of the complex change enormously, and Cu(II) in tetb is soft[9] in HSAB, as seen in Table 6.1.

Table 6.1. Formation constants[3,11-13] for $Cu^{2+}(aq)$ and $[Cu(tetb)]^{2+}$ with a selection of ligands, illustrating the softness[10] of $[Cu(tetb)]^{2+}$ in the HSAB classification,[5] which goes hand-in-hand with increased tolerance for bulky ligands, whether with large donor atoms, such as I^-, or bulky substituents, such as $NH(CH_3)_2$.

log K_1 for	Cl^-	Br^-	I^-	NH_3	$NH_2CH_2CH_3$	$NH(CH_3)_2$
$Cu^{2+}(aq)$	0.4	-0.03	<-1	4.04	3.2^a	no complex
$[Cu(tetb)]^{2+}$	1.19	1.26	1.35	1.95	2.84	2.69

^{a} Estimated by comparison with Ni^{2+} complex.

The change of Lewis acidity of $[Cu(tetb)]^{2+}$ relative to $Cu^{2+}(aq)$ resembles the change in Lewis acidity observed for passing from Zn^{2+}(aq) to Zn(II) in carbonic anhydrase. The D parameters in equation 2.8 have been interpreted[14] as steric effects on ligands with large donor atoms in tight coordination spheres of metal ions such as $Cu^{2+}(aq)$.   An observation regarding HSAB theory in section 2.4 is that metal ions that are soft in aqueous solution must be able to form covalent M-L bonds, and also have a loose enough coordination sphere to tolerate potentially adverse steric effects such as bulky donor atoms or substituents on a ligand. Thus, in common with only very large metal ions such as Ag^+ and Pb^{2+}, $[Cu(tetb)]^{2+}$ as a Lewis acid complexes the large I^- ion, and also complexes the very sterically hindering $N(CH_3)_3$ ligand strongly. Small metal ions such as $Zn^{2+}(aq)$ have almost no affinity for either I^- or $N(CH_3)_3$ in aqueous solution.[3] To produce a soft Zn(II) ion in zinc metalloenzymes, the coordination number of Zn(II) must drop from six in the aquo ion[15] to four in alcohol dehydrogenase or carbonic anhydrase. Lowered coordination number firstly decreases steric crowding around the metal ion, and makes way for bulky donors such as I. Also, tolerance for bulky substituents, like the ethyl group on ethanol, is increased. This could be an important aspect of increasing strength of binding of alcohols, whether as neutral molecules or alkoxides, to Zn(II) in alcohol dehydrogenase. Figure 2.4 shows that in the gas phase alcohols are stronger Lewis bases than the water molecule. In water they are weaker bases than water itself because bulky alkyl groups cause steric problems with most Lewis acids. By adopting tetrahedral geometry in metalloenzymes, Zn(II) can coordinate more effectively with alcohols, and other bulky substrates, than most ordinary metal aquo ions. This is due to lowered steric crowding, and also, as a second contribution, the ability to form covalent bonds should be higher for four coordinate than six coordinate zinc.

Vallee and Williams[16] proposed that in the "entatic state" metal ions are forced to adopt unusual coordination numbers and geometries to produce properties required for specific functions of enzymes in which they occur. This appears to be particularly true for Zn(II). There are several examples[4] where Zn(II) ions are forced into low coordination numbers where they occur in metalloenzymes, with geometries approximating to tetrahedral, or trigonal bipyramidal, rather than octahedral. Achievement of stronger Lewis acidity, as well as softer acid character, and hence reduction in steric hindrance at the coordination site on the metal, may be important in binding bulky organic substrates. Some examples of this type of entatic state are shown in Figure 6.1. Enzymes containing

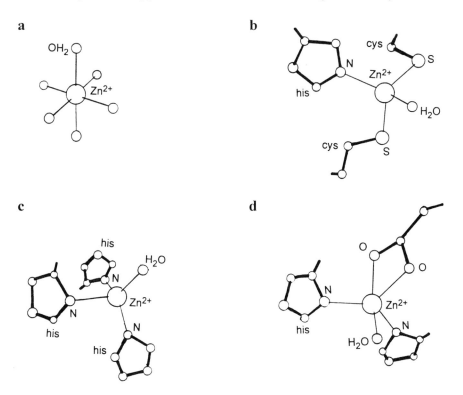

Figure 6.1. Examples of coordination geometry of metal ions in the entatic state[4,16] where the Lewis acidity of the metal ion in its metalloprotein is greatly altered as compared with that for the aquo ion. a) the Zn^{2+} aquo ion b) Zn(II) in alcohol dehydrogenase c) Zn(II) in carbonic anhydrase d) carboxypeptidase A. Redrawn after Ref. 4.

Zn(II) in a low coordination number to enhance Lewis acidity occur in alcohol dehydrogenase, carbonic anhydrase, and in various peptidases, ribonucleases, and snake venoms. This type of entatic state may be more difficult to achieve for other metal ions, possibly accounting for how often Zn(II) occurs in metalloenzymes. In addition, that octahedral Zn(II) is a weaker and harder Lewis acid than tetrahedral Zn(II) means that the Lewis acid Zn(II) in alcohol dehydrogenase can be "switched off"[17] when the required coenzyme NAD^+ is not bound to the protein. Binding of NAD^+ to the protein causes a conformational change that promotes the octahedral-tetrahedral transition for Zn(II) in the metalloenzyme.

It is not clear how important assumption of low coordination number is for other metal ions in metalloenzymes in order to promote catalytic reactions. Copper proteins like hemocyanin,[18] and possibly iron non-heme proteins like nitrogenase[19] appear to achieve unusual Lewis acidity by adopting unusual coordination geometries (Figure 6.2). Copper hemocyanin has been successfully modeled by Kitajima et al.[20] Cu(II) in a tetrahedral environment appears reluctant to stay that way. Thus, the tetrahedral structure of [Cu(II)(TPB)Cl] (TPB = trispyrazolyl borate) is maintained[20] only in the solid state or solvents of low dielectric constant. In a polar solvent like DMF, a five coordinate square pyramidal structure is assumed.[20] This may be one reason why Cu(II) is not widely used in a strong Lewis acid role as is Zn(II).

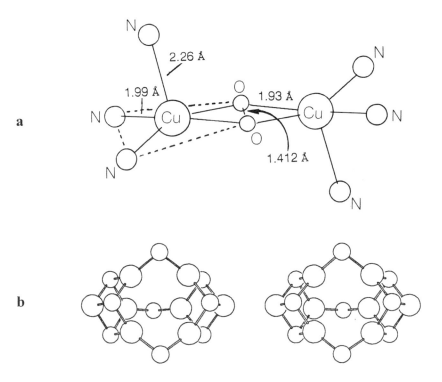

Figure 6.2. Structure of a) the bis[copper(II) tris-pyrazolylborate] complex of Kitajima et al.[20] which appears to be a good model for the coordination of dioxygen in hemocyanin. b) Stereoview of the FeMo cluster[19] which is the proposed binding site for dinitrogen in nitrogenase. Small circles are S atoms, large are Fe, except for terminal atom on right of drawing which is Mo. Redrawn after Refs. 19 and 20.

6.2 Chelate Ring Size and Metal Ion Selectivity

Chelate ring size and metal ion selectivity has been discussed in Chapters 3 and 4. As pointed out, the geometry of chelate rings is such that (Figure 6.3) five membered chelate rings incorporate large metal ions (M-L length ~ 2.5 Å) with small L-M-L angles (~70°) with least steric strain, while six membered chelate rings best incorporate small metal ions (M-L ~1.6 Å) with large (~109.5°) L-M-L angles. As M-L bond length and coordination geometry of the metal ion deviates from the requirements of these two different size chelate rings, so steric strain rises, and complexes are destabilized. Figure 6.3 shows ideal geometry for four membered chelate rings involving naphthyridine and carbonate.[30] Four membered chelate rings require very large metal ions, with even higher coordination number to produce the small O-M-O bond angles required.

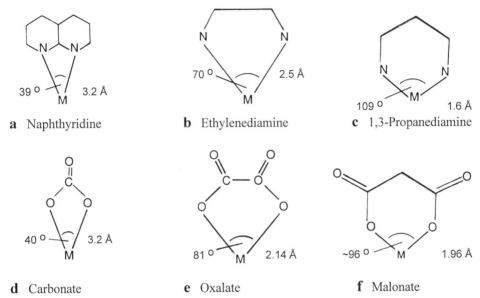

a Naphthyridine **b** Ethylenediamine **c** 1,3-Propanediamine

d Carbonate **e** Oxalate **f** Malonate

Figure 6.3. The size of metal ion, and L-M-L bond angles, that produce minimum strain when coordinated in chelate rings of different sizes, as calculated from molecular mechanics.[30] The calculations show that four membered chelate rings for both a) nitrogen donors and d) oxygen donors, require longer M-L bond lengths than observed in any real metal ion, and so the values given refer to the longest M-O or M-N bond lengths observed in complexes, which are about 3.2 Å. For six membered chelate rings with c) nitrogen donors or negatively charged oxygen donors (f), metal ions with shorter bond lengths and much larger L-M-L angles produce complexes of lower strain than is the case for the analogous five membered chelate rings (b) and (e).

Structures of coordination compounds suggest that smaller chelate rings promote higher coordination numbers. Twelve coordination[21] is always associated with four membered chelate rings, as in the complexes $[Ce(NO_3)_6]^{2-}$ and $[Pr(naphth)_6]^{3+}$, where naphth is 1,8-naphthyridine. Part of what determines coordination number may be steric strain that would result from mismatch between metal ion size and coordination number on the one hand, and chelate ring size on the other. With six membered chelate rings a metal ion should adopt its lower accessible coordination numbers, while with four membered chelate rings higher accessible coordination numbers should be preferred.

The effect of chelate ring size on metal ion selectivity may be discerned in biology with the Ca^{2+} ion. Selective binding of Ca^{2+} rather than Mg^{2+} may be essential in calcium proteins such as annexin[22] parvalbumin[23] or calmodulin.[24] In the binding sites in annexin and calmodulin (Figure 6.4) Ca(II) is coordinated to several neutral oxygens, and a bidentate carboxylate from an aspartate. The Ca^{2+} occupying the "EF hand" site in parvalbumin is bound to two[23] bidentate carboxylates. The four membered ring in the bound carboxylate should favor binding of the large Ca^{2+} over the small Mg^{2+} ion. How the neutral oxygen donors in these sites produce selectivity for large relative to small metal ions is discussed in section 2.5.1.

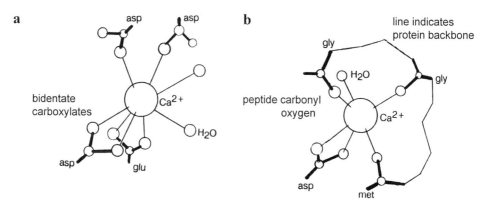

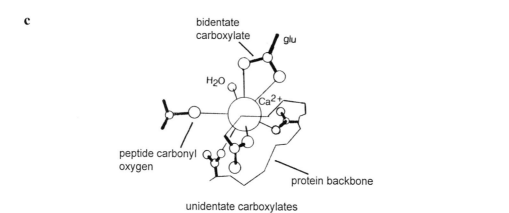

Figure 6.4. Drawings of the binding sites of calcium in a) parvalbumin,[23] b) annexin[22] and c) calmodulin.[24] Note bidentate coordination of carboxylates to give four membered chelate rings. All the donor atoms coordinated to the calcium are oxygens. Asp = aspartate, glu = glutamate. Redrawn after Refs. 22-24.

Proteins, as opposed to simple chelating ligands, are well suited to orient carboxylates to favor chelation. Carboxylates need long connecting chains to other donor atoms in a ligand if they are to twist around enough to be correctly oriented for binding in the chelating mode. Usually, long flexible connecting chains lead to weak complexes, but proteins can provide rigid structures where carboxylates can be well spaced but correctly oriented. In simple chelating ligands, occurrence of carboxylate as part of a small chelate ring, e.g. carboxylates in EDTA, strongly favors only unidentate coordination of carboxylate.

The chelate ring size rule can have structural effects as well as effects on thermodynamic stability in aqueous solution. An example is coordination of metal ions by sugars.[25] The cyclic polyol *cis*-inositol can coordinate metal ions in two distinct ways (Figure 6.5).[26] In *ax-ax-ax* bonding (Figure 6.5), the metal ion is part of three fused six-membered chelate rings. Alternatively, in *ax-eq-ax* coordination, the metal ion is part of two fused five membered and one six membered chelate rings. Angyal has noted[25] that metal ions of radius more than 0.8 Å adopt the *ax-eq-ax* structure, while with an ionic radius less than 0.8 Å the *ax-ax-ax* structure is adopted. The chelate ring size rule indicates why this should occur.[26] Large metal ions chelate with least steric strain as part of five membered chelate rings, and the *ax-eq-ax* structure maximizes the number of five membered chelate rings of which the large metal ion is a part. Small metal ions adopt the *ax-ax-ax* structure, as this maximizes the number of six membered chelate rings involving the metal ion. Other examples of the chelate ring size rule occur in structures[27,28] of complexes of citrate ion. Tridentate coordination of citrate (Figure 6.6) with metal ions can lead to two six membered chelate rings, or one five membered and one six membered chelate ring. The Cambridge crystallographic data base reveals that nearly all metal ions adopt the structure that gives one five membered and one six membered chelate ring, shown for the large Bi(III) ion[27] in Figure 6.6. Only the very small Al(III) shows a structure with two six membered chelate rings[28] (Figure 6.6), in accord with the chelate ring size rule.[29,30] Unpublished MM calculations by the present authors show that only metal ions of M-O bond length about 1.90 Å or less should favor the form of citrate complex with two six membered chelate rings, compared to the structure with one five and one six membered chelate ring. The average Al-O bond length in the Al(III) citrate complex is[28] 1.91 Å, and in its structure are present examples of coordination of Al(III) to give both six,six and five,six chelate ring sizes.

The enzyme aconitase[31,32] contains a Fe_4S_4 cluster which catalyzes the conversion of citrate to *iso*-citrate[1] via aconitic acid. The crystal structure of aconitase[31] with *iso*-citrate bound to an Fe of the Fe_4S_4 cluster leads to the geometry around Fe shown in Figure 6.6. The *iso*-citrate is bound so as to form a five membered chelate ring. The Fe in aconitase has long Fe-O bonds to citrate, averaging[31] 2.3 Å in length, so that a five membered chelate ring would be expected. One might speculate on how the enzyme might drive the

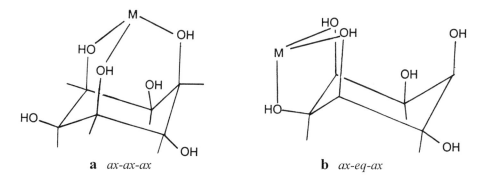

Figure 6.5. a) the *ax-ax-ax* structure adopted by small metal ions, and b) the *ax-eq-ax* type of structure adopted by large metal ions in their complexes with cis-inositol.[25,26]

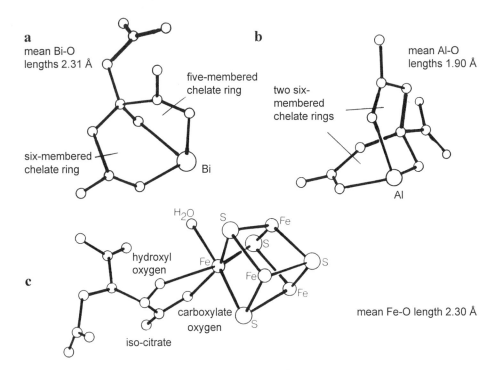

Figure 6.6. Crystal structures of citric acid complexes of a) Bi(III)[27] and b) Al(III).[28] The structures show[26] that, in accord with chelate ring size rules,[30] the large Bi(III) ion adopts the structure with one five membered and one six membered chelate ring, while the small Al(III) ion adopts a structure which gives it two six membered chelate rings. In c) is shown isocitrate bound to iron in the active site of aconitase,[31] with formation of a five membered chelate ring with the iron.

switch from citrate to *iso*-citrate, which might involve the formation of a six membered chelate ring in the product. Possibly[10] an electron shift within the Fe_4S_4 cluster could occur, which would make the Fe to which the citrate binds more like an Fe(III) ion. This would decrease the Fe-O bond lengths, thus favoring the formation of the six membered chelate ring. Thus the sequence of reactions indicated by Figure 6.7 would take place. The deprotonated hydroxyl group in 6.7c would be protonated again in the Fe(II) bound isocitrate, illustrated by Figure 6.6c.

The basis of the chelate ring size rule is that orientation of lone pairs on donor atoms is important. The entatic state hypothesis[16] proposes that the coordination geometry of metal ions may be distorted in metalloenzymes to produce required properties. MM calculations[29] suggest that bond length deformation by external steric factors is more difficult than bond angle deformation. The observation[33] that bond lengths in carboxypeptidase A substituted with the metal ions Co(II), Mn(II), Cd(II) or Cu(II) all appear normal is not unexpected. For example, the M-O bond length to the same glutamate carboxylate oxygen donor varies from 2.17 to 2.61 Å with these different metal ions, depending on the bond lengths required by the metal ion. This shows how the metal ion can distort its environment in the protein to suit its bond length needs. However, all the metal ions studied[33] have greatly distorted coordination geometry, showing that bond angles around metal ions are more easily distorted than bond lengths. This indicates that the entatic state should more readily involve distortion of bond angles around metal ions than bond lengths.

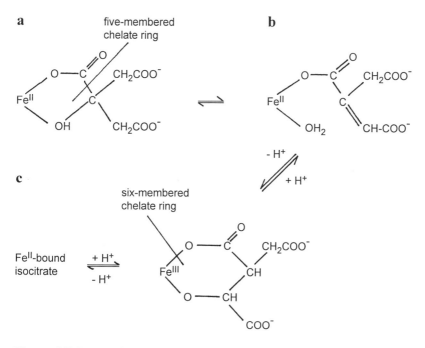

Figure 6.7. Proposed mechanism of conversion of citrate to isocitrate by aconitase. a) citrate coordinated to Fe in aconitase, b) conversion to aconitate by removal of water, and c) conversion to isocitrate by hydrolysis.

6.3   The Neutral Oxygen Donor

Coordinating properties of neutral oxygen donors have been discussed in chapter 2.5.1. Here the coordinating properties are related to the occurrence of this donor group in systems of biological interest. The neutral oxygen donor is of particular interest, since it is the donor atom of water, the solvent in biological systems. Figure 2.4 shows that ligands containing neutral oxygen donors are not all of equal donor strength in the gas phase. Firstly, there is an increase in basicity as alkyl substitution increases along the series H_2O < CH_3OH < $(CH_3)_2O$. Ether oxygens are, for example, of considerable importance in monensin type antiobiotics[34] (Figure 6.8), which selectively complex Na^+. The other neutral oxygen donor of biological interest is the carbonyl group as found in esters or amides. The carbonyl oxygen of the amide group is commonly found as a donor for binding calcium in proteins,[35] and potassium in enniatin,[36] while carbonyls of ester groups commonly occur in small biological molecules such as valinomycin[36] that binds K^+ (Figure 6.8). Studies of crown ethers, and many ligands of biological interest, show that some metal ion types complex well with ligands containing neutral oxygen donors, while others do not. As discussed in sections 2.5.1 and 4.6, the determining factor on whether a metal ion will complex well with ligands containing neutral oxygen donors such as crown ethers, is metal ion size. Only metal ions with an ionic radius (r^+) above 1.0 Å [37], whether these are alkali metal ions (e.g. Na^+, r^+ = 1.00 Å,[37] K^+, 1.38 Å), alkali earth metal ions (Ca^{2+}, $r^+ = 1.00$ Å,[37] Sr^{2+}, 1.18 Å. Ba^{2+}, 1.36 Å), or heavy post-transition metal ions (Pb^{2+}, $r^+ = 1.19$ Å,[37] Hg^{2+}, 1.13 Å, Tl^+, 1.50 Å), bind significantly with crown ethers in aqueous solution. What may partly cause neutral oxygen donors to promote complex stability of larger metal ions, and destabilize complexes of smaller metal ions, is[30] that in virtually all ligands in which neutral oxygen donors are present, they are part of five membered chelate rings. A further factor that may be important for neutral oxygen donors that are alcohols, ethers, or derived from carbonyl groups, may be the reduced ability, compared with water, to hydrogen bond with solvent water, and so disperse charge from the metal ion to the solvent, which should be more important for smaller more highly charged metal ions. Thus, a trivalent metal ion such as La^{3+}, which has the required ionic radius of 1.03 Å,[37] does not[38] appear to form complexes with crown ethers of any stability in aqueous solution. It seems likely that an even larger trivalent metal ion would form stable complexes with crown ethers in aqueous solution, and this hypothesis could be tested with Ac^{3+}, which is the largest trivalent metal ion with an ionic radius of 1.12 Å.[37]

a   Monensin

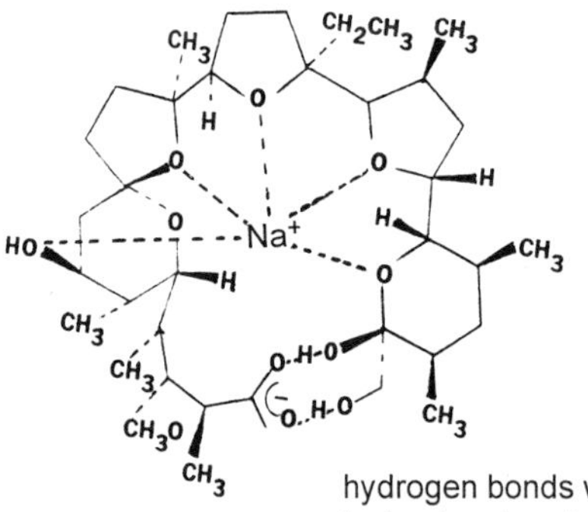

donor atoms are ethereal and alcoholic oxygens. The ionized carboxylate does not coordinate

hydrogen bonds with ionized carboxylate form a pseudo-macrocyclic ring

b   Valinomycin 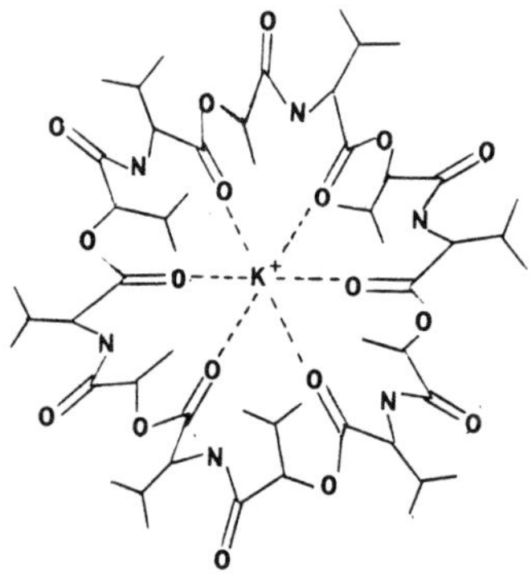   **c**   Enniatin-B

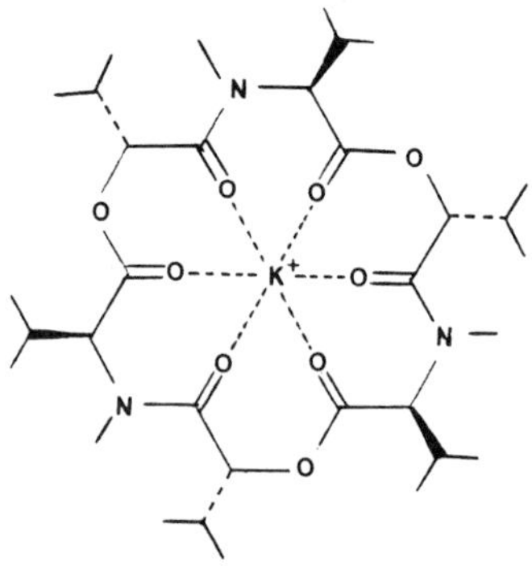

donor atoms are ester carbonyl oxygens   donor groups are alternating amide and ester carbonyl oxygens

Figure 6.8. The ligands a) monensin, which binds Na^+ selectively, and b) enniatin, and c) valinomycin, which bind K^+.

Selectivity of ion channels for metal ions is currently of considerable interest,[39] since an understanding of the factors controlling selectivity would aid in understanding conduction of nervous signals, and drug action. The fact that peptides that act as potassium channels appear[40-43] to contain few polar side groups is perhaps surprising. It has been suggested[40-43] that the selectivity that potassium ion channels display toward K^+ derives from a π-interaction between the phenyl side chains present in the region of the protein that generates the selectivity, and the K^+. This suggestion would need a crystal structure of a channel-forming protein to investigate it further, to see whether these phenyl groups were directed into the ion channel. However, Figure 6.9 suggests an alternate explanation for the potassium selectivity. The ability of metal ions to permeate through potassium channels is shown as a function of log K_1 with 18-crown-6[3] in Figure 6.9. Figure 6.9 suggests that the groups that metal ions might bind to in potassium channels have some resemblance to ethereal oxygens of 18-crown-6. As indicated above, the sequences of peptides from potassium ion channels have few oxygen donor side groups.[40-43] It seems possible that the neutral oxygen donors in the ion channels derive from carbonyl oxygens of peptide amide groups, oriented into the channel so as to coordinate selectively with ions passing through the channel. Figure 6.9 offers an explanation of the strong selectivity of potassium ion channels against Ca^{2+}, which has a low log K_1 with 18-crown-6. An interesting aspect of Figure 6.9 is the high log K_1 which

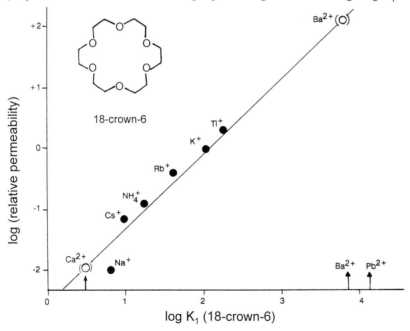

Figure 6.9. Correlation between permeability of the potassium ion channels to univalent metal ions and formation constant of the same metal ions with the crown ether 18-crown-6 (shown on Diagram). Data for permeability from Ref. 39 for potassium channels of giant squid axon. The low log K_1 for Ca^{2+} with 18-crown-6 suggests an origin of the selectivity of potassium ion channels against Ca^{2+}. The high log K_1 of 18-crown-6 with Ba^{2+} suggests why Ba^{2+} is able to block potassium ion channels. The high log K_1 for Pb^{2+} with 18-crown-6 suggests that Pb^{2+} should also be able to block potassium ion channels. Redrawn after Ref. 10.

Ba^{2+} has for 18-crown-6. For optimal transport the distribution coefficient of the metal ion into the organic phase should not be too low, since this leads to slow rates of transfer from the aqueous to the organic phase. The coefficient of distribution of the metal ion into the organic phase should also not be too high, as this leads to the metal ion sticking in the organic phase, with passage through the membrane being again slow. Thus, in line with the high log K_1 of Ba^{2+} with 18-crown-6, a single Ba^{2+} ion can block the passage of K^+ through potassium ion channels for several seconds at a time.[40-43] Figure 6.9 suggests, from the high[3] log K_1 of Pb^{2+} with 18-crown-6, that Pb^{2+} should also be able to block the passage of K^+ ions through potassium ion channels. It would be interesting to see whether this is so.

6.4   The Negative Oxygen Donor

The negative oxygen donor occurs in a number of donor groups of biological interest, such as carboxylates, hydroxamates, phosphates, and phenolates. What controls[10] thermodynamic stability of complexes of metal ions with ligands with negative oxygen donors, as discussed in section 2.5.2, is affinity for the archetypal negative oxygen donor, hydroxide ion. Figure 6.10 shows the relationship between log K_1 for the siderophore[44] DFO (desferriferrioxamine-B) and log $K_1(OH^-)$ for metal ions. Highest complex stability for DFO occurs with the most acidic metal ions. Acidity of metal ions is related to metal ion size and charge, and also covalence in the M-O bond. Two important factors[10] control correlations such as Figure 6.10. Firstly, more basic donor atoms on the ligand lead to steeper slopes for such correlations. Steeper slopes lead to higher selectivity for more acidic metal ions such as Fe(III). Accordingly, donor atoms in siderophores[44] contain more basic negative oxygen donor groups such as hydroxamates and phenolic oxygens, which produce greater selectivity for Fe(III). Secondly, size of intercept in correlations such as Figure 6.10 depends on level of preorganization of the ligand. Analysis of the chelate effect[45] (section 3.3) suggests that it should give a stabilization of (n - 1) log 55.5, where n is denticity of the multidentate ligand in equation 6.1:

$$\log K_1 \text{ (multidentate)} = \log \beta_n \text{ (unidentate)} + (n - 1) \log 55.5 \qquad 6.1$$

log β_n refers to the cumulative stability constant of the complex containing n unidentate analogues of the multidentate ligand bound to the metal ion. The (n - 1) log 55.5 term is an entropy contribution to complex stability. More flexible ligands give smaller intercepts than expected from equation 6.1. Thus, a ligand such as DFO, with its long connecting bridges between the donor groups is of low preorganization, and the intercept for DFO in Figure 6.10 is smaller than 5 log 55.5, or 8.7 log units, expected[45] for a hexadentate ligand. For smaller ligands with short connecting bridges between donor groups, such as citric acid in Figure 6.10, which has n = 2, the intercept is close to the theoretical[45] value of 2 log 55.5 for a tridentate ligand. The intercept on a correlation such as Figure 6.10 is an indicator of the level of preorganization of the ligand.

Greater basicity favors complexation of more acidic metal ions, while a large number of highly preorganized donor groups will produce a large intercept on correlations such as Figure 6.10. Thus,[44] Fe(III) is complexed in biological systems by flexible ligands with low levels of preorganization, such as DFO and enterobactin. The high basicity of the hydroxamate or phenolate oxygens of these ligands favors more acidic Fe(III), while the flexibility of the ligand leads to low intercepts in correlations such as Figure 6.10, and

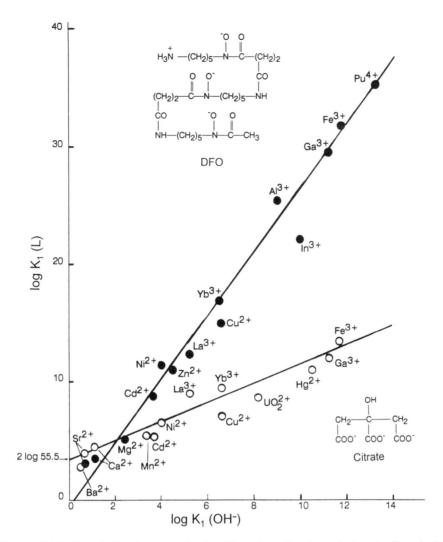

Figure 6.10. Correlation between the log K_1 values for the polydentate ligands DFO (desferriferrioxamine-B) and log K_1 (OH⁻) (●), and for citrate and log K_1 (OH⁻) (○). The sizes of the intercepts are indications of the level of preorganization of polydentate ligands. Formation constants from Ref. 3.

poor affinity for metal ions of low acidity such as Ca(II) or Zn(II). An additional factor here is that the low level of preorganization will lead to more rapid metallation reactions. The price of high preorganization is slow metallation and demetallation kinetics.[10] In contrast, to complex ions of low acidity such as Ca(II), without the ligand being preferentially complexed by more acidic metal ions, the basicity of the negative oxygen donor must be low, and so in calcium binding proteins such as calmodulin[24] the negative oxygen donor atoms are low basicity carboxylate groups. To maximize affinity of the ligand for the low acidity Ca(II) ion, several well preorganized carboxylates or other donor groups could be used, which would promote complex stability for the Ca(II) by maximizing the entropy contribution to the chelate effect.

6.5 The Neutral Nitrogen Donor

The nitrogen donor commonly found coordinated to metal ions in biology is the imidazole group from histidine. This is found coordinated to metal ions such as Cu(II) in hemocyanin,[18] and Zn(II) in carbonic anhydrase,[4] which metal ions have an extensive and well studied[3] aqueous chemistry with nitrogen donor ligands. However, transferrin has an imidazole coordinated to Fe(III), in addition to two phenolates.[46] One expects that negative oxygen donors would be used to complex the acidic Fe(III), but what affinity does Fe(III) have for imidazole nitrogen? A variety of procedures have been used (section 2.5.3) to estimate[47] the formation constants of metal ions such as Fe(III) for nitrogen donor ligands such as ammonia, which suggest $\log K_1 = 3.8$ for the Fe(III) complex of ammonia. In Table 6.2 are values of $\log K_1$ with imidazole for metal ions of interest in biology, both experimental[3] and predicted by equation 2.8. As expected from its high stability constant with ammonia, Fe(III) forms a strong complex with imidazole, accounting for its use in transferrin. Imidazole is a more effective donor group than a saturated amine, because it has a lower protonation constant[3] of 7.02, as compared to protonation constants for primary amines, for example, of about 10.6. This means that at biological pH of 7.4 imidazole groups will be largely unprotonated, whereas with primary amines, complexation of metal ions would have to occur in competition with the proton. Cobalt(III) has a very high affinity for nitrogen donor ligands, relating to its occurrence[48,49] in vitamin B_{12} where it is held by four nitrogen donors.

Table 6.2. The stability of complexes of imidazole with metal ions of biological and biomedical interest.[a]

metal ion	$\log K_1$ (imidazole)	metal ion	$\log K_1$ (imidazole)
metal ions of biological interest			
Ca(II)	-0.1^b	Mg(II)	0.0^b
Zn(II)	2.56	Cu(II)	4.18
Cu(I)	6.8	Fe(II)	1.4
Mn(II)	0.8^b	Ni(II)	3.02
Fe(III)	3.5^b	Co(III)	6.4^b
VO^{2+}	2.5^b		
metal ions of biomedical interest			
Gd(III)	0.5^b	Y(III)	0.4^b
Ga(III)	3.5^b	In(III)	3.8^b
Bi(III)	5.0^b	Pb(II)	1.6
Cd(II)	2.80	Hg(II)	9.2

[a] Experimental data from Ref. 3. [b] Estimated data from equation 2.8 and parameters in Tables 2.4 and for imidazole $E_B = -1.1$, $C_B = 12.3$, and $D_B = 0.1$.

The porphyrins[50,51] are amongst the most highly preorganized ligands. A recent study[52] of the formation constants of metal ions with $TSPP^{6-}$ (tetrasulfophenylporphyrin) in 80:20 $DMSO/H_2O$ shows that even with metal ions such as Mg(II) which has only low affinity for nitrogen donor ligands (Table 6.2), the formation constant is extremely high. An advantage of the solvent used is that, unlike water, a high enough pH can be attained to measure the high first and second protonation constants. In water these can only be estimated. The results in DMSO/water should otherwise not be too different from those in water. One can contrast (Table 6.3) formation constants of the porphyrin TSPP^{6-} with the tetraaza macrocycles cyclam and THEC [tetrakis(2-hydroxyethyl)cyclam] for the same metal ions. Mg(II) has only a low affinity for THEC, but a high affinity for TSPP^{6-}, showing the high preorganization of porphyrins. The high levels of preorganization of porphyrins leads to high kinetic inertness to metallation and demetallation, as seen in the fact that equilibration of the Cu(II)/TSPP^{6-} solutions with the proton took[52] two years.

Table 6.3. Stability of complexes of the porphyrin TSPP compared with the nitrogen donor macrocycles cyclam and THEC, with metal ions of biological interest.[a]

	log K$_1$[b]		
metal ion	TSPP^{6-}	cyclam	THEC
Cu(II)	38.1	27.2	15.7
Zn(II)	34.6	15.5	6.4
Mg(II)	28.8	(~2?)	1.9
	protonation constants		
$L^{6-} + 2H^+ = LH_2^{4-}$	32.8	22.0	17.0
$LH_2^{4-} + 2H^+ = LH_4^{2-}$	3.6	3.9	3.9

[a] abbreviations: TSPP = tetrakis(p-sulfophenyl)porphyrin, cyclam = 1,4,8,11-tetraaza-cyclo-tetradecane, THEC = N,N',N'',N'''-tetrakis(2-hydroxyethyl)cyclam. [b] Formation constants for TSPP^{6-} in 80:20 DMSO/water,[52] for cyclam and THEC in water[3] at ionic strength 0.1.

6.6   Sulfur Donors

The sulfur donor atom in biology occurs as mercapto groups from cysteine, and complexes, for example, metal ions such as Zn(II) in metalloenzymes where it has a catalytic role,[4] and in "zinc fingers" where it has[53] a structural role. Sulfur is also involved in complexing iron in numerous iron/sulfur proteins.[54] Metallothionein[54] is a protein containing cysteines which protects against toxic levels of metal ions by binding Cu(I), Zn(II), Cd(II), Pb(II), and Hg(II). Our interest here is the affinity which the mercapto group has for different metal ions, as a guide to the distribution of metal ions on mercapto sites of proteins. As a start, the formation constants for metal ions of biological and biomedical interest with simple mercaptans such as mercaptoethanol[3] can serve as a guide to relative affinity of metal ions for cysteine sites on proteins. Unfortunately, not a great number of such systems have been studied,[3] and in Table 6.4 many of the values shown have had to be estimated from equation 2.8 and parameters in Table 2.3, and with E_B = -3.78, C_B = 39.1, and D_B = 0.9 for mercaptoethanol.[10] Some confidence in the estimated values can be derived from the good agreement between values predicted by equation 2.8, and observed values shown in Table 6.4. The I_B value for mercaptoethanol[10] of -0.097 corresponds to a ligand that is intermediate in HSAB, not soft (see section 2.4 for a discussion of I_B values). In agreement with this, it is found that metal ions such as Ga(III), or In(III), which are classified[5] as hard, form complexes of great stability with ligands with mercapto groups,[55] which they do not do with the truly soft cyanide ion. What one can say about mercapto sulfur is that it can form bonds of great covalent strength to metal ions, but that there is also an ability to forms bonds of some ionicity. Thus, Cd(II) is classified as soft, and Zn(II) and Pb(II) are classified as intermediate in

HSAB, but, as seen in Table 6.4, mercapto groups have very similar formation constants with all three metal ions. Thus, use of RS⁻ type ligands for removal of Pb(II) and Cd(II) from the body[56] produces very useful ligands, but it is unlikely that these ligands will display selectivity for the poisonous Pb(II) and Cd(II) over the essential Zn(II) ion.

Table 6.4. Formation constants of mercaptoethanol with metal ions of medical and biomedical interest.[a]

metal ions of biological interest			
Ca(II)	-0.55[b]	Mg(II)	-1.42[b]
Zn(II)	5.7[b]	Cu(II)	8.1[b,c]
Cu(I)	16.7[b]	Ni(II)	3.9[b]
Mn(II)	1.8[b]	Fe(II)	2.5[b]
Fe(III)	8.6[b,c]	VO^{2+}	5.0[b]
metal ions of biomedical interest			
Gd(III)	0.1[b]	Y(III)	-0.5[b]
Ga(III)	8.7[b]	In(III)	9.1 (9.6[b])
Bi(III)	13.8 (13.4[b])	Pb(II)	6.6 (5.7[b])
Cd(II)	6.1 (7.4[b])	CH_3Hg^+	15.9 (15.6[b])

[a] From reference 3, and estimated[b] using equation 2.8. [c] These ions tend to be reduced by thiols ($Fe^{III} \rightarrow Fe^{II}$, and $Cu^{II} \rightarrow Cu^{I}$), unless the complex formed in the higher oxidation state is sufficiently more stable than that in the lower oxidation state to reduce the reduction potential sufficiently. Thus Fe^{III} is stable in Fe_4S_4 clusters, since the four sulfurs produce a very stable complex.

Metallothioneins (MT) in mammals usually bind seven metal ions in cluster structures, with bridging sulfur groups, as seen in the X-ray structure[57] of the Cd_5Zn_2MT complex. This makes it difficult to develop a simple formation constant description of the binding of metal ions to MT.[58] The log K_1 values with mercaptoethanol (ME) in Table 6.4 are, however, a useful guide to the complexing strength[58,59] of metal ions with MT. Thus, both Co(II) and Ni(II) in Table 6.4 have lower log K_1 values with ME than Zn(II), and so cannot displace Zn(II) from MT. The Pb(II) ion has log K_1 with ME similar to that with Zn(II), and so Pb(II) only partially displaces Zn(II) from MT. The metal ions Bi(III), In(III), Cu(I), and Hg(II) all have log K_1 values with ME higher than do Zn(II) and Cd(II), and so completely displace Zn(II) and Cd(II) from MT.[60]

References

1. J. J. R. Frausto da Silva and R. J. P. Williams, *The Biological Chemistry of the Elements*, Clarendon Press, Oxford, 1991.
2. S. J. Lippard and J. M. Berg, *Principles of Bioinorganic Chemistry*, University Science Books, Mill Valley, California, 1994.
3. A. E. Martell and R. M. Smith, *Critical Stability Constants*, Vols. 1-6, Plenum Press, New York, 1974, 1975, 1976, 1977, 1982, 1989.
4. I. Bertini, C. Luchinat, W. Maret and M. Zeppezauer, Eds., *Progress in Inorganic Biochemistry and Biophysics*, Vol. 1, *Zinc Enzymes*, Birkhauser Verlag, Basel, 1986.
5. R. G. Pearson, *J. Am. Chem. Soc.*, **1963**, *85*, 3533: *Coord. Chem. Rev.*, **1990**, *100*, 403.
6. Y. Pocker and J. T. Stone, *Biochemistry*, **1968**, *7*, 2936.
7. a) P. Woolley, *Nature (London)*, **1975**, *258*, 677. b) S. H. Gellma, R. Petter and R. J. Breslow, *J. Am. Chem. Soc.*, **1986** , *108*, 2388. c) J. T. Groves and R. R. Chambers, *J. Am. Chem. Soc.*, **1984**, *106*, 630.

8. E. Kimura, T. Shiota, T. Koike, M. Shiro and M. Kodama, *J. Am. Chem. Soc.*, **1990**, *112*, 5805.

9. G. Pettersen and J. Kvassman, *Eur. J. Biochem.*, **1980**, *103*, 565.

10. R. D. Hancock and A. E. Martell, *Adv. Inorg. Chem.*, **1995**, *42*, 89.

11. S.-H. Wu, D.-S. Lee and C.-S. Chung, *Inorg. Chem.*, **1984**, *23*, 2548.

12. C.-C. Chang and C.-S. Chung, *J. Chem. Soc., Dalton Trans.*, **1991**, 1685.

13. H.-R. Sheu, T.-J. Lee, T.-H. Lu, B.-F. Liang and C.-S. Chung, *Proc. Natl. Sci. Council R.O.C.*, **1983**, *7*, 113.

14. R. D. Hancock and F. Marsicano, *Inorg. Chem.*, **1978**, *17*, 560; **1980**, *19*, 2709.

15. H. Ohtaki and T. Radnai, *Chem. Rev.*, **1993**, *93*, 1157.

16. B. L. Vallee and R. J. P. Williams, *Proc. Nat. Acad. Sci.*, **1968**, *59*, 498.

17. E. S. Cedergren-Zeppezauer, J.-P. Samama and H. Eklund, *Biochemistry*, **1982**, *21*, 4895.

18. R. Huber, *Angew. Chemie., Intnl. Edn.*, **1989**, *28*, 848.

19. a) M. M. Georgiadas, H. Komiya, P. Chakrabarti, D. Woo, J. J. Kornuc and D. C. Rees, *Science*, **1992**, *257*, 1653. b) J. Kim and D. C. Rees, *Science*, **1992**, *257*, 1677. c) J. Kim and D. C. Rees, *Nature*, **1992**, *360*, 553.

20. a) N. Kitajima, K. Fujisawa, C. Fujimoto, Y. Moro-oka, S. Hashimoto, T. Kitagawa, K. Toriumi, K. Tatsumi and A. Nakamura, *J. Am. Chem. Soc.*, **1992**, *114*, 1277. b) N. Kitajima, K. Fujisawa, M. Tanaka and Y. Moro-oka, *J. Am. Chem. Soc.*, **1992**, *114*, 9232. c) N. Kitajima, *Adv. Inorg. Chem.*, in press.

21. M. G. B. Drew, *Coord. Chem. Rev.*, **1977**, *24*, 179.

22. R. Huber, M. Schneider, I. Mayr, J. Romisch and E.-P. Paques, *FEBS Lett.*, **1990**, *275*, 15.

23. P. C. Moews and R. H. Kretzinger, *J. Mol. Biol.*, **1975**, *91*, 201.

24. Y. S. Babu, C. E. Bugg and W. J. Cook, *J. Mol. Biol.*, **1988**, *204*, 191.

25. a) S. J. Angyal, *Chem. Soc. Rev.*, **1981**, *415*. b) K. Burger and L. Nagy, in *Biocoordination Chemistry*, Ed. K. Burger, Ellis Horwood, New York, 1990, p 236. c) D. M. Whitfield, S. Stojkovski and B. Sarkar, *Coord. Chem. Rev.*, **1993**, *122*, 171.

26. R. D. Hancock and K Hegetschweiler, *J. Chem. Soc., Dalton Trans.*, **1993**, 2137.

27. E. Asato, W. L. Driessen, R. A. G. de Graff, F. B. Hulsbergen and J. Reedijk, *Inorg. Chem.*, **1991**, *30*, 4210.

28. T. L. Feng, P. L. Gurian, M. D. Healy and A. R. Barron, *Inorg. Chem.*, **1990**, *29*, 408.

29. R. D. Hancock, *J. Chem. Educ.*, **1992**, *69*, 615.

30. R. D. Hancock, *Progr. Inorg. Chem.*, **1989**, *37*, 187.

31. H. Lauble, M. C. Kennedy, H. Beinert and C. D. Stout, *Biochemistry*, **1992**, *31*, 2735.

32. M. C. Kennedy and C. D. Stout, *Adv. Inorg. Chem.*, **1992**, *38*, 323.

33. D. C. Rees, J. B. Howard, P. Chakrabarti, T. Yeates, B. T. Hsu, K. D. Hardman and W. N. Lipscomb, in *Progress in Inorganic Biochemistry and Biophysics, Vol 1. Zinc Enzymes*, Eds. I. Bertini, C. Luchinat, W. Maret and M. Zeppezauer, Birkhauser, Basel, l986, p 155.

34. J. W. Westley, Ed., *Polyether Antibiotics: Naturally Occurring Ionophores*, Vols. 1 and 2, Dekker, New York, 1982.

35. H. Einspahr and C. E. Bugg, in *Metal Ions in Biological Systems*, H. Sigel, Ed., Marcel Dekker, New York, 1984, p 51.

36. M. Dobler, *Ionophores and their Structures*, Wiley, New York, 1981.

37. R. D. Shannon, *Acta Crystallogr., Sect. A*, **1976**, *A32*, 751.

38. H.-Y. An, J. S. Bradshaw and R. M. Izatt, *Chem. Rev.*, **1992**, *92*, 543.

39. B. Hille, *Ionic Channels of Excitable Membranes*, 2nd Ed., Sinauer Associates, Sunderland MA, 1992.

40. C. Miller, *Science*, **1993**, *261*, 1692.

41. R. A. Kumpf and D. A. Dougherty, *Science*, **1993**, *261*, 1708.

42. J. Neyton and C. Miller, *J. Gen. Physiol.*, **1988**, *92*, 549, 569.

43. C. Miller, *Biophys. J.*, **1987**, *52*, 123.

44. R. J. Bergeron and G. M. Brittenden, Eds., *The Development of Iron Chelators for Clinical Use*, CRC Press, Boca Raton, 1994.

45. R. D. Hancock and F. Marsicano, *J. Chem. Soc., Dalton Trans.*, **1976**, 1096.

46. E. N. Baker and T. Blundell, *Proc. Nat. Acad. Sci., USA*, **1987**, *84*, 1769.

47. F. Mulla, F. Marsicano, B. S. Nakani and R. D. Hancock, *Inorg. Chem.*, **1985**, *24*, 3076.

48. J. M. Pratt, *Inorganic Chemistry of vitamin B12*, Academic Press, New York, l972.

49. D. Dolphin, Ed. "*B12*", Vols. 1 and 2, John Wiley, New York, l982.

50. W. R. Scheidt and C. A. Reid, *Chem. Rev.*, **1981**, *81* 543.

51. W. R. Scheidt and Y. J. Lee, *Struct. Bonding.*, **1987**, *64*, 1.

52. H. R. Jimenez, M. Julve and J. Faus, *J. Chem. Soc., Dalton Trans.*, **1991**, 1945.
53. A Klug and D. Rhodes, *Trends Biochem. Sci.*, **1987**, *12*, 464.
54. A. G. Sykes, Ed., *Advances in Inorganic Chemistry*, Vol. *38*, Academic Press, San Diego, CA, 1992.
55. K. Tanaboylu and G. Schwarzenbach, *Helv. Chim. Acta.*, **1972**, *55*, 2065.
56. J. S. Casas, A. Sanchez, J. Bravo, S. Garcia-Fontan, E. E. Castellano and M. M. Jones, *Inorg. Chim. Acta*, **1989**, *158*, 119.
57. W. F. Furey, A. H. Robbins, L. L. Clancy, D. R. Wings, B. C. Wang and C. A. Stout, *Science*, **1986**, *231*, 704.
58. M. Vasak and J. H. R. Kagi, in *Metal Ions in Biology*, Ed., H. Sigel, Vol. 15, p.213, Marcel Dekker, New York, 1983.
59. M. Vasak, J. H. R. Kagi, B. Holmquist and B. L. Vallee, *Biochemistry*, **1981**, *20*, 6659.
60 K. B. Nielsen, C. L. Atkin and D. P. Winger, *J. Biol. Chem.*, **1985**, *260*, 5342.

CHAPTER 7

STABILITY CONSTANTS AND THEIR MEASUREMENT

7.1   Introduction

The equilibrium constant involving the formation of a metal complex from the aquo metal ion and the most basic form of the ligand is a standard measure of the effectiveness of the ligand in coordinating metal ions. The constants involved are called stability constants or formation constants. Most complex formation reactions are measured in aqueous medium under controlled conditions, and the formation constants generally apply to that medium. However, ligands that are not soluble in water but are soluble in organic solvents are frequently employed and their formation constants with metal ions are often determined in mixed solvents such as dioxane-water (up to 70% dioxane by volume); ethanol- and methanol-water systems are also quite common. For completely organic systems such as acetonitrile or tetrahydrofuran, metal complexes can be formed quite readily but their formation constants are generally not known. Approximate values have been used occasionally for such systems but there is no mathematical expression that can relate the equilibrium constants in such systems to the formation constants or stability constants in water or water/organic mixtures.

7.2   Early Work

The development of the measurement of stability constants and their use paralleled the development of the instrumentation for their measurement, as well as the theory of electrolyte solutions on which such measurements are based. Potentiometric measurements were first used for the measurement of stability constants by Arrhenius, Ostwald and Nernst, who provided the basis for the introduction of electrodes responding reversibly and selectively to only one species present in solution. The potentials of such electrodes are shown below to provide sufficient information for the determination of stability constants of complex formation reactions. At that early period only ligands which were not protonated could be measured. Also determinations were limited to metal ions for which metal electrodes were characterized by potentials which corresponded reversibly to the concentrations of their cations. The investigations of Bjerrum, Bronsted and McGuinness on activity coefficients and the development of the theory of strong electrolytes in solution by Debye and Huckel in 1923, formed the basis for exact studies of metal ions and of anions in solution. The determination of empirical formulas and overall formation constants were pioneered by workers such as von Euler[1] and Bodlander.[2] Stepwise formation of complexes was first demonstrated for the system Hg^{2+}, Cl^- by Abegg and coworkers.[3] Also Bodlander[2] and Grossman[4] were the first to apply the idea of an ionic medium to control the ionic strength of the solution, in studying the formation of Hg^{2+} thiocyanate complexes. The stepwise hydrolysis constants of

Cr(III) were described by Bjerrum[5] and later Bjerrum described the thiocyanate complexes of Cr(III) and calculated the six-step stability constants by taking advantage of the inert nature of the complexes formed.[6] Subsequently work on the glass electrode, which is very sensitive to the hydrogen ion activity or concentration, and appropriate potentiometric apparatus for the measurement of the hydrogen ion with considerable accuracy was developed. The work on the glass electrode is summarized by Dole[7] in the book *The Glass Electrode*. The prolific work of R. Bates[8] on the measurement with the glass electrode of hydrogen ion activities and concentrations gave precise affinities of various ligands for the hydrogen ion and the results which have been published represent the most careful work that has been done in this field to date.

The introduction of general methods for computing stepwise stability constants was developed by I. Leden[9] and J. Bjerrum.[10] The thesis by J. Bjerrum on the stepwise ammonia formation constants of a number of metal ions received considerable attention. He used a large excess of the ligand (ammonia) to prevent hydrolysis and precipitation and developed approximation methods for the calculation of the metal ammonia stability constants. In 1945 there appeared a classic paper by Calvin and Wilson[11] in which the stability constants of a number of complexes were calculated without the Bjerrum simplifications by the use of exact algebraic treatment of equilibrium constants and mass balance equations. Because of the power of these methods and the relatively inexpensive instruments required, the work in this field rapidly developed. A large part of the work was carried out by three research groups, those of Sillen *et al.* in Stockholm, Bjerrum's group in Copenhagen, and Schwarzenbach and coworkers in Zurich. Subsequently a group involving Martell and coworkers developed in the United States and the methods used are described in a book on The Chemistry of the Metal Chelate Compounds.[12] The properties of metal complexes and chelates are also described in a book by Chaberek and Martell,[13] and by Dwyer and Mellor.[14] The extensive treatment of the Determination of Stability Constants by Rossotti and Rossotti[15] still remains the major text on the methods employed for stability constant determinations prior to the use of computers. Other major works that should be mentioned are a book by Bailar and his students,[16] in which many reaction mechanisms are described and a more modern text by Lewis and Wilkins[17] in which the use of ligand field theory is introduced.

7.3   Recent Work

The ubiquitous availability of computers has expedited and changed the methods of calculation of stability constants from equilibrium data so that now very complex systems may be handled with relative ease and the approximations used previously by the workers described in the books mentioned above have been eliminated. The use of computers has also resulted in the development of various ways of displaying experimental results, such as the production of species distribution diagrams. Such graphical illustrations may be used to test the calculated results and the validity of the data used for computer calculations, as well as the methodology employed for defining the chemical system under investigation. A number of computer programs have been published[18-20] which use various methods[21,22] and new programs for this purpose are being described in still more recent publications. The programs for the calculation of stability constants have been compared in a number of publications.[23-25] The programs employed in our research group handles any number of ligands and metal ions and is the general procedure described as program BEST.[26] Determinations of pK's of ligands are also described.[27] These programs for the potentiometric determination of stability constants are summarized in a book by

Motekaitis and Martell on The Determination and Use of Stability Constants[28] which contains, in addition the experimental methods employed, the determination of stability constants in complex systems and the displaying of the results of stability constant calculations by species distribution diagrams. The main advantage of these computer programs is the processing of potentiometric pH data and calculation of measured pH directly for comparison with experimental results, as well as a description of the precautions to be used in the implementation of such methods.

Because of the relative ease of measurement and the simple equipment involved there has been a large proliferation of stability constants of metal ion complexes reported in the literature. These data are collected in a series of non-critical volumes in which every reaction resulting in the formation of a complex is described.[29-33] These compilations are sponsored by the Subcommission V.6 on Equilibrium Data of the Analytical Section, IUPAC. The non-expert faces considerable difficulty in selecting the correct constants from among a number of values differing even in orders of magnitude, that might be offered for a metal complex equilibrium in the literature. For this reason Martell and Smith have provided a series of critical compilations of stability constants.[34] The result is the listing of a recommended value for the formation constant of a metal complex when enough information is supplied in the publications involved. The problem has been recognized by the Subcommission on Equilibrium Data and has resulted in the development of in-depth critical reviews of stability constants of individual ligands or of groups of related ligands or metal ions. These compilations give the reasons for the choices and give a detailed account of the problems faced by the compiler. A number of these critical surveys have been published.[35] The difficulties of access to the volumes of critical stability constants is made much simpler by the development of a computerized database of critical constants developed by Smith, Motekaitis, and Martell.[36] This work is being supported by The National Institute of Standards and Technology (NIST) and is now available for distribution. There is also a non-critical Stability Constants Database compiled by Pettit and Powell and sponsored by IUPAC.

The growth of stability data has been extensive and a large number of publications are now available for critical compilations. Much of the data published in the literature is rather hastily developed, inaccurately described, and much of the data that has been published should be looked upon with considerable doubt. Because of the proliferation of unreliable data, however, the field has suffered considerably with respect to its evaluation by non-experts. This is especially true and can be seen in the non-critical compilations whereby stability constants are reported which differ by an order of magnitude or even more for the same reaction. In the early days such problems were often the result of graphical methods where perhaps unwarranted assumptions and simplifications had to be made to make the problem tractable. In modern days the careless use of computer programs allows for the mindless modeling of systems whereby defective experimental data can be actually made to fit exactly by invoking numerous untenable minor species. In practice, the most common errors center around very large protonation constants, whose accurate values are nearly impossible to obtain by direct potentiometry, and large stability constants which render the measurements insensitive to the exact values of such large constants. The results have been the reporting of sometime meaningless equilibrium formation constants of metal complexes.

Further expansion of this field is inevitable, not only because of the large number of metal ions involved but also because the ligands have taken on considerable importance and their number and variety has been recently expanded. For example, the development of macrocyclic and macrobicyclic (cryptand) complexing agents has added new dimensions to the previous concepts of complexes and chelate compounds. Also, a large number of ligands of biological significance have now been defined and are the subject of rather detailed investigations. In addition the many ligands that are present in the environment can also now be considered. These ligands may be very complex in nature and a considerable number of them may be involved simultaneously. Computer methods have made metal complexation in such systems quite feasible, and activity in both the biological and environmental systems with respect to metal ion complexation has greatly expanded in recent years. Because of the many-component systems involved and the ability of the investigator to assume the presence of unusual or meaningless species in order to fit the data, it is important the investigator exercise known principles of coordination chemistry in order to avoid over-interpretation of experimental data. The various ways that this kind of difficulty can be avoided are indicated in Reference 28. The previous proliferation of poor data in the field and the possibilities for over interpretation of the data in multicomponent systems has resulted in the development of a set of standards, given in Section 7.14.1, for the measurement and reporting of stability constants of metal ion complexes.

7.4. The Stability Constant

An equilibrium constant is a quotient consisting of the products of the activities of the products of the reaction raised to the appropriate power divided by the products of the reactants also raised to the appropriate power in accordance with the following reaction and equation.

$$aA + bB \quad \rightleftharpoons \quad cC + dD \qquad K_{eq} \quad = \quad \frac{a_C^c \cdot a_D^d}{a_A^a \cdot a_B^b} \qquad \qquad 7.1$$

The logarithm of the equilibrium constant is directly related to the Gibbs free energies of the products of the reaction minus the Gibbs free energies of the reactants in their standard states and it is therefore a measure of the difference in reactivities of the products and reactants. If heat of reaction, ΔH^o, is also known, measured directly or by the temperature coefficient of the equilibrium constant calculated with the use of the Van't Hoff expression (equation 7.2) then the Gibbs free energy, ΔG^o can be broken down into heats and entropies of reaction as indicated by equation 7.3.

$$\Delta H^o/2.303 \ RT^2 \ = \ d \ logK/dT \qquad \qquad 7.2$$

where T is absolute temperature

$$\Delta G^o \ = \ \Delta H^o \ - \ T\Delta S^o \qquad \qquad 7.3$$

Further insights into the nature of the reactions are provided by ΔH^o which is related to the heats of solvation and bond energies, and ΔS^o which is related to the freedom of motion of the species involved in the reaction.

The determination of the activities of complex ionic species under real conditions and at infinite dilution is a complex and time-consuming operation which is beyond the interests and capabilities of most coordination chemists. These techniques are illustrated by the work of Bates[8] who has determined thermodynamic equilibrium constants at infinite dilution for many organic acids. However, because concentrations closely parallel activities under carefully controlled conditions involving both temperature and ionic strength it is the practice to determine equilibrium concentration constants in place of activity constants. The ionic strength is controlled by a non-reacting ionic species such as KNO_3 or $(Bu)_4N \cdot NO_3$ in concentrations in large excess over the ionic species involved in the reaction under consideration. This practice may be traced back to the time of Grossman, who employed KNO_3 to keep ionic strength constant, although the formal concept of ionic strength did not appear until its introduction by Lewis and Randall almost twenty years later. The equilibrium concentration constant would then be represented by equation 7.4.

$$aA + bB \rightleftharpoons cC + dD \qquad K'_{eq} = \frac{[C]^c[D]^d}{[A]^a[B]^b} \qquad 7.4$$

The use of concentrations of species indicated by equation 7.4 for the equilibrium constant of the reaction has several advantages. First, the concentration of the species involved can be substituted directly into the mass balance equations used in solving the equations for the equilibrium constant or formation constant of the metal ion complexes that are formed in the reaction. The choice of the supporting electrolyte has not been standardized. At the present time it is essential that the supporting electrolyte not interfere appreciably with the reaction under consideration. The species having the least interaction with the ionic media under consideration are the tetraalkylammonium salts which do not interact appreciably with any anions or with any chelating agents. The choice of supporting electrolyte and the interactions of various electrolytes with the reference electrode in potentiometric studies has been discussed in detail in the book by Motekaitis and Martell.[28] It is further pointed out that the so-called concentration constants are very close to the true thermodynamic constants of the reaction in the electrolyte medium chosen, because the activity coefficients change very little in going from finite concentrations, used in the measurement of the equilibrium constants, to infinite dilution in the supporting electrolyte. Therefore very little error is involved in the assumption that the concentration constants measured under controlled conditions and under a controlled ionic atmosphere are very close to the true thermodynamic constants of the reaction and therefore the terms $-\Delta G^o$, $-\Delta H^o$, and ΔS^o are determined from the concentration equilibrium constants of the reaction.

The following types of equilibria may be involved in the formation of a metal complex, ML, from the metal ion, M^{n+}, and a ligand, HL^{n-}:

1. Protonation of the ligand

$$H^+ + L^{n-} \rightleftharpoons HL^{(n-1)-} \qquad K_1^H = \frac{[HL]}{[H][L]} \qquad 7.5$$

$$H^+ + HL^{(n-1)-} \rightleftharpoons H_2L^{(n-2)-} \qquad K_2^H = \frac{[H_2L]}{[H][HL]} \qquad 7.6$$

$$xH^+ + L^{n-} \rightleftharpoons H_xL^{(n-x)-} \qquad \beta_x^H = \frac{[H_xL]}{[H]^x[L]} \qquad 7.7$$

2. Formation constants

$$M^{m+} + L^{n-} \rightleftharpoons ML^{(n-m)-} \qquad K_1 = \frac{[ML]}{[M][L]} \qquad 7.8$$

$$ML^{(n-m)-} + L^{n-} \rightleftharpoons ML_2^{(2n-m)-} \qquad K_2 = \frac{[ML_2]}{[ML][L]} \qquad 7.9$$

$$M^{m+} + NL^{n-} \rightleftharpoons ML_N^{(Nn-m)-} \qquad \beta_N = \frac{[ML_N]}{[M][L]^N} \qquad 7.10$$

3. Formation of hydrogen complexes

$$ML^{(m-n)-} + H^+ \rightleftharpoons MHL^{n-m+1} \qquad K^{MHL} = \frac{[MHL]}{[H^+][ML]} \qquad 7.11$$

4. Formation of hydroxyl complexes

$$ML^{(n-m)-} + OH^- \rightleftharpoons MOHL^{(n-m+1)-} \qquad K^{MLOH} = \frac{[M(OH)L]}{[ML][OH]} \qquad 7.12$$

5. Ionic equilibrium of water

$$H_2O \rightleftharpoons H^+ + OH^- \qquad K_w = [H][OH] \qquad 7.13$$

6. Hydrolysis of the metal ion

$$M^{m+} + H_2O \rightleftharpoons M(OH)^{(m-1)+} + H^+ \qquad K_M^{MOH} = \frac{[MOH][H]}{[M]} \qquad 7.14$$

7.5   pH and p[H]

In conformity with the general practice described above regarding concentration constants, hydrogen ion concentrations are employed exclusively in place of hydrogen ion activities. In order to make this distinction clear, the generally used term pH is replaced by p[H], with the brackets indicating concentration. Because of the widespread use of the term pH in potentiometric studies of metal complex stabilities, it is important to comment on the definition of this quantity. The term defined as -log a_{H+}, is not a measurable quantity, because it is not possible to determine the activity of a single ion. The potential of the hydrogen electrode in the cell $H_2|H^+,Cl^-|AgCl|Ag$ is RT/F log $(a_{H+} \cdot \gamma_{Cl^-})$ where γ_{Cl^-} is the activity coefficient of the chloride counter ion. In spite of this built-in restriction, the pH scale is nevertheless widely used as a measure of acidity or alkalinity. This should be done, however, with the understanding that the values cannot be precisely measured or defined. Under conditions of constant ionic strength maintained by an inert supporting electrolyte, activity coefficients are essentially constant, and the potential of the hydrogen electrode in a cell without liquid junction (e.g., $H_2|H^+,Cl^-|AgCl|Ag$; electrolyte = HCl, KCl) varies linearly with hydrogen ion concentration as well as with hydrogen ion activity. When the hydrogen electrode is replaced by a glass electrode (an ion-selective electrode for H^+), the parallel relationship remains. Below pH 2 and above pH 12, hydrogen ion and hydroxide ions begin to be responsible for appreciable fractions of the conductance, so that liquid junction potentials change as the pH is lowered and raised below and above these limits. Accurate measurements of hydrogen ion concentration (or activity) with a glass electrode-reference electrode system are restricted therefore to the p[H] range 2-12.

7.6 Ca(II)-EDTA Complexes

7.6.1 Stability constants for Ca(II)-EDTA

The formation of the Ca(II)-EDTA chelates is an example which illustrates the kind of reasoning possible in terms of species present and the corresponding stability constants. EDTA may be represented by the conventional symbolic formula H_4L showing the presence of four ionizable hydrogens. In the absence of metal ions, depending on the p[H], there are five possible forms of EDTA: H_4L, H_3L^-, H_2L^{2-}, HL^{3-}, and L^{4-}. These forms are related by the four protonation constants β_{H_nL}; $K_1^H = \beta_{HL}$, $\beta_{H_2L} = K_1^H K_2^H$, $\beta_{H_3L} = K_1^H K_2^H K_3^H$ and $\beta_{H_4L} = K_1^H K_2^H K_3^H K_4^H$.

The titration curve of EDTA with standard KOH (Figure 7.1) shows a low pH buffer region (log K_4^H = 2.0 and log K_3^H = 2.68) terminating in a break after the first two equivalents. This is followed by a single equivalent buffer region near pH 6 (log K_2^H = 6.11) terminating with a break at 3 equivalents. A one equivalent buffer region near pH 10 represents the final neutralization (log K_1^H = 10.17).

There are two calcium containing EDTA species possible: CaL^{2-} and $CaHL^-$. The calcium(II) formation constant is β_{CaL} and the overall protonated chelate formation constant is β_{CaHL}. The titration curve of 1:1 Ca:EDTA solution with KOH shows that

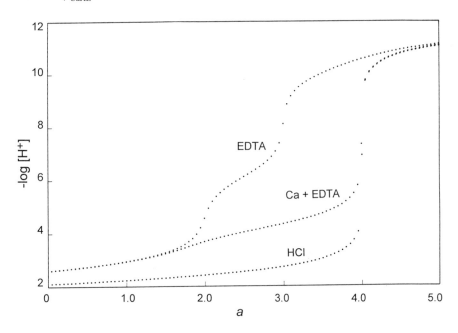

Figure 7.1. Potentiometric titration curves for EDTA, Ca-EDTA and 4HCl. The upper curve is a measure of the ligand protonation constants. The middle curve is a measure of the displacement of protons from EDTA during the course of complex formation. The bottom curve is a strong acid curve showing what to expect if all of the titratable protons were displaced by the metal. T_L = 2.00 mM, T_M = 2.00 mM, μ = 0.100 M (KCl), t = 25.0 °C, a = moles of KOH/mole of ligand.

Ca^{2+} begins to interact with EDTA just above pH 3. This is revealed by the so-called formation curve which is a measure of the stability constants. This curve is intermediate between the two extremes of no formation and 100% formation. If no complex formation was to take place, the EDTA-Ca curve would be coincident with the ligand curve. At 100% formation the curve would be coincident with the 4-mole equivalent strong acid curve shown on the bottom of the graph (i.e. each mole of complex formed would release 4 moles of acid). All of the species appearing in equations (7.15)-(7.20) are present in the system over the entire pH range, but not all of these species are present in significant concentrations. A complete set of equations is needed to describe all the equilibria in the pH range over which hydrogen ion concentrations are measured. It is imperative to keep in mind that in addition to the above species, also H^+, Ca^{2+}, as well as OH^- may be present.

$$H^+ + L^{4-} \rightleftharpoons HL^{3-} \qquad \beta_{HL} = \frac{[HL^{3-}]}{[H^+][L^{4-}]} \qquad \qquad 7.15$$

$$2H^+ + L^{4-} \rightleftharpoons H_2L^{2-} \qquad \beta_{H_2L} = \frac{[H_2L^{2-}]}{[H^+]^2[L^{4-}]} \qquad \qquad 7.16$$

$$3H^+ + L^{4-} \rightleftharpoons H_3L^{3-} \qquad \beta_{H_3L} = \frac{[H_3L^-]}{[H^+]^3[L^{4-}]} \qquad \qquad 7.17$$

$$4H^+ + L^{4-} \rightleftharpoons H_4L \qquad \beta_{H_4L} = \frac{[H_4L]}{[H^+]^4[L^{4-}]} \qquad \qquad 7.18$$

$$Ca^{2+} + L^{4-} \rightleftharpoons CaL^{2-} \qquad \beta_{CaL} = \frac{[CaL^{2-}]}{[Ca^{2+}][L^{4-}]} \qquad \qquad 7.19$$

$$H^+ + Ca^{2+} + L^{4-} \rightleftharpoons CaHL^- \qquad \beta_{CaHL} = \frac{[CaHL^-]}{[H^+][Ca^{2+}][L^{4-}]} \qquad \qquad 7.20$$

The EDTA-Ca system is considered as consisting of three components: $EDTA^{4-}$ (L), Ca^{2+} (M), and H^+. The species possible are $EDTA^{4-}$, $HEDTA^{3-}$, H_2EDTA^{2-}, H_3EDTA^-, H_4EDTA, $CaEDTA^{2-}$, $CaHEDTA^-$, H^+, Ca^{2+}, and OH^-. There would be three mass balance equations in terms of total ligand, total metal ion, and total initial hydrogen ion concentrations: T_L, T_M, T_H, respectively.

$$T_L = [L^{4-}] + [HL^{3-}] + [H_2L^{2-}] + [H_3L^-] + [H_4L] + [CaL^{2-}] + [CaHL^-] \qquad 7.21$$

$$T_M = [Ca^{2+}] + [CaL^{2-}] + [CaHL^-] \qquad 7.22$$

$$T_H = [HL^{3-}] + 2[H_2L^{2-}] + 3[H_3L^-] + 4[H_4L] + [CaHL^-] + [BASE] + [H^+] - [OH^-] \qquad 7.23$$

In (7.23) T_H represents the amount of H^+ initially present and [BASE] that which has been removed by the added base (e.g., KOH). The internal computer representation of the equilibrium constants is set up in terms of β's, and in terms of the concentrations of the individual species, as expressed by the following:

$$T_L = [L^{4-}] + \beta_{HL}[H^+][L^{4-}] + \beta_{H_2L}[H^+]^2[L^{4-}] + \beta_{H_3L}[H^+]^3[L^{4-}] + \beta_{H_4L}[H^+]^4[L^{4-}] \qquad 7.24$$

$$+ \beta_{ML}[M^{2+}][L^{4-}] + \beta_{MHL}[M^{2+}][H^+][L^{4-}]$$

$$T_M = [M^{2+}] + \beta_{ML}[M^{2+}][L^{4-}] + \beta_{MHL}[M^{2+}][H^+][L^{4-}] \qquad 7.25$$

$$T_H = \beta_{HL}[H^+][L^{4-}] + 2\beta_{H_2L}[H^+]^2[L^{4-}] + 3\beta_{H_3L}[H^+]^3[L^{4-}] + 4\beta_{H_4L}[H^+]^4[L^{4-}] + [BASE] \qquad 7.26$$

$$+ [H^+] - \beta_{OH}[H^+]^{-1} + \beta_{MHL}[M^{2+}][H^+][L^{4-}]$$

The set of simultaneous equations is solved for each component (i.e. $[L^{4-}]$, $[M^{2+}]$, and $[H^+]$). The value of the calculated concentration of H^+, is then compared with the measured hydrogen ion concentration. This calculation process is repeated at all measured equilibrium points. In any calculation based on a p[H] profile there will be some known, previously calculated, β values as well as the unknown values to be determined. The first pass of the calculation procedure uses both the known and the estimated values of the unknown constants. Refinement of the unknown constants follows.

Thus the use of the algorithm for computing equilibrium constants in BEST involves the following sequence: 1, start with a set of known and estimated overall stability constants (β's) and compute $[H^+]$ at all equilibrium points; 2, compute the weighted sum of the squares of the deviations in p[H] as in (7.27).

$$U = \Sigma w\,(p[H]_{obs} - p[H]_{calcd})^2 \qquad 7.27$$

where $w = 1/(p[H]_{i+1} - p[H]_{i-1})^2$, a weighting factor which serves to lessen the influence on the calculation of the less accurate p[H] values in the steeply sloped regions of the p[H] profile; 3, change the unknown stability constants and repeat the calculations until no further minimization of U can be obtained, thus providing the final calculated β values. The standard deviation in pH units is obtained by the use of equation (7.28).

$$\sigma_{fit} = (U/N)^{1/2} \qquad 7.28$$

where $N = \Sigma w$.

The program PKAS is a special case of the more general algorithm found in BEST. Since only two components (L and H) are present when metal ions are absent, much simpler equations become available without the need to solve for more than two simultaneous equations (T_L and T_H). It is convenient when the ligand is of known exact molecular weight and the T_H is an exact integral multiple of T_L.

7.7   Species Distribution Diagrams

A species distribution diagram is a powerful tool for the assessment of the concentrations of the species present as a function of p[H]. The construction of a species diagram always starts with the total concentration of each component and its accuracy depends on the quality of all of the equilibrium constants. Shown in Figure 7.2 is the species distribution diagram for 1.00×10^{-3} M EDTA and 1.00×10^{-3} M Ca(II) drawn up using Program SPE[28] and the six overall equilibrium formation constants whose log values are listed in the left-hand table in the Figure. As the p[H] is varied from 2 to 12, the solution changes from one containing uncomplexed Ca^{2+}, H_4L^o, H_3L^-, and a small but significant

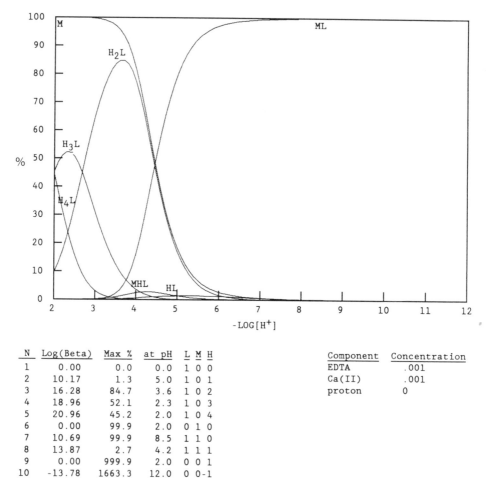

N	Log(Beta)	Max %	at pH	L	M	H
1	0.00	0.0	0.0	1	0	0
2	10.17	1.3	5.0	1	0	1
3	16.28	84.7	3.6	1	0	2
4	18.96	52.1	2.3	1	0	3
5	20.96	45.2	2.0	1	0	4
6	0.00	99.9	2.0	0	1	0
7	10.69	99.9	8.5	1	1	0
8	13.87	2.7	4.2	1	1	1
9	0.00	999.9	2.0	0	0	1
10	-13.78	1663.3	12.0	0	0	-1

Component	Concentration
EDTA	.001
Ca(II)	.001
proton	0

Figure 7.2. Species Distribution Diagram for 0.0010 M Ca(II) and 0.0010 M EDTA at 0.100 M Ionic Strength and 25.0 °C

concentration of H_2L^{2-} to solutions which ultimately contain only CaL^{2-}. At pH *ca* 4.5, the major species are about 0.50×10^{-3} M Ca^{2+}, CaL^{2-}, and H_2L^{2-} while in this region one also finds minor concentrations of $CaHL^-$ and HL^{3-}. Such diagrams are very useful in raising questions about the logic of a published result. The fact that $CaHL^-$ appears with a maximum concentration of only 2.7% of the total Ca(II) species present casts some doubt on its existence or, at the very least, indicates that its concentration cannot be accurately known under the conditions employed, and that the β value calculated for this species may have a relatively large error. The species HL, however, is not in doubt since it is a major species (as is L) when the metal ion is absent.

7.8   Experimental Methods for Measuring Complex Equilibria

Table 7.1 is a partial list of the methods available for measuring equilibrium constants or complex equilibria. In general, any method can be used if it can measure the concentration of at least one of the species in equilibrium in which a metal complex is formed. The concentration of that species plus the stoichiometry of the solution provides the information necessary to calculate the concentration of all species present at equilibrium. If a sufficient number of such equilibrium measurements is made over a range of conditions in which the concentrations of the species vary considerably, then accurate calculations can be carried out for the equilibrium constant of the reaction. Except in the special case of metal electrodes discussed below, the concentration of each species involved must be sufficient to allow accurate calculations; in other words, the complex must be appreciably dissociated under the reaction conditions chosen for the measurements.

Table 7.1. Methods Available for Determining Complex Equilibrium Constants

---

Standard Methods
      Potentiometry
      Spectrophotometry
      Specific metal ion electrodes
      Nuclear magnetic resonance spectroscopy
      Polarography
      Ion exchange
      Colorimetry
      Ionic conductivity
      Distribution between two phases
      Reaction kinetics
      Partial pressure measurements
      Solubility measurements
Competition Methods for Strong Complexes
      Ligand-ligand competition measured potentiometrically
      Metal-metal competition measured spectrophotometrically
      Ligand-ligand competition measured spectrophotometrically
Amphoteric Metal Ions

Of the methods listed in Table 7.1 potentiometry accounts for approximately 80% of the stability constants in the literature. This may be supplemented by spectrophotometry under special conditions where the pK's cannot be determined otherwise. Another method which has been employed to a considerable extent, and which may be developed further in the future, involves the use of specific metal ion electrodes. The first three methods therefore constitute the bulk of the experimental determinations of stability constants, and if one remembers that the competition methods for strong complexes, also listed in Table 7.1, are variations of the standard methods, one can list potentiometry, spectrophotometry, and specific electrode potentials as being the basis for about ninety five per cent of the stability constant data in the literature.

The use of the potentiometric method for determination of hydrogen ion concentration has been described above (Sections 7.2, 7.3, 7.5, 7.6). Early treatises on this subject have described the standard methods for stability constant determination in some detail.[12-15] A complete discussion of the methods available for stability constant determination may be found in a recent review by Anderegg.[37] Since most ligands have considerable basicity and control the hydrogen ion concentration as well as the metal ion concentration, potentiometry has and will continue to be the method of choice.

Therefore, a determination of the hydrogen ion concentration provides a measure of the concentration of the ligand at equilibrium. Such measurements, with and without the metal ion under investigation, provide enough data to determine the affinity constant between the metal ion and the ligand. In this type of measurement the glass electrode is very convenient and is in effect a single ion electrode whose potential is sensitive to the concentration of the free hydrogen ion. In general if the pH at equilibrium is between 2 and 12 the hydrogen ion concentration can be determined accurately enough to carry out the stability constant calculations. Figure 7.1 illustrates such titration curves.

A problem that is frequently encountered with very stable complexes of multidentate ligands, is the inability to form appreciable concentrations of the most basic form of the ligand without raising the pH above 12. This does not affect the accuracy of measurements made between 2 and 12 and the equilibrium constants thus obtained. The constant calculated, however, involves the displacement of hydrogen ions from the ligand by the metal ion according to equation 7.29. This constant can be measured reasonably accurately and provides the concentration of the ligand and the metal ion in the pH range for which the measurement is valid. However, it is not the stability of the reaction between the metal ion and the most basic form of the ligand L^{n-} and comparisons between the constants determined and the stability constants of other metal complexes cannot then be made.

$$M^{n+} + H_nL \rightleftharpoons ML^{m-n} + nH^+ \qquad K_M^{ML} = \frac{[ML][H^+]^n}{[M^{m+}][H_nL]} \qquad 7.29$$

7.8.1 Absorbance methods

For the above reasons attempts have been made to measure the hydrogen ion association constants of basic ligands at pH values above 12, if the hydrogen ion concentration (actually the hydroxide ion concentration) can be determined from the stoichiometry of the solution, while making a reasonable attempt to keep the ionic strength at a constant value. When the concentration of free ligand and its acid forms are measured by spectrophotometry and the hydroxide ion concentration can be determined from the stoichiometry of the solution, approximate pK's can be calculated for the ligand. Potentiometry does not provide microscopic information on the protonation and metal coordination reactions of ligands. For such microscopic information, spectroscopic measurements (e.g. NMR) or spectrophotometric absorbance studies are required. It should be noted however, that while these measurements are microscopic in nature the reaction involved and the constant determined may indeed be a macroscopic constant because of the presence of non-absorbing or other species in the equilibrium reaction. For a detailed description of the other methods of stability constant determination and their limitations the reader is referred to the review by Anderegg.[37] In special cases where participation of such colorless species can be ruled out the equilibrium constants obtained are both microscopic and macroscopic constants.

The use of optical methods can become quite complicated, however, if more than one species is involved in the absorbance, and this is usually the case. It is usually found that the metal complex also absorbs in a region which may or may not coincide with that of the protonated form of the ligand. If the absorbances occur with peaks at different wavelengths and can be separated, then both the metal complex and its protonated form can be determined individually. Also, if the absorbance of the free ligand can be determined separately then the concentrations of the metal complex and the protonated form can sometimes be determined. The quantity measured spectrophotometrically is the absorbance A expressed by equation 7.30.

$$A = \log I_0/I = \ell \sum \varepsilon_i C_i \tag{7.30}$$

where ℓ is the path length of the light, ε_i is the extinction coefficient of species i, C_i is the concentration, I_0 is the intensity of the incident light and I is the intensity of the transmitted light. Solution of equation 7.30 is fairly simple when there is only one absorbing species or when the absorbing species absorb at different wavelengths and are easily separated. In the general case in which two or more species have overlapping absorption bands the expression for A (7.31) must take into account all the absorbing species, which may include the metal ion, the various complexes of the metal ion, and the absorbance of the free ligand. The solution of 7.31 for the β value of the complex or complexes involved may be simple or difficult depending on whether the value of the extinction coefficient may be obtained for each individual species. Those of the free ligand and of the metal ion may, of course, be determined separately. The values of E for the individual complex species sometimes cannot be obtained directly from the data available if the maximum concentration of the species involved cannot be obtained under the experimental conditions that prevail. In such cases the various extinction coefficients can be left as unknown quantities and the absorbance data be taken over sufficient change in concentration that the unknown extinction coefficients can be calculated as well as the β values. For each set of absorbance curves the hydrogen ion concentration is important and must be determined separately by the potentiometric (glass electrode) method. This is particularly true when the ligand exists in several protonated forms over the experimental conditions employed.

$$A = \ell \left(\sum_{0}^{n} \varepsilon_n[ML_n] + \sum \varepsilon_{H_iL}(H_iL) \right) \tag{7.31}$$

$$= \ell \left(\varepsilon_M[M] + \sum_{1}^{n} \varepsilon_n \beta_n[M][L]^n + \sum \varepsilon_{H_iL}[H_iL] \right)$$

An example of the use of spectrophotometry in the determination of a metal chelate stability constant can be found in a recent paper by Clarke on the determination of the 1:1 stability constant of 1,2-dimethyl-3-hydroxy-4-pyridinone iron(III) chelates.[38] Although this ligand forms 3:1, 2:1 and 1:1 complexes with iron(III) it was found that in the lowest pH region only the 1:1 complex was formed. However, insufficient dissociation of this complex was observed around pH 2 by the potentiometric method. Therefore, the complex dissociation was measured spectrophotometrically at 10^{-4} M concentration by observing the variation in the 568 nm absorbance band of the 1:1 chelate. The ionic strength was controlled at 0.100 M by using mixtures of HCl and KCl, the total of which equaled one tenth molar concentration.

7.9   Specific Metal Ion Electrodes

For certain metal ions the complexes are so stable that the potentiometric method cannot be applied since the complex is completely formed at the lower limit of the pH measurement, which is around pH 2. Schwarzenbach used the mercury electrode for determining among other things the stability constant of the Hg(II)-EDTA complex.[39] In this case the metal ion concentration [Hg^{2+}] may be determined by the use of the mercury electrode, in the presence of a known excess of the ligand. Since the ligand has an affinity for hydrogen ions it is necessary to know the hydrogen ion concentration. Also it is desirable to take the potential measurements over a pH range. The concentration of free mercury ion is then expressed by equation 7.32.

$$Hg^{2+} + 2e^- \rightleftharpoons Hg(\ell) \qquad E = E_o + RT/2F \ Ln[Hg^{2+}] \qquad 7.32$$

The mercury electrode is a very sensitive and reversible electrode and is sensitive to the concentration of mercuric ion down to 10^{-40} molar. This property allows the determination of the stabilities of the most stable mercury complexes through the use of equation 7.32. The electrode simply consists of a pool of mercury exposed to the experimental solution.

A simple variation of this procedure can be used for determining the stability constants of complexes that are weaker than those of the mercuric ion. Cell 7.33 provides a system in which the mercury electrode is sensitive to the concentration of the magnesium ion, which in turn regulates the concentration of EDTA in solution and thus the potential of the mercury electrode which is in equilibrium with the mercuric-EDTA complex. Another variation of the mercury electrode is a mercury lead amalgam coated with lead oxalate indicated by 7.34. The surface of the lead oxalate also contains calcium lead oxalate, and this electrode is sensitive to the concentration of the calcium ion in solution. The potential of the electrode is sensitive to the calcium ion through the regulation of the oxalate concentration on the surface of the electrode.

$$Hg(\ell), \text{ test solution with HgEDTA, MgEDTA, } Mg^{2+} \qquad 7.33$$

$$Hg\text{-Pb, } PbC_2O_4(S), CaC_2O_4(S), Ca^{2+} \qquad 7.34$$

Still another variation of the metal electrode system is the very familiar silver-silver chloride electrode indicated by 7.35. The potential of this electrode is sensitive to the chloride ion concentration in the solution. This type of electrode is often used as a reference electrode by keeping the chloride ion concentration constant in the solution in contact with the silver-silver chloride electrode.

$$Ag(S), AgCl(S), Cl^- \qquad 7.35$$

Equation 7.36 indicates the potential of an inert electrode in equilibrium with an oxidized and reduced metal ion or metal complex species in solution. The electrode consists of an inert metal foil, usually platinum, immersed in the experimental solution. Examples of the oxidation-reduction pairs that can be measured are Fe(III)/Fe(II), Hg(II)/Hg(I) and Tl(III)/Tl(I).[39] The variation of the potential of cell 7.36 with the free ligand concentration is often simplified by the fact that the lower valence form of the metal is usually much weaker than the higher valence form and its stability can be determined by conventional methods such as the potentiometric or spectrophotometric techniques. Thus the stability of the ferric EDTA complex was determined by Schwarzenbach[40] by the use of the reduction potential of the ferric to the ferrous form of EDTA. This was carried out over a variation of pH so that the stability constant of the hydroxo form of the ferric EDTA complex was also determined. Since the stability of the ferrous complex of EDTA was known it was a simple matter to calculate the stability constant for the ferric EDTA complex.

$$Ox + ne^- \rightleftharpoons Red \qquad E = E_0 + RT/nF\ Ln[Ox]/[Red] \qquad 7.36$$

In recent years a number of specific metal ion electrodes have been developed which allow the direct determination of the concentration of the metal ion. Also a number of specific anion electrodes are now available. Thus it is possible to measure directly the concentrations of the alkali metal ions and the alkaline earth metal ions and a number of other ions by determining the potential of the electrode that is specific for that ion. However, in most cases the specific ion electrodes have not been accurate enough for the determination of stability constants as described above, and are characterized not only by this inherent inaccuracy but by drifting of the electrode. This can be partially overcome by frequent calibration of the electrode, preferably before and after each measurement, with a solution containing the specific ion involved.

7.10   Polarography and the Study of Solution Equilibria

Polarography has been in some ways the poor relation of glass electrode potentiometry as far as the study of solution equilibria is concerned. This is reflected in the fact that texts dealing with the study of metal ion complexation by polarography are now becoming quite dated.[42-45]   Reluctance to use polarography relates to the greater ease of interpretation of the results of glass electrode potentiometry, particularly with the advent of computer programs that greatly facilitate analysis of the potentiometric data. However, polarography can have definite advantages, and sometimes can provide answers that are inaccessible by other means. Polarography can operate at total metal ion concentrations of as low as 10^{-6} M making it advantageous in situations of low complex solubility and when dealing with easily hydrolyzable ions to form hydroxide precipitates. The ability of polarography to work at very low total metal ion concentrations means that precipitation of solid hydroxides will thermodynamically occur at much higher pH values.   Further, the slow formation of solid hydroxides at the very low total metal concentrations accessible to polarography means that for kinetic reasons one can study complexes of metal ions that should not exist at all from the thermodynamic point of view because of hydrolysis. The consequence of this is that for complexes of some metal ions such as, for example, Pb^{2+} with cryptand-222, in addition to the ML species observed by glass electrode potentiometry, MLOH and $ML(OH)_2$ species were observed[46] by polarography. For the EN complexes of Pb^{2+}, only a ML complex was observed by glass electrode potentiometry. In a polarographic study, the additional complexes, MLH, MLOH, ML_2, and ML_2OH species were observed as well. Even more important, very acidic metal ions such as Bi^{3+} form complexes even in 0.5M HNO_3 which can usually only be studied by polarography.[47]   This has revealed a rich chemistry of Bi(III) with polyamines such as TRIEN and 15-aneN_4 which were previously unsuspected. The combined use of polarography and glass electrode potentiometry is recommended for very complex systems with many species present, as aspects of polarography discussed below provide for an excellent analysis of speciation.

The essence of the polarographic technique is the production of a polarographic wave of current as a function of applied potential as electroactive species are reduced at the mercury drop electrode. The most convenient technique for studying speciation and determining formation constants is DPP (Differential Pulse Polarography). In conventional DC (Direct Current) polarography, a dropping mercury electrode is used in which a succession of mercury drops form at the end of a capillary as a result of pressure applied to the Hg reservoir. The increase in surface area as the drop grows causes the current to increase, and produces a large fluctuation in current with time, making the polarographic trace oscillate around a mean current. In addition, the charging of the electrical double layer around the drop causes a current which limits the detection of species present in very low concentration, which contribute only a very small current. In DPP the current is sampled only as a short pulse when the drop has reached almost full size, just prior to and at the end of a short voltage pulse imposed on a voltage step, so that charging of the electrical double layer is already complete, and the change in size of the drop during the process of applying the potential is minimal. This produces much smoother polarograms, which are presented in differential form ($i_{pulse} - i_{step}$), so that peaks of current as a function of applied potential, rather than waves, are obtained, which makes the detection of minor species much easier (see Figure 7.3).

The response of the polarographic peaks to variation in ligand concentration is of two kinds. In the first kind, known as *labile* systems, the free metal, the free ligand, and metal-ligand species present at the electrode surface are in rapid equilibrium. The variation of peak potential as a function of concentration of free ligand is given by the Lingane equation.[42,43]

$$E = (E)_s - (E)_c = 0.0591/n \log_{10} \beta_j + j\ 0.0591/n\ \log_{10} C_x \qquad 7.37$$

In equation 7.37, $(E)_s$ and $(E)_c$ are the half-wave potentials (or in the case of DPP, the peak potentials) of the free metal ion and the complexed metal ion respectively. C_x is the concentration of the free ligand, and j and β_j are the coordination number and stability constant of the complex. This equation as it stands is best suited to the study of complexes of non-protic ligands such as chloride. Use has been made of this equation where the total ligand concentration of a protic ligand such as EDTA is increased at constant pH. This is a potentially misleading approach, because at constant pH there is no proton dependency that might allow one to infer the number of protons involved in the metal species. It is usually assumed that the species present is the ML species, rather than MLH or MLOH, for example, and it is this kind of approach that may have led to inaccurate formation constants reported by polarography. Equation 7.37 needs to be expanded[45] to include the effect of pH in the form of equation 7.38.

$$E_{peak} - RT/nF \ln (i_c/i_s) = RT/nF \ln \beta_{ML\ (j)} + j\ RT/nF \ln [L] + b\ RT/nF \ln [OH^-] \qquad 7.38$$

Where (i_c/i_s) is fraction of free metal ion in the presence of ligand and M and L denote the metal ion and ligand in the overall reaction:

$$M + jL + b\,(OH^-) \longrightarrow ML_j\,(OH^-)_b \qquad 7.39$$

where [L] is calculated from the known pH, total ligand concentration ($C_{L(t)}$), and protonation constants ($K_1, K_2, K_3, ...$) of the ligand:

$$[L] = C_{L(t)} /\{ 1 + K_1[H^+] + K_1 K_2\,[H^+]^2 + K_1 K_2 K_3 [H^+]^3 + \ . . . . . \} \qquad 7.40$$

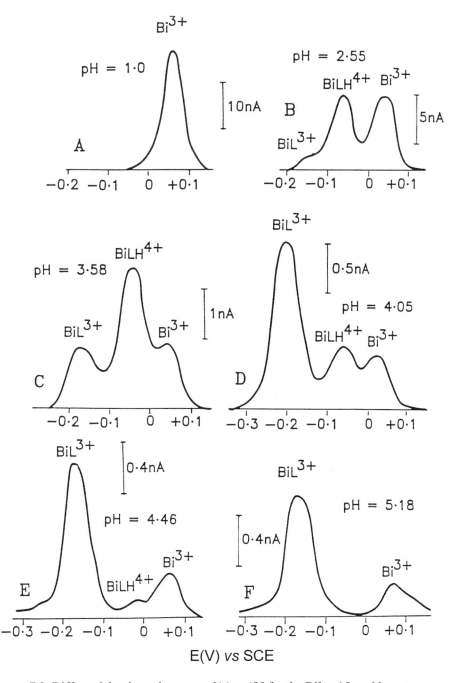

Figure 7.3 Differential pulse polarograms [(a) to (f)] for the Bi^{3+} - 15-aneN$_4$ system, as a function of pH, illustrating the analysis of non-labile metal-ligand systems in polarography. The vertical scale indicated by a bar on each polarogram corresponds to the current shown. The peaks correspond to the solution species Bi^{3+}, [Bi(15-aneN$_4$)]$^{3+}$, and [Bi(15-aneN$_4$)H]$^{4+}$, as indicated, although,[47] the peak labelled Bi^{3+} represents bismuth(III) hydroxide species as well. The pH values of the solutions at which the polarograms were recorded are indicated. Redrawn after reference 47.

The approach adopted with protic ligands follows that of glass electrode potentiometry. A glass electrode is used to record the pH, and at regular pH intervals, adjusted by titration with accurately standardized base, a polarogram is recorded. Shifts in peak position as a function of pH are obtained, and the shifts can be analyzed using equation 7.38 to obtain values of $\beta_{ML_jH_b}$. The slope of E_{peak} as a function of pH gives a clear indication of the composition of the species involved. Thus, for example, for the ligand[46] N,N'-dipicolylethylenediamine, DPA2, with Pb[II], the peak potential varies with pH as shown in Figure 7.4. One can analyze Figure 7.4 as follows.

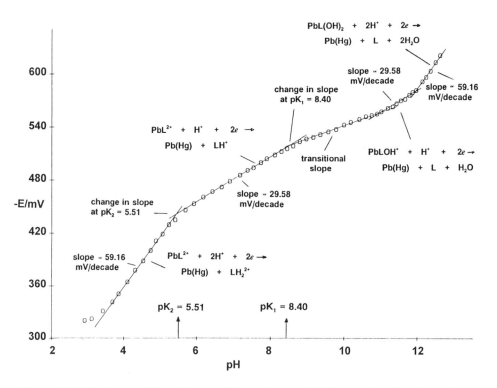

Figure 7.4. Variation of the polarographic peak potential (E) as a function of pH for the Pb(II) complex of DPA2, illustrating the analysis of these shifts for labile metal-ligand systems in polarography. The slopes of E versus pH are in accord with the electrode processes indicated. The changes in slope at the pH values corresponding to the protonation constants, pK_1 and pK_2, correspond to a change in the number of protons involved in the reduction step at the Hg electrode. The transitional slope corresponds to a region where both ML and MLOH are present, leading to a 'mixed' slope. The total Pb concentration is approximately 10^{-4} M, and the ligand 10^{-3} M. Redrawn after reference 46.

The protonation constants of DPA2 in the pH range covered in Figure 7.4 are[46] 8.40 and 5.51. One sees in Figure 7.4 that the potential increases with increasing pH with a slope of close to 59.16 mV per pH unit between pH 3.0 and 5.5. This slope is consistent with the reduction process at the Hg electrode:

$$ML + 2H^+ + 2e = Pb(Hg) + LH_2^{2+} \qquad\qquad 7.41$$

Equation 7.38 indicates that the slope at 25 °C is given by $59.16\{mH^+/ne\}$ where mH^+ is the number of protons involved in the reduction process, and ne is the number of electrons. Thus, in [5] mH^+ is 2 and ne is 2, which suggests a slope of 59.16 mV per pH unit, as observed. The change in slope at pH 5.5 corresponds to the pK_2 of 5.5 for DPA2, so that there is only one proton in the reduction process above pH 5.5, which gives a slope of close to 29.58 mV per pH unit. Above pH 8.4, where the last proton on L is lost, the slope should now be independent of pH. The slope does initially drop down quite considerably, but rises to a steady slope of close to 29.58 mV per pH unit, up to a pH of 11.8, which corresponds to the process:

$$MLOH^+ + H^+ + 2e = Pb(Hg) + L + H_2O \qquad\qquad 7.42$$

Above a pH value of about 11.8 the slope reaches a constant value close to 59.16 mV per pH unit, which suggests the involvement of two protons in the electrode reduction process, according to:

$$ML(OH)_2 + 2H^+ + 2e = Pb(Hg) + L + 2H_2O \qquad\qquad 7.43$$

In *non-labile* systems, the rate at which the ligand, metal ion, and complex species is in equilibrium is slower than the time constant of the polarographic reduction process. Under these circumstances, separate peaks for the different species present are observed. This is a great aid to the analysis of speciation. The peak height for each species present is proportional to the diffusion coefficient for each species. One can usually quite easily obtain the diffusion coefficient for each species, and it is then a simple matter to calculate the concentration of each species present at each pH, and so obtain the formation constants. A selection of polarograms for the Bi(III) complex of 15-aneN$_4$ at different pH values is seen in Figure 7.3. One sees that at (a) in Figure 7.3, where the pH is 1.0, only a peak due to Bi^{3+} is present. As the pH is raised to 2.55 in (b), a peak due to $BiHL^{4+}$ appears, with a small peak due to BiL^{3+}. In (c) the pH is now 3.58, the BiL^{3+} peak has grown further, and is the dominant peak in (d) with a pH of 4.05. As the pH is raised further in (e) and (f), the "BiL^{3+}" peak shifts with a slope of 59 mV per decade. This is typical of most ligands where at least some of the steps will be labile, and at higher pH the electrode process corresponding to a slope of 59 mV is:

$$BiLOH^{2+} + 3H^+ + 3e = Bi(Hg) + LH_2^{2+} + H_2O \qquad\qquad 7.44$$

The occurrence of labile and non-labile behavior in polarography resembles the familiar behavior of NMR spectra, where averaged peaks, or separate peaks, are observed depending on whether the chemical processes are occurring more slowly or faster than the NMR time scale. One notes that where complex formation involves the formation of more than three Bi-N bonds, non-labile behavior occurs, while Bi-O bonds appear to lead to labile polarographic behavior. Steps involving protonation and deprotonation equilibria of the complex (e.g. $ML^{3+} + OH^- = MLOH^{2+}$) are invariably labile, as would be expected, except for steps such as the conversion of $BiLH^{4+}$ to BiL^{3+} ($L = 15\text{-aneN}_4$) in Figure 7.2, where one supposes that there is a slow conformational change on loss of a proton from $BiLH^{4+}$ to give BiL^{3+}. The complexes of the divalent Pb^{2+} ion appear to be invariably labile, as seen in Figure 7.4, except with ligands such as tetraaza macrocycles, where the slowness is related to the macrocyclic structure of the ligand.

Polarography works best for studying complexation of heavy post-transition metal ions such as Cd^{2+}, Tl^+, Pb^{2+}, Sb^{3+}, or Bi^{3+}, as well as Cu^{2+}, Cu^+, Co^{2+}, Ni^{2+}, and Zn^{2+}. Its very specific advantages for the study of complexes at low pH and low total metal concentration mean that it can be invaluable in the study of complexes of such very acidic metal ions as Bi^{III}, and the more complete study of complexes of other metal ions. This is important, since metal ions in biomedical applications and in the environment, are more often than not in the very low concentration ranges amenable to study by polarography.

7.11   Other Methods

While there are many other methods for determining protonation constants and metal ion equilibrium constants, such as ion exchange, NMR, etc. as listed in Table 7.1, they are usually useful under special circumstances which arise when potentiometry, spectrophotometry or specific metal ion concentration electrodes cannot be used. For example, ion exchange methods may be employed for studying complex formation by trace amounts (carrier free) of radioactive metal ions which have concentrations so low as to not be detected by any means except radioactivity.

7.12   Competition Methods

When the stability of a complex is so high that an appreciable amount of the complex is not dissociated at the lowest pH available (pH ~2), competition methods are almost always used for determining the stability constants involved. These methods may involve two ligands competing for the same metal ion, or two metal ions competing for the same ligand. The equilibria established in this manner may be detected by potentiometric or spectrophotometric methods. The following are a few examples of the competition reactions that have been investigated in this manner.

7.12.1 Ligand-ligand competition by potentiometric methods

This is by far the most common of the competition methods employed in the determination of high stability constants, and was first used by Schwarzenbach to determine the stabilities of the EDTA complexes of the transition metals which were not dissociated at the lowest pH's obtained by direct potentiometric titration of the metal vs EDTA.[48] The competing ligand in this case is TREN (tris(aminoethyl)amine) which has a higher affinity for transition metals than does EDTA but also has much higher pK_a's and therefore at low pH TREN is present in the protonated form while the EDTA complexes of the transition metals are completely formed. The equilibria investigated are represented by equation 7.45 in which the EDTA complexes of the transition metal ions (represented here by Cu^{2+} ion) are converted to the TREN complex at higher pH. Schwarzenbach added calcium ion to the experimental solution in order to form a complex with EDTA and simplify the reaction mixture, which would otherwise contain protonated forms of EDTA. While the addition of the second metal ion greatly simplified Schwarzenbach's calculations, such procedures are no longer necessary for today's computers because it makes no difference to the computer whether the protonated forms of EDTA are formed or whether a calcium complex is formed. In any case, the stability of the TREN-copper complexes and the pK's of the ligand can be obtained separately, as can the pK's of EDTA, by potentiometric methods. Thus only the competition constant is unknown in equation 7.45. From this constant, if the stability constant of the copper-TREN complex is known, then that of the copper-EDTA complex can be calculated.

$$CuEDTA^{2-} + H_3TREN^{3+} + Ca^{2+} \rightleftharpoons CuTREN^{2+} + CaEDTA^{2-} + 3H^+ \qquad 7.45$$

$$K_X = \frac{[CuTREN^{2+}][CaEDTA^{2-}][H^+]^3}{[CuEDTA^{2-}][H_3TREN^{3+}][Ca^{2+}]}$$

$$\log \frac{K^{CuEDTA}}{K^{CuTREN}} = \log K^{Ca}_{CaEDTA} - \log K_X - \log \beta_3^H \qquad 7.46$$

Further examples of the use of the potentiometric competition method for the determination of high stability constants can be found in the paper by Harris and Martell.[49] In this work potentiometric pH profiles were obtained for the systems containing gallium complexes of SEDDA-HBIDA, EDTA-HBIDA, NTA-HBIDA, HEDTA-HBIDA, CDTA-NTA, and EDTA vs CDTA, where SEDDA is symmetrical ethylenediaminediacetic acid, HBIDA is o-hydroxybenzyliminodiacetic acid, DTPA is di-ethylenetriaminepentaacetic acid, NTA is nitrilotriacetic acid, CDTA is cyclohexanedi-aminetetraacetic acid, HEDTA is N-hydroxyethylethylenediaminetriacetic acid, and EDTA is ethylenediaminetetraacetic acid. Another use of the application of the potentiometric competition method can be found in papers by Raymond in which he, for example, used CDTA as a competing ligand for such catechol-like hexadentate ligands as TRENCAM.[50]

7.12.2 Spectrophotometric determinations of competition constants

Because gallium(III) ion has a completely filled d-shell, the absorption spectra of the gallium ion and its complexes is very low and for all practical purposes zero. Harris took advantage of this by competing the ferric ion with gallium ion for the formation of complexes of HBED (hydroxybenzylethylenediaminediacetic acid), EHPG (ethylenebis-hydroxybenzylglycine), and HBIDA.[49] Also experiments were run with NTA complexes in which the strongly absorbing Cu(II) ion was used as the competing metal ion. In no case did the gallium(III) chelate of the free gallium ion as well as its complexes have appreciable absorbance in the wavelength region studied. The literature values were used for the stability constants of the ferric chelates of EHPG,[51] HBED,[52] and HBIDA[53] and the cupric complex of NTA.[52]

The spectrophotometric method may also be used for detection of the absorbance of the free ligand and its metal complexes. For example, the gallium(III) ion or the aluminum(III) ion, which have no effective absorbance over most of the measurable absorbance range in the uv-visible region, can be examined with respect to their competition for a highly absorbing anion such as a catecholate vs a non-absorbing or very low absorbing anion such as EDTA or CDTA. The protonated forms of the catecholate type ligands have absorbances at different wavelengths from those of their metal chelates and the degree of formation of the metal chelate can therefore be obtained unambiguously from the absorption spectrum. Corrections, if necessary, are made for the absorbances of the protonated forms of the ligand.

7.13 Amphoteric Metal Ions

An amphoteric metal ion is one which forms very stable hydroxo complexes which are soluble in excess base. Gallium(III) is a metal ion which has been the most investigated in this respect but there are other metal ions which fit these requirements such as Pb(II) and Al(III). The tendency of the hydroxo complexes of gallium(III) to form the tetrahydroxogallium complex has been reported by Baes and Mesmer.[54] This hydroxo complex has been used by Motekaitis and Martell[55] for the determination of the stabilities of many gallium complexes which are completely formed at very low pH (~2) so that their stabilities cannot be determined by the customary potentiometric methods. The technique involves the formation of the soluble tetrahydroxogallate complex in the presence of excess alkali hydroxide ion and the addition of standard HCl to form the gallium complexes of the ligand under investigation while the pH is directly determined in the standard way with potentiometric apparatus. Thus the method is very similar to the potentiometric pH determination except that the titration is begun on the alkaline side rather than with the acidic form of the ligand. In this manner the stabilities of over twenty gallium(III) complexes were measured and reported,[55] most of which were fully formed at low p[H] (~2). The general reaction involved was:

$$Ga(OH)_4^- + L^{n-} + 4H^+ \rightleftharpoons GaL^{(n-3)-} + 4H_2O \qquad\qquad 7.47$$

Other, more recent, examples of the use of the $Ga(OH)_4^-$ complex for the determination of high stability constants of Ga(III) chelates are the reports of the stability of the Ga(III)-N,N',N''-tris(5-methyl-3-hydroxy-2-pyridylmethyl)-1,3,7-triazacyclononane complex by Martell et al.[56] and of the tetraazacyclododecanetetraacetate complexes by Clarke and Martell.[57]

7.14 Critical Stability Constants and Their Selection

As noted in 7.3, stability constants of metal ion complexes in the literature have frequently been published by several investigators who have reported values that differ by up to one or two orders of magnitude. These variations found in the published literature may be due to several factors including impure ligand, poor experimental design, faulty experimental conditions, inaccurate measurements and erroneous calculations. For this reason there has been developed by Smith and Martell[34,35] a compilation of critically selected stability constants such that every complex formation reaction is indicated by one and only one equilibrium constant for a specific set of conditions. The selection of critical (recommended) data from series of published constants differing by a considerable amount may at first seem difficult but the application of well established criteria eliminates most of the problems in coming to an appropriate decision. One of the prime factors is the carrying out of essential experimental control, in which the care with which the reaction conditions and measurements are made and clearly explained. There are many papers which are deficient in specifying essential reaction conditions such as temperature, ionic strength, nature of supporting electrolyte and these are excluded from the compilation. An adequate description of the measurements, apparatus and its calibration is also necessary. Also a basis for disqualification of data is the lack of information on the purity of the ligand and failure to define the equilibrium quotients that are reported. When several investigators are in close agreement on a particular constant the average of the results is selected. In cases where the agreement is poor and few results are available for comparison the selection may be guided by comparison with values obtained for other metal ions with the same ligand and with values obtained for the same metal ion with similar ligands. While established trends among similar metal ions and among similar ligands are valuable in deciding between widely varying data, such guidelines are used cautiously so as not to overlook occasional real examples of specificity or anomalous behavior. When there is poor agreement between published values, and comparisons with other metals or ligands do not suggest the best value, the results of the more experienced research groups who have supplied reliable values for other stability constants are selected. The values reported by only one investigator are included in the database unless there is some reason to doubt their validity. Generally the values are accepted if the paper meets the criteria described below which have been established as minimum standards for obtaining accurate values of stability constants. A complete bibliography for each ligand is also included in the database so that the user may determine the completeness of the literature search employed in the selection of critical values. This means that many papers not used in compilation are also included in the literature survey. The user may thus employ all the references to make an independent evaluation if there are any questions or doubts concerning the selection of the critical constant.

7.14.1 Minimum requirements for equilibrium data

An indication of how the selection of critical constants from non-critical data works is provided by the following example of DTPA complexes. Table 7.2 gives a complete list of the stability constants for the copper-DTPA complexes that have been published up to the present time. The constants reported differ by as much as one and a half log units. While such data may be of some value they are useless for the non-specialist who needs an equilibrium constant for work in his field or for teaching purposes. The critical value for the Cu(II) complex at 25.0 °C was derived from the constants listed in Table 7.2 at 25.0 °C and with the help of ΔH for those listed at 20.0 °C and 0.10 M ionic strength. Thus examination of the data in Table 7.2 resulted in the recommendation of a single critical constant 21.40 ± 0.05 which may be found in the tables of critical constants published by Smith and Martell.[34,36]

Table 7.2. Complete List of Stability Constants for $Cu^{2+} + DTPA$

Method[a]	Temperature	Ionic Strength	Background Electrolyte	Reported Log K	Ref.
ise	25°	0.10	KNO_3	21.45	59
Hg	20°	0.10	$NaNO_3$	21.6	60
gl	20°	0.10	KCl	21.53	61
gl	25°	0.10	KNO_3	21.1	61
gl	20°	0.10	KCl	21.03	62
Hg	25°	0.10	$NaClO_4$	20.5	63
gl	20°	0.10	NaOH	20.1	61

[a] ise = ion selective electrode; Hg = mercury electrode; gl = glass electrode with competition from another ligand as pH is varied.

In order to develop a systematic way of evaluating new data on stability constants, a set of minimum criteria has been developed for the reporting of such data in publications. They are the following:

1. The definition of equilibrium quotients reported, and each term in the equilibrium quotient should be carefully described.

2. The purity of ligands, reagents and solvents, and the procedures followed in solvent and reagent purification should be described. The standardization of metal solutions as well as the determination of the exact molecular weight of the ligand must be indicated.

3. The composition of the solution, especially its ionic strength, and any other relevant factors including the range of metal and ligand concentrations investigated. For mixed solvent systems, the solvent composition must be defined.

4. When appropriate, the pH range over which measurements have been made, the titrant used, and the K_w value determined or used should be given.

5. The instrument(s) (e.g. pH meter, electrode, spectrophotometer, etc.) used in the experimental studies, and an explicit description of the method of calibration.

6. The temperature and temperature range.

7. The number of data points recorded in a titration (or elsewhere as appropriate), and the number of replicate measurements.

8. The computer program, or any other method of calculation used to derive final results from the experimental values, and a literature reference if the programs are the work of others; new programs should be shown in terms of the stepwise logic involved. Any discussion of the reasons for choosing a given program is considered appropriate.

9. The range of the results, the standard deviation of the final results, the sources of error, and the methods used in establishing parameters.

10. Any assumptions made in working up or modeling the data should be clearly set forth, including any problems encountered during the determinations or the calculations.

This list is similar to the one recommended in Appendix 5 by Motekaitis and Martell[28] and is also similar to the minimum criteria set forth by the Equilibrium Data Commission of the Analytical Division of IUPAC, described in a paper by Tuck.[58]

7.15 Development of a Complete Metal Complex Database

There are many fields in which the number of metal ions and ligands encountered is quite extensive and the probability that most of them appear in the critical compilation is quite small. Examples of such fields are the environment, which contains many toxic as well as essential metal ions and a great many ligands some of which are known and others are unknown, and biological systems, for example, the blood serum containing many amino acid ligands and several metal ions. In order to determine the speciation of metal complexes in such complicated systems it is necessary to estimate many of the stability constants since the possibility that they will be determined in the near future is very remote. For this purpose linear correlations have been very useful (see Section 2.2). Table

7.3 gives examples of early publications of linear free energy relationships for stability constants. Correlations with pK's of the ligands or ionization potential of the metal and electronegativity of the metal ion, or of one metal ion with another or between closely related ligands are very helpful in determining the values of unknown constants.

Table 7.3. Examples of Early Linear Free Energy Relationships

Authors	Relationships	Ref.
Larsson (1934)	$\log K_{ML}$ vs ligand basicity	64
Bjerrum (1950)	$\log K_{ML}$ vs $\log K_{HL}$	65
Calvin, Wilson (1945)	$\log K_{ML}$ vs $\log K_{HL}$ (graphical)	11
Calvin, Melchior (1948)	$\log K_{ML}$ vs ionization potential (graphical)	66
Irving, Williams (1948) (1953)	$\log K_{ML}$ vs atomic number (graphical)	67,68
Davies (1951)	$\log K_{ML}$ vs e^2/r	69
Martell, Calvin (1952)	$\log K_{ML}$ vs e^2/r	12
Van Uitert, Fernelius, Douglas (1953)	$\log K_{ML}$ vs electronegativity (graphical)	70
Irving, Rossotti (1956)	$\log K_{ML} = a \log K_{HL} + b$	71
	$\log K_{ML}$ vs $\log K_{ML'}$ (graphical)	
	$\log K_{ML}$ vs $\log K_{M'L}$ (graphical)	
Nieboer, McBryde (1970)	$\log K_{ML} = b \log K_{M'L} + (\log K_{ML'} - b \log K_{M'L'})$	72
	$\log K_{ML} = c \log K_{ML'} + (\log K_{M'L} - c \log K_{M'L'})$	

Also, careful inspection of stability constants compiled in the critical tables indicates that many constants may be divided into groups of related ligands which show little variation in the their stability constants. The sample data listed in Table 7.4 differ in the stability constants or protonation constants by as little as one tenth of a log unit. For such types of ligands other members of the group can be assigned the same stability constant with some degree of confidence. For ligands which show greater variation, close approximations may be made if the effects of variation in structure on stability constants are studied and the observed trends are taken into consideration. This process is aided by the use of structure stability constant relationships which have been described in the literature compiled in Table 7.3.

Stability constants also vary with ionic strength and the charge on the complex as well as on the metal ion and the ligand and have considerable influence on the values obtained. A series of corrections for stability constants is listed for metal ions with charges of +1 to +4 and for ligands with charges of -1 to -3. The corrections listed in Table 7.5 are based on known constants in most cases. The correction values in brackets are based on the trends observed in the known constants.

With the resulting values of critical constants supplemented by the estimated constants many metal ions and ligands can be covered. Such an expanded database of stability constants and protonated constants of ligands may be used for the speciation calculations of many complex systems such as blood serum and environmental systems. For further details on how these calculations may be used, the reader is referred to the book by Motekaitis and Martell.[28]

Table 7.4. Critical Protonation and Formation Constants[a]

Ligand Type	Log Protonation Constant		Log Formation Constants, Cu(II)		Log Formation Constants, Ag(I)	
	NR_3	RCO_2^-	K_2	β_2	K_1	β_2
Primary amine	10.6 ± 0.1		(4)		3.5 ± 0.1	7.4 ± 0.2
Carboxylic acid		4.7 ± 0.1	1.8 ± 0.1	3.2 ± 0.5	0.7	0.6
2-Amino acid	9.6 ± 0.1	2.3 ± 0.1	8.1 ± 0.1	14.9 ± 0.1	3.3 ± 0.3	6.7 ± 0.4
2-Hydroxylamine	9.6 ± 0.1		5.7	9.8	3.1 ± 0.1	6.7 ± 0.0
(Glycyl)$_n$amide	7.9 ± 0.1		5.0 ± 0.2	5.2 ± 0.2^b		
Dipeptide	8.1 ± 0.1	3.1 ± 0.1	5.5 ± 0.3	4.2 ± 0.7^b		
Secondary amine	11.1 ± 0.2				3.5 ± 0.5	7.0 ± 0.7
Tertiary amine	10.5 ± 0.4				3.8 ± 0.3	4.5 ± 0.7

[a] 25.0 °C, 0.100 M ionic strength. [b] Amide protonation constants.

Table 7.5. Variation of Stability Constants with Ionic Strength and Charge Relative to 0.1 M Ionic Strength. Values in [ ] are Estimated Based on Observed Trends.

Ionic Strength		0.0	0.5	1.0	2.0	3.0
				M$^+$		
L$^-$	log K_1	+0.2	-0.1	-0.1	0.0	+0.2
	log β_2	+0.3	-0.1	0.0	+0.3	+0.6
L^{2-}	log K_1	+0.4	-0.2	-0.2	-0.1	0.0
	log β_2	+0.6	-0.4	-0.4	-0.3	0.0
L^{3-}	log K_1	+0.6	-0.3	-0.3	-0.3	-0.2
	log β_2	+1.0	-0.5	-0.5	[-0.4]	[-0.2]
				M^{2+}		
L$^-$	log K_1	+0.4	-0.2	-0.2	-0.1	0.0
	log β_2	+0.6	-0.4	-0.4	-0.3	0.0
L^{2-}	log K_1	+0.8	-0.4	-0.4	-0.4	[-0.3]
	log β_2	+1.2	-0.8	-0.8	[-0.7]	[-0.5]
L^{3-}	log K_1	+1.2	-0.6	-0.6	[-0.7]	[-0.6]
	log β_2	+1.8	[-1.0]	[-1.0]	[-1.0]	[-0.9]
				M^{3+}		
L$^-$	log K_1	+0.6	-0.3	-0.3	-0.3	-0.2
	log β_2	+1.0	-0.5	-0.5	[-0.4]	[-0.2]
L^{2-}	log K_1	+1.2	-0.6	-0.6	[-0.7]	[-0.6]
	log β_2	+1.8	[-1.0]	[-1.0]	[-1.0]	[-0.9]
L^{3-}	log K_1	[+1.8]	[-0.9]	-0.9	[-1.1]	[-1.0]
	log β_2	[+2.6]	[-1.5]	[-1.5]	[-1.6]	[-1.5]
				M^{4+}		
L$^-$	log K_1	[+0.8]	[-0.4]	[-0.4]	[-0.5]	[-0.4]
	log β_2	[+1.6]	[-0.6]	[-0.6]	[-0.6]	[-0.5]
L^{2-}	log K_1	[+1.6]	[-0.8]	[-0.8]	[-1.0]	[-0.9]
	log β	[+2.4]	[-1.4]	[-1.4]	[-1.5]	[-1.4]
L^{3-}	log K_1	[+2.4]	[-1.2]	[-1.2]	[-1.5]	[-1.4]
	log β	[+3.4]	[-2.0]	[-2.0]	[-2.2]	[-2.1]

References

1. H. von Euler, *Ber.* **1903**, 1854.
2. G. Bodlander and O. Z. Storbeck, *Anorg. Chem.* **1902**, *31*, 1.
3. M. S. Sherrill and R. Z. Abegg, *Elektrochem.* **1903**, *9*, 549.
4. H. Z. Grossman, *Anorg. Chem.* **1905**, *43*, 356.
5. N. Bjerrum, Ph.D. Dissertation, Copenhagen, 1908.
6. N. Bjerrum, *Kgl. Danski Videnskab Selskab Naturvidenskab Math. Afdel* **1915**, 12 (7), 147.
7. M. Dole, *The Glass Electrode*; Chapman & Hall: New York, 1941.
8. R. G. Bates, *Determination of pH*, 2nd Ed; Wiley: New York, 1973.
9. I. Leden, *Z. Physik. Chem.* **1941**, *A188*, 160.
10. J. Bjerrum, *Metal Amine Formation in Aqueous Solution*, Thesis, Copenhagen, 1941; reprinted 1957, P. Haase and Son, Copenhagen.
11. M. Calvin and K. W. Wilson, *J. Am. Chem. Soc.* **1945**, *67*, 2003.
12. A. E. Martell and M. Calvin, *Chemistry of the Metal Chelate Compounds*; Prentice-Hall: New York, 1952.
13. S. Chaberek and A. E. Martell, *Organic Sequestering Agents*; John Wiley: New York, 1959.
14. *Chelating Agents and Metal Chelates*, Eds. F. P. Dwyer and D. P. Mellor, Academic Press: New York 1964.
15. F. C. Rossotti and H. Rossotti, *The Determination of Stability Constants*; McGraw-Hill: New York, 1961.
16. *Chemistry of the Coordination Compounds*, Eds. J. Bailar and D. C. Busch, ACS Monograph No.131; Reinhold: New York, 1956
17. J. Lewis and R. G. Wilkins, *Modern Coordination Chemistry*; Interscience: New York, 1960.
18. F. Gaizer, *Coord. Chem. Revs.* **1979**, *27 (3)*, 195.
19. F. Gans, *Coord. Chem. Revs.* **1976**, *19 (2)*, 99.
20. F. J. C. Rossotti, H. S. Rossotti and R. J. Whewell, *J. Inorg. Nucl. Chem.* **1971**, *33*, 2051.
21. A. Izquierdo and J. L. Beltran, *Anal. Chim. Acta*, **1986**, *181*, 87.
22. T. Hofman and M. Krzyzanowska, *Talanta*, **1986**, *33*, 851.
23. T. B. Field and W. A. E. McBryde, *Can. J. Chem.* **1978**, *56*, 1202.
24. D. J. Leggett, *Talanta*, **1977**, *24*, 535.
25. M. Meloun, J. Havel and E. Hogfeldt, *Computation of Solution Equilibria: A guide to Methods in Potentiometry, Extraction and Spectrophotometry*; Halsted Press: New York, 1988.
26. R. J. Motekaitis and A. E. Martell, *Can. J. Chem.* **1982**, *60*, 2403.
27. R. J. Motekaitis and A. E. Martell, *Can. J. Chem.* **1982**, *60*, 168.
28. A. E. Martell and R. J. Motekaitis, *Determination and Use of Stability Constants*; VCH Publishers: New York, **1989**.
29. *Stability Constants. Part I. Organic Ligands; Part II. Inorganic Ligands*, Eds. G. Schwarzenbach and L. G. Sillen, Chemical Society: London, 1957, 1958.
30. *Stability Constants, Special Publication No.17*, Eds. L. G. Sillen and A. E. Martell, Chemical Society: London, 1964.
31. *Stability Constants, Supplement No.1, Special Publication No.25*, Eds. L. G. Sillen and A. E. Martell, Chemical Society: London, 1971.
32. *Stability Constants of Metal-ion Complexes. Part B: Organic Ligands*, Ed. D. D. Perrin; Pergamon Press: Oxford, 1979.
33. *Stability Constants of Metal-ion Complexes. Part A. Inorganic Ligands*, Ed. E. Hogfeldt; Pergamon Press: Oxford, 1982.
34. R. M. Smith and A. E. Martell, *Critical Stability Constants*; Plenum Press: New York, 1974, 1975, 1976, 1977, 1982, 1989; Vols. 1-6.
35. a) G. Anderegg, *Critical Survey of Stability Constants of EDTA Complexes*; Pergamon Press: Oxford, 1977; b) W. A. E. McBryde, *A Critical Review of Equilibrium Data for Proton and Metal Compounds of 1,10-Phenanthroline, 2,2'-bipyridyl and Related Compounds*; Pergamon Press: Oxford, 1978; c) J. Stary, Yu. A. Zolotov, O. M. Petrukhin, *Critical Evaluation of Equilibrium Constants involving 8-Hydroxyquinoline and its Metal Chelates*; Pergamon Press: Oxford, 1979; d) A. M. Bond and G. T. Hefter, *Critical Survey of Stability Constants and Related Thermodynamic Data of Fluoride Complexes in Aqueous Solution*; Pergamon Press: Oxford, 1980; e) J. Stary and J. D. Liljenzin, *Pure & Appld.*

Chem. **1982**, 54, 2557. f) G. Anderegg, *Pure & Appld. Chem.* **1982**, *54*, 2693. g) D. G. Tuck, *Pure & Appld. Chem.* **1983**, *55*, 1477. h) L. D. Pettit, *Pure & Appld. Chem.* **1984**, *56*, 247. i) P. Paoletti, *Pure & Appld. Chem.* **1984**, *56*, 491. j) M. R. Beck, *Pure & Appld. Chem.* **1987**, *59*, 1703. k) R. M. Smith, A. E. Martell and Y. Chen, *Pure & Appld. Chem.* **1991**, *63*, 1015.

36. R. M. Smith, A. E. Martell and R. J. Motekaitis, *Critical Stability Constants Database*, 46, NIST, Gaithersburg, MD, USA, 1993.
37. G. Anderegg, *Pure & Appld. Chem.* **1982**, *54*, 2693.
38. E. T. Clarke and A. E. Martell, *Inorg. Chim. Acta.* **1992**, *191 (1)*, 57.
39. G. Schwarzenbach and J. Heller, *Helv. Chim. Acta*, **1951**, *34*, 1876.
40. G. Anderegg and E. Bottari, *Helv. Chim. Acta* **1967**, *50*, 2341
41. G. Schwarzenbach and J. Heller, J. *Helv. Chim. Acta* **1951**, *34*, 576.
42. D. R. Crow, *Polarography of Metal Complexes*, Academic Press, New York, 1969.
43. J. J. Lingane, *Chem. Rev.*, 1, 29, (1941).
44. D. D. de Ford and D. N. Hume, *J. Am. Chem. Soc.*, 73, 53 (1951).
45. See also: a) I. M. Kolthoff and J. J. Lingane, *Polarography*, Interscience, New York, 1952. b) D. R. Crow and J. V. Westwood, *Polarography*, Methuen, London, 1968. b) J. Heyrovsky and I. Kuta, *Principles of Polarography*, Academic Press, 1966.
46. I. Cukrowski, F. Marsicano, R. D. Hancock, and E. Cukrowska, *Electroanal.*, in the press (1995).
47. R. D. Hancock, I. Cukrowski, J. Baloyi, and J. Mashishi, *J. Chem. Soc., Dalton Trans.*, 2895 (1993).
48. G. Schwarzenbach and E. Freitag, *Helv. Chim. Acta* **1951**, *34*, 1503.
49. W. R. Harris and A. E. Martell, *Inorg. Chem.* **1976**, *15*, 713.
50. L. D. Loomis and K. N. Raymond, *Inorg. Chem.* **1991**, *30*, 906.
51. R. M. Smith and A. E. Martell, A. E. *Critical Stability Constants*, Vol.1; Plenum: New York, **1974**.
52. F. L'Eplattenier; I. Murase and A. E. Martell, *J. Am. Chem. Soc.* **1967**, *89*, 837.
53. W. R. Harris, R. J. Motekaitis and A. E. Martell, *Inorg. Chem.* **1975**, *14*, 974.
54. C. J. Baes, Jr. and R. E. Mesmer, *The Hydrolysis of Cations*; Wiley: New York, 1976.
55. R. J. Motekaitis and A. E. Martell, *Inorg. Chem.* **1980**, *19*, 1646.
56. R. J. Motekaitis, Y. Sun and A. E. Martell, *Inorg. Chim. Acta* **1992**, *198-200*, 421.
57. E. T. Clarke and A. E. Martell, *Inorg. Chim. Acta* **1992**, *190*, 37..
58. D. G. Tuck, *Pure & Appld. Chem.* **1989**, *61*, 1162.
59. E. W. J. Baumann, *Inorg. Nucl. Chem.* **1974**, *36*, 1827.
60. G. Anderegg, P. Nageli, F. Muller, F. and G. Schwarzenbach, *Helv. Chim. Acta* **1959**, *42*, 827.
61. S. Chaberek, A. E. Frost, M. A. Doran and N. J. Bicknell, *J. Inorg. Nucl. Chem.* **1959**, *11*, 184.
62. E. J. Durham and D. P. Ruskiewick, D. P. *J. Am. Chem. Soc.* **1958**, *80*, 4812.
63. E. Wanninen, *Acta Acad. Aboensis, Math. et Phys.* **1960**, *21*, 17.
64. E. Z. Larsson, *Phys. Chem.* **1934**, *13*, 16.
65. J. Bjerrum, *Chem. Revs.* **1934**, *A169*, 208.
66. M. Calvin and N. C. Melchior, N. C. *J. Am. Chem. Soc.* **1948**, *70*, 3270.
67. H. Irving and R. J. P. Williams, *Nature (London)* **1948**, *162*, 746.
68. H. Irving and R. J. P. Williams, *J. Chem. Soc.* **1953**, 3192.
69. C. W. Davies, *J. Chem. Soc.* **1951**, 1256.
70. L. G. Van Uitert, W. C. Fernelius and B. E. Douglas, *J. Am. Chem. Soc.* **1954**, *75*, 2736.
71. H. Irving and H. Rossotti, *Acta Chem. Scand.* **1956**, *10*, 72.
72. E. Nieboer and W. A. E. McBryde, *Can. J. Chem.* **1970**, *48*, 2549; 2565.

INDEX

Ethylenediamine, 1, 58, 203
Ethylenediaminediacetic acid, 58, 157, 237
Expanded porphyrin, 137, 139

Fe(III)
    chelate of EHPG, 191
    complex of NOTA, 130
    complex with TACN-MHP, 130
Fe_4S_4 cluster, 205
Ferritin, 155, 161
First row transition metals, 31
Five-membered chelate ring, 104, 118
Force constant, 13
Force field, 73
Formation constants, 1, 46, 76, 77, 222
    ammonia complexes, 48
    for polyamines, 65
    hydroxide complexes, 48
    of complexes of Cu(II), 33
Free energy, 57
    of complex formation, 56
    of protonation, 56
Fuoss equation, 9, 86

Gallic acid, 163
Gas phase, 15-18, 109
    basicities, 16
    enthalpies
        of formation, 22
        of protonation, 22
Gem dimethyl group, 138
Geometry of chelate ring, 77
Glass electrode, 218, 222, 228
GLY, 71

ΔH^o
Half-buckled conformer, 122, 124
Hammett σ-function, 19
Hard metal ions, 151, 153
Hardness, 10, 39
    parameter, 56
HBED, 84, 153, 156, 158, 237
    alkyl derivatives of, 168
HBIDA, 237
Heats of solvation, 5
HEDTA, 237
Hemocyanin, 212

Hg(II) complexes, 57
HIDA, 47, 51
    complex, 52
Hidden inductive effects, 52
High-spin complexes, 31
    octahedral, 32
Highly preorganized ligands, 87
    macrocycles, 138
HPED, 173, 174
HSAB, 199
    trends, 22
Hydrogen
    bond, 16, 17, 18, 19
    electrode, 222
    ion, 218
Hydroxamate, 84
Hydroxide complex, 45
Hydroxo complexes of gallium(III), 238
Hydroxybenzylethylendiaminediacetic acid, 237
o-Hydroxybenzyliminodiacetic acid, 237
1-Hydroxycyclohepta-3,5,7-trien-2-one (tropolonate), 59
Hydroxyethyl group, deprotonation of, 47
N-Hydroxyethylethylenediaminetriacetic acid, 237
2-Hydroxyethyliminodiacetic acid, 59
5-Hydroxy-2-hydroxymethyl-4-pyrone (kojate), 59
Hydroxyl complexes, 222
3-Hydroxy-2-methyl-4-pyrone (maltol), 59
Hydroxypyridinones, 159, 174
1-Hydroxy-2-pyridinone, 174
3-Hydroxy-2-pyridinone, 159, 174
3-Hydroxy-4-pyridinone, 159, 174

I_A parameters, 37
I_B parameters, 38
IDA
    complex, 49
    ligand, 49
Imidazole, 55, 212
Iminodiacetic acid (IDA), 59
Inductive effect, 18, 19, 52, 54, 94, 108, 109, 143
    of methyl groups, 16

ΔS°

Saturated polyamines, 33
Schiff base, 55
Secondary coordination sphere, 7
SEDDA, 237
Selectivity patterns, 125
Selectivity, 112
Sepulchrate, 3, 128, 136
Siderophores, 84
Silver-silver chloride electrode, 230
Six-membered, chelate rings, 104, 118
Size match selectivity, 97, 112, 120
Size of chelate ring, 73, 83
Sodium diethyldithiocarbamate, 163
    derivative, 163
Sodium-2,3-dimercaptopropane-1-
    sulfonate, 163, 165
Soft metal ions, 154
Softness, 10, 39
    series, 22
Solubility measurements, 227
Solvation, 17, 25
Species distribution diagram, 225
Specific electrode potentials, 227
Specific metal ion electrodes, 227, 229,
    231
Spectrochemical series, 29
Spectrophotometric determination of
    competition constants, 237
Spectrophotometry, 227, 228
Spherand, 3
Spin pairing, 31
    energy, 29
Stabilities of EDTA complexes, 236
Stabilities of complexes of macrocyclic
    ligands, 42
Stability constant, 217, 220
    determination, 227
    of Ca(II)-EDTA, 223
Standard reference state, asymmetry, 64
Stepwise stability constants, 218
Steric effects, 19, 54, 94
Steric hindrance, 17, 39, 109, 143
    to solvation, 16, 18
Steric strain, 12, 73, 100
Strain energy, 73, 75, 122, 124
    of the EN chelate ring, 78
    of the six-membered chelate ring, 79

of the five-membered chelate ring,
    79
Strength of the M-L bond, 34
Structure of Ni(DME-DAC), 58
Sulfur donors, 213

TACN-Me2HB, 158, 169
TACN-MeHP, 158, 170, 171
TACN-MHP, 128, 129
TACN-TM, 168, 169
TACN-TX, 128, 129
Taft equation, 19
Taft σ-function, 19,
Technetium, 177
Technetium(I), 184
Technetium(II), 184
Technetium(III), 182
Technetium(IV), 181
Technetium(V), 180
Technetium(VI), 178
Technetium(VII), 178
2,2,2-Tet, 92, 98
2,3,2-Tet, 92, 98, 100, 101, 107
3,2,3-Tet, 92
3,3,3-Tet, 92
TETA, 128, 131, 172
TETB, 200
1,4,7,10-Tetraazacyclododecane-
    1,4,7,10-tetraacetic acid, 189, 190
1,4,8,11-Tetraazacyclotetradecane
    (cyclam), 59, 213
Tetradecane, rac and meso-
    5,5,7,12,12,14-hexamethyl-1,4,8,11-
    tetraazacyclo-, 59
Tetrahedral coordination geometry, 119
[Tetrakis(2-hydroxyethyl)cyclam, 213
Tetrakis(2-sulfophenyl)porphyrin, 213
Tetrakis-(2-hydroxy-3,5-dimethyl-
    benzyl)ethylenediamine, 169
Tetrakis-(2-hydroxy-5-methylbenzyl)-
    ethylenediamine, 169
Tetrakis-(1-sulfophenyl)porphyrin, 138,
    212
Tetrasubstituted tetraazacycloalkane,
    172
TETREN, 63, 65, 66, 98, 127
T_H, 224
Thalassemia, 84, 155